PE POWER ELECTRICAL ENGINEERING
License Review Study Guide

Howard Smolleck, Ph.D., PE

This publication is designed to provide accurate and authoritative information in regard to the subject matter covered. It is sold with the understanding that the publisher is not engaged in rendering legal, accounting, or other professional service. If legal advice or other expert assistance is required, the services of a competent professional person should be sought.

President: Dr. Andrew Temte

Chief Learning Officer: Dr. Tim Smaby

Vice President of Engineering Education: Dr. Jeffrey Manzi, PE

PE POWER ELECTRICAL ENGINEERING LICENSE REVIEW STUDY GUIDE

© 2009 by Dearborn Financial Publishing, Inc.®

Published by Kaplan Engineering Education

1905 Palace Street

La Crosse, WI 54603

800-420-1432

www.kaplanengineering.com

All rights reserved. The text of this publication, or any part thereof, may not be reproduced in any manner whatsoever without permission in writing from the publisher.

Printed in the United States of America.

09 10 11   10 9 8 7 6 5 4 3 2 1

ISBN: 1-4277-9106-6
PPN: 1570-9500

CONTENTS

Introduction   VII
HOW TO USE THIS BOOK   VII
BECOMING A PROFESSIONAL ENGINEER   VIII
ELECTRICAL ENGINEERING PROFESSIONAL ENGINEER EXAM   IX
ACKNOWLEDGMENTS   XIV

CHAPTER 1 — Mathematics Review   1
LOGARITHMS   2
CIRCULAR AND HYPERBOLIC TRIG FUNCTIONS AND THEIR EQUIVALENTS   2
CALCULUS, DIFFERENTIAL EQUATIONS, AND INTEGRALS   4
COMPLEX NUMBERS   5
MATRICES AND SIMULTANEOUS LINEAR EQUATIONS   6
DETERMINANTS   8

CHAPTER 2 — Electric and Magnetic Fields   11
RELATION OF ELECTROMAGNETICS TO POWER ENGINEERING   12
ELECTRIC FIELDS   12
MAGNETIC FIELDS   15
DYNAMIC FIELDS   16
RECOMMENDED REFERENCES   17

CHAPTER 3 — Basic Circuit Concepts and DC Circuit Analysis   19
NOTATION   20
DIAGRAMS AND SYMBOLS   21
BASIC CIRCUIT LAWS   21
BASIC CIRCUIT-ANALYSIS TECHNIQUES   22
POWER   30
DEPENDENT SOURCES AND MAXIMUM POWER TRANSFER   30
DELTA-WYE TRANSFORMATION   33
SUMMARY   35
RECOMMENDED REFERENCES   35

CHAPTER 4 — Single-Phase AC Circuits   37
THE SINUSOID   38
THE IMPORTANCE OF CIRCUIT MODELS AND SSAC ANALYSIS   39
LINEAR, BILATERAL, PASSIVE CIRCUIT ELEMENTS   39

EFFECTIVE (RMS) VALUES 41
THE PHASOR 43
IMPEDANCE AND ADMITTANCE 45
RESONANCE 47
POWER TO A SINGLE-PHASE LOAD 47
SERIES AND PARALLEL EQUIVALENTS 51
POWER FACTOR CORRECTION 53
THE NODAL PROBLEM AND THE BUS IMPEDANCE AND ADMITTANCE MATRICES 59
RECOMMENDED REFERENCES 62

CHAPTER 5 — Balanced Three-Phase Circuits 63
THE BALANCED THREE-PHASE SOURCE 64
PHASE SEQUENCES: abc AND acb 65
THE THREE-PHASE SYNCHRONOUS GENERATOR 66
THREE-PHASE CONNECTIONS AND THE PER-PHASE EQUIVALENT 67
RECOMMENDED REFERENCE 77

CHAPTER 6 — Per-Unit Methods and Calculations 79
PER-UNIT, PERCENT, AND BASE QUANTITIES 79
PRINCIPLE ADVANTAGES OF PER-UNIT REPRESENTATION 83
OTHER NUMERICAL EXAMPLES 83
RECOMMENDED REFERENCES 83

CHAPTER 7 — Transformers 85
ELECTROMAGNETIC FUNDAMENTALS OF TRANSFORMERS 85
PER-UNIT CALCULATIONS AND BASE QUANTITIES 89
MULTI-WINDING, BUCK-BOOST, AND AUTOTRANSFORMER CONNECTIONS 92
REAL TRANSFORMERS: LOSSES, IMPEDANCES, AND MODELS 96
TRANSFORMER TESTS TO DETERMINE MODEL PARAMETERS 100
BALANCED THREE-PHASE TRANSFORMER CONNECTIONS 102
RECOMMENDED REFERENCES 104

CHAPTER 8 — Transmission and Distribution Lines 105
LUMPED PARAMETER MODEL OF LINE 106
CONDUCTOR TYPES AND CONFIGURATIONS 107
UNITS, MEASUREMENTS, AND CONVERSION FACTORS 108
OBTAINING THE PARAMETER VALUES FOR THE PER-PHASE MODEL 108
SHORT AND MEDIUM-LENGTH MODELS 112
ABCD PARAMETERS 112
LONG LINE EQUATIONS 114
WAVELENGTH 114
THE LOSSLESS MODEL AND SURGE-IMPEDANCE LOADING 115
FULL-LOAD AND NO-LOAD PERFORMANCE CALCULATIONS 115
RECOMMENDED REFERENCES 117

CHAPTER 9 — Direct-Current Machines and Machine Basics 119
BASIC CONCEPTS AND TERMINOLOGY 120
THE SEPARATELY EXCITED GENERATOR AND MAGNETIC SATURATION 121
SELF-EXCITED DC GENERATOR AND DC MOTOR CONFIGURATIONS 124
RECOMMENDED REFERENCES 124

CHAPTER 10 — Alternating-Current Machines 125
AC MACHINE TYPES AND GENERAL CHARACTERISTICS 126
SYNCHRONOUS MACHINES 127
POLYPHASE INDUCTION MOTORS 138
SINGLE-PHASE SYNCHRONOUS AND INDUCTION MOTORS 154
SOME OTHER MACHINE TYPES 155
RECOMMENDED REFERENCES 156

CHAPTER 11 — Power Flow Study 157
FUNDAMENTALS OF POWER FLOW STUDY 158
CIRCUIT THEORY 160
CALCULATION OF LINE FLOWS 163
RECOMMENDED REFERENCE 164

CHAPTER 12 — Symmetrical Components and Unbalance Factors 165
THE METHOD OF SYMMETRICAL COMPONENTS 165
PHASE UNBALANCE FACTORS 170
RECOMMENDED REFERENCES 171

CHAPTER 13 — Generator Transient Behavior, Short Circuit Study, and Power System Protection 173
FAULTS AND FAULT PROTECTION 174
POWER SYSTEM PROTECTION 185
RECOMMENDED REFERENCES 187

CHAPTER 14 — Power Quality and Power Electronics 189
ELECTRIC POWER QUALITY 190
POWER ELECTRONICS 197
RECOMMENDED REFERENCES 204

CHAPTER 15 — Measurement, Reliability, and Lighting 205
FUNDAMENTALS OF MEASUREMENT AND INSTRUMENTATION 206
METERS AND WAVEFORMS 206
POWER SYSTEM RELIABILITY 217
FUNDAMENTALS OF LIGHTING (ILLUMINATION ENGINEERING) 219
RECOMMENDED REFERENCES 222

CHAPTER 16 — Codes, Standards, and Safety 223

CODE-RELATED EXAM QUESTIONS 224
AN OVERVIEW OF ELECTRICAL CODES AND STANDARDS 224
THE NATIONAL ELECTRICAL CODE 226
THE NATIONAL ELECTRICAL SAFETY CODE 229
ELECTRICAL SAFETY 230
ELECTRICAL SHOCK AND OTHER INJURY 231
RECOMMENDED REFERENCES 236

CHAPTER 17 — Engineering Economics 237

CASH FLOW 239
TIME VALUE OF MONEY 240
EQUIVALENCE 241
COMPOUND INTEREST 241
NOMINAL AND EFFECTIVE INTEREST 248
SOLVING ENGINEERING ECONOMICS PROBLEMS 250
PRESENT WORTH 251
FUTURE WORTH OR VALUE 253
ANNUAL COST 254
RATE OF RETURN ANALYSIS 256
BENEFIT-COST ANALYSIS 260
BREAKEVEN ANALYSIS 261
OPTIMIZATION 262
VALUATION AND DEPRECIATION 264
TAX CONSEQUENCES 269
INFLATION 270
RISK ANALYSIS 272
REFERENCE 274
INTEREST TABLES 275

CHAPTER 18 — Final Tips for the Power Exam 285

DON'T COMPLICATE 286
BE AWARE OF UNITS 286
USE COMMON SENSE 286
BE PREPARED FOR CODE QUESTIONS 286
LOOK FOR THE FAMILIAR IN THE UNFAMILIAR 287
DON'T TRIP OVER TRICKY QUESTIONS 287
SUMMARY 288

Introduction

OUTLINE

HOW TO USE THIS BOOK   VII

BECOMING A PROFESSIONAL ENGINEER   VIII
Education ■ Fundamentals of Engineering (FE/EIT) Exam ■ Experience ■ Professional Engineer Exam

ELECTRICAL ENGINEERING PROFESSIONAL ENGINEER EXAM   IX
Examination Development ■ Examination Structure ■ Exam Dates ■ Exam Procedure ■ Preparing for and Taking the Exam ■ Exam Day Preparations ■ What to Take to the Exam ■ Examination Scoring and Results

ACKNOWLEDGMENTS   XIV

HOW TO USE THIS BOOK

- *PE Power Electrical Engineering License Review Study Guide* contains conceptual review of topics for the power subdiscipline of the PE exam, including key terms, equations, analytical methods and reference data. In addition to reviewing concepts, the book provides solved Examples to help you apply your understanding of equations and techniques as you proceed. Because this book does not contain exam-type problems and solutions, it can be brought into the open-book PE exam as one of your references.

- Chapters 1 through 5 review fundamental material from undergraduate engineering courses. In fact, most of the information in the book (with the exception of the chapter on economics) is material that is either presented or reviewed in a typical electric power undergraduate course sequence. We have tried to zero in on the highlights of this background material to make it easy for you to review without having to read a lot of proofs and background theory. If this material is no longer fresh in your mind, you probably will want to review it before proceeding to more exam-specific chapters. Alternatively, you can use these chapters as reference if you find you need some background review while studying more advanced topics.

- Some questions on the exam can probably be answered from the background material in Chapters 2 through 5 alone. For instance, a question involving power-factor correction or formulation of a nodal-analysis problem might be expected since these topics are so fundamental to electric power engineering.

- Chapters 4 through 17 primarily review exam-specific topics. The assumption throughout is that exam candidates already have some knowledge of these topics from both their education and professional practice. Therefore, the approach is not to teach the material "from scratch," but to refresh your understanding of important, exam-specific concepts. You will notice that some chapters contain both background review and more advanced topical content, in order to move more logically from the former to the latter.

- Chapter 18 provides some final tips and perspectives that should help you optimize your performance on the exam. These insights should help you solve some specific problems that may at first appear to be difficult or outside your area of study.

- The goal of this book is to provide a focused review on critical topics for busy engineers. No single text can provide exhaustive coverage of exam topics given the scope of the exam and the unique strengths and weaknesses of individual candidates. Therefore, this book also provides references to texts you may want to consult for additional review.

- Finally, we recommend that you pair this book with Kaplan's *PE Power Electrical Engineering Sample Exam* to optimize your preparation. The Sample Exam provides exam-style questions to give you practice in problem-solving. These practice problems test your mastery of exam topics and simulate what you will experience in the actual exam, including the tight time constraints. On the actual PE exam you will have on average six minutes per problem, so practicing your pace as well as problem-solving skills is important.

BECOMING A PROFESSIONAL ENGINEER

To achieve registration as a professional engineer there are four distinct steps: (1) education, (2) the Fundamentals of Engineering/Engineer-In-Training (FE/EIT) exam, (3) professional experience, and (4) the professional engineer (PE) exam, more formally known as the Principles and Practice of Engineering Exam. These steps are described in the following sections.

Education

The obvious appropriate education is a B.S. degree in electrical engineering from an ABET accredited college or university. This is not an absolute requirement. Alternative, but less acceptable, education is a B.S. degree in something other than electrical engineering, or a degree from a non-accredited institution, or four years of education but no degree.

Fundamentals of Engineering (FE/EIT) Exam

Most people are required to take and pass this eight-hour multiple-choice examination. Different states call it by different names (Fundamentals of Engineering, EIT, or Intern Engineer), but the exam is the same in all states. It is prepared and graded by the National Council of Examiners for Engineering and Surveying (NCEES). Review materials for this exam are found in other Kaplan AE books, such as *Fundamentals of Engineering FE/EIT Exam Preparation*.

Experience

Typically one must have four years of acceptable experience before being permitted to take the Professional Engineer exam (California requires only two years). Both the length and character of the experience will be examined. It may, of course, take more than four years to acquire four years of acceptable experience.

Professional Engineer Exam

The second national exam is called Principles and Practice of Engineering by NCEES, but just about everyone else calls it the Professional Engineer or PE exam. All states, plus Guam, the District of Columbia, and Puerto Rico, use the same NCEES exam.

ELECTRICAL ENGINEERING PROFESSIONAL ENGINEER EXAM

The reason for passing laws regulating the practice of electrical engineering is to protect the public from incompetent practitioners. Most states require engineers working on projects involving public safety to be registered, or to work under the supervision of a registered professional engineer. In addition, many private companies encourage or require engineers in their employ to pursue registration as a matter of professional development. Engineers in private practice, who wish to consult or serve as expert witnesses, typically also must be registered. There is no national registration law; registration is based on individual state laws and is administered by boards of registration in each of the states. You can find a list of contact information for and links to the various state boards of registration at the Kaplan AE Education Web site: *www.kaplanae.com*. This list also shows the exam registration deadline for each state.

Examination Development

Initially the states wrote their own examinations, but beginning in 1966 the NCEES took over the task for some of the states. Now the NCEES exams are used by all states. This greatly eases the ability of an engineer to move from one state to another and achieve registration in the new state.

The development of the engineering exams is the responsibility of the NCEES Committee on Examinations for Professional Engineers. The committee is composed of people from industry, consulting, and education, plus consultants and subject matter experts. The starting point for the exam is a task analysis survey, which NCEES does at roughly 5- to 10-year intervals. People in industry, consulting, and education are surveyed to determine what electrical engineers do and what knowledge is needed. From this NCEES develops what it calls a "matrix of knowledge" that forms the basis for the exam structure described in the next section.

The actual exam questions are prepared by the NCEES committee members, subject matter experts, and other volunteers. All people participating must hold professional registration. Using workshop meetings and correspondence by mail, the questions are written and circulated for review. Although based on an understanding of engineering fundamentals, the problems require the application of practical professional judgment and insight.

Examination Structure

The Electrical and Computer Engineering PE exam is an 8-hour exam consisting of 40 morning questions and 40 afternoon questions. All questions are multiple-choice format with four possible answer choices each.

Beginning in 2009, exam candidates must select one of three subdiscipline topics for their exam at the time they register with their state board: 1) Power; 2) Electrical and Electronics; or 3) Computer.

Table 1 summarizes the topics covered in the Power examination and their relative emphasis as specified by NCEES. It also points to the sections of this book where you will find review coverage of these topics. For more information on the topics and subtopics on the exam, visit the NCEES Web site at *www.ncees.org*.

Table 1 PE Power Electrical Engineering License Review Topics

	Review Chapter(s) in this Text
General Power Engineering (30%)	
Measurement and Instrumentation (7.5%)	Chapters 14, 15
Special Applications (10%)	Chapter 14
Codes and Standards (12.5%)	Chapter 16
Circuit Analysis (20%)	Chapters 3-5
Analysis (11%)	Chapters 5, 6, 12
Devices and Power Electronic Circuits (9%)	Chapter 14
Rotating Machines and Electromagnetic Devices (20%)	
Rotating Machines (12.5%)	Chapters 9, 10
Electromagnetic Devices (7.5%)	Chapter 7
Transmission and Distribution (High, Medium, and Low Voltage) (30%)	Chapters 4, 5, 6, 7, 8
System Analysis (12.5%)	Chapters 11, 12, 13, 14, 15
Power System Performance (7.5%)	Chapters 11, 14, 16
Protection (10%)	Chapter 15

Exam Dates

NCEES prepares Professional Engineer exams for use on a Friday in April and October of each year. Some state boards administer the exam twice a year in their state, whereas others offer the exam once a year. The scheduled exam dates for the next ten years can be found on the NCEES Web site (*www.ncees.org/exams/schedules*).

People seeking to take a particular exam must apply to their state board several months in advance.

Exam Procedure

Before the morning four-hour session begins, proctors will pass out an exam booklet, answer sheet, and mechanical pencil with eraser to each examinee. The provided pencil is the only writing instrument you are permitted to use during the exam. If you need an additional pencil during the exam, a proctor will supply one.

Fill in the answer bubbles neatly and completely. Questions with two or more bubbles filled in will be marked as incorrect, so if you decide to change an answer, be sure to erase your original answer completely.

The afternoon session will begin following a one-hour lunch break. In both the morning and afternoon sessions, if you finish more than 15 minutes early you may turn in your booklet and answer sheet and leave. In the last 15 minutes, however, you must remain to the end of the exam in order to ensure a quiet environment for those still working and an orderly collection of materials.

Preparing for and Taking the Exam

Give yourself time to prepare for the exam in a calm and unhurried way. Many candidates like to begin several months before the actual exam. Target a number of hours per day or week that you will study, and reserve blocks of time for doing so. Creating a review schedule on a topic-by-topic basis is a good idea. Remember to allow time for both reviewing concepts and solving practice problems.

In addition to review work that you do on your own, you may want to join a study group or take a review course. A group study environment might help you stay committed to a study plan and schedule. Group members can create additional practice problems for one another and share tips and tricks.

You may want to prioritize the time you spend reviewing specific topics according to their relative weight on the exam, as identified by NCEES, or by your areas of relative strength and weakness. When you find a problem or topic during your study that needs more attention, make a note of it or look it up immediately. Don't wait; you may forget to address something important.

People familiar with the psychology of exam taking have several suggestions for people as they prepare to take an exam.

1. Exam taking involves, really, two skills. One is the skill of demonstrating knowledge that you know. The other is the skill of exam taking. The first may be enhanced by a systematic review of the technical material. Exam-taking skills, on the other hand, may be improved by practice with similar problems presented in the exam format.

2. Since there is no deduction for guessing on the multiple choice problems, answers should be given for all of them. Even when one is going to guess, a logical approach is to attempt to first eliminate one or two of the four alternatives. If this can be done, the chance of selecting a correct answer obviously improves. In fact, careful attention to this procedure, including a moment's thought on each alternative answer, can significantly improve your probability of correct answers.

3. Problem statements may seem confusing at first glance. They may be designed to see if you can isolate critical data from extraneous information and make good typical assumptions. Read the problems carefully, make good assumptions, and select the best answer.

4. Plan ahead with a strategy for working through the exam. For example, you may want to take a two-pass approach. In the first pass, answer those problems that you find you can solve quickly and easily with reasonable certainty. Mark or list these problems so you don't waste time looking at them again. Then make a second pass to address those problems you find especially challenging or feel uncertain about. During the test, a busy hour or two away from a difficult problem might allow you a fresh insight into the solution. Even if you have answered every question before the exam session ends, don't walk out early. Use this time to go back and check your work, as well as revisit any problems about which you are still uncertain.

5. Read all four multiple-choice answers before making a selection. An answer in a multiple-choice question is sometimes a plausible decoy—not the best answer.

6. Do not change an answer unless you are absolutely certain you have made a mistake. Your first reaction is likely to be correct.

7. Do not sit next to a friend, a window, or other potential distractions.

Exam Day Preparations

The exam day will be a stressful and tiring one. This will be no day to have unpleasant surprises. For this reason we suggest that an advance visit be made to the examination site. Try to determine such items as the following:

1. How much time should I allow for travel to the exam on that day? Plan to arrive about 15 minutes early. That way you will have ample time, but not too much time. Arriving too early, and mingling with others who also are anxious, will increase your anxiety and nervousness.

2. Where will I park?

3. How does the exam site look? Will I have ample workspace? Where will I stack my reference materials? Will it be overly bright (sunglasses), cold (sweater), or noisy (earplugs)? Would a cushion make the chair more comfortable?

4. Where are the drinking fountain, lavatory facilities, pay phone?

5. What about food? Should I take something along for energy in the exam? A bag lunch during the break probably makes sense.

6. As with college exams, avoid cramming late the night before. Do your more intensive study days or weeks earlier. By the day before the exam, you should have your study items well marked and well organized, possibly in a carton that you can use as a portable bookcase. Be sure to put together everything you will need on the exam (including all reference materials and calculator(s) early the night before the text, and then get a good rest in order to be refreshed for the next day.

What to Take to the Exam

The NCEES guidelines say you may bring only the following reference materials and aids into the examination room for your personal use:

1. Handbooks and textbooks, including the applicable design standards.

2. Bound reference materials, provided the materials remain bound during the entire examination. The NCEES defines "bound" as books or materials fastened securely in their covers by fasteners that penetrate all papers. Examples are ring binders, spiral binders and notebooks, plastic snap binders, brads, screw posts, and so on.

3. A battery-operated, silent, nonprinting, noncommunicating calculator from the NCEES list of approved calculators. For the most current list, see the NCEES Web site (*www.ncees.org*). You also need to determine whether or not your state permits preprogrammed calculators. Bring extra batteries for your calculator just in case; many people feel that bringing a second calculator is also a very good idea.

At one time NCEES had a rule barring "review publications directed principally toward sample questions and their solutions" in the exam room. This set the stage for restricting some kinds of publications from the exam. *State boards may adopt the NCEES guidelines, or adopt either more or less restrictive rules.* Thus an important step in preparing for the exam is to know what will—and will not—be permitted. We suggest that if possible you obtain a written copy of your state's policy for the specific exam you will be taking. Occasionally there has been confusion at individual examination sites, so a copy of the exact applicable policy will not only allow you to carefully and correctly prepare your materials, but will also ensure that the exam proctors will allow all proper materials that you bring to the exam.

As a general rule we recommend that you plan well in advance what books and materials you want to take to the exam. Then, use the same materials in your review that you will have in the exam. Being able to find the appropriate information quickly presupposes a lot of advance organization on your part, and familiarity with your reference materials. Catalog your reference material by tabbing or marking important topics, equations, tables, and procedures. Keep everything well organized as you prepare these materials during the weeks before the exam, and be sure to insert new items in their correct places and catalog them properly.

License Review Books
The review books you use to prepare for the exam are good choices to bring to the exam itself. After weeks or months of studying, you will be very familiar with their organization and content, so you'll be able to quickly locate the material you want to reference during the exam. Keep in mind the caveat just discussed—some state boards will not permit you to bring in review books that consist largely of sample questions and answers.

Textbooks
If you still have your university textbooks, they are the ones you should use in the exam, unless they are too out of date. To a great extent, the books will be like old friends with familiar notation, and perhaps some of your own notes.

Bound Reference Materials
The NCEES guidelines suggest that you can take any reference materials you wish, so long as you prepare them properly. You could, for example, prepare several volumes of bound reference materials, with each volume intended to cover a particular category of problem. Maybe the most efficient way to use this book would be to cut it up and insert portions of it in your individually prepared bound materials. Use tabs so that specific material can be located quickly. If you do a careful and systematic review of electrical engineering, and prepare a lot of well-organized materials, you just may find that you are so well prepared that you will not have left anything of value at home.

Calculators
Be sure you can perform needed calculations on your calculators prior to the test. You may need to work with exponentials and logarithms, trigonometrics and their inverses, complex numbers (including polar-rectangular conversions, and complex trigonometric and hyperbolic numbers), and perhaps some simple matrix operations.

Calculators are usually kept in soft cases; if they are thrown together with books and other materials, buttons can be pressed and held unintentionally. Be sure you understand your calculator well enough to get out of unusual modes (such as graphics or programming) if you accidentally get into them. The exam is no time to find you have been inadvertently prevented from using your calculator.

Other Items

In addition to the reference materials just mentioned, you should consider bringing the following to the exam:

- *Clock*—You must have a time plan and a clock or wristwatch.

- *Exam assignment paperwork*—Take along the letter assigning you to the exam at the specified location. To prove you are the correct person, also bring something with your name and picture.

- *Items suggested by advance visit*—If you visit the exam site, you probably will discover an item or two that you need to add to your list.

- *Clothes*—Plan to wear comfortable clothes. You probably will do better if you are slightly cool.

- *Box for everything*—You need to be able to carry all your materials to the exam and have them conveniently organized at your side. Probably a cardboard box is the answer.

Examination Scoring and Results

The questions are machine-scored by scanning. The answers sheets are checked for errors by computer. Marking two answers to a question, for example, will be detected and no credit will be given.

Your state board will notify you whether you have passed or failed roughly three months after the exam. Candidates who do not pass the exam the first time may take it again. If you do not pass you will receive a report listing the percentages of questions you answered correctly for each topic area. This information can help focus the review efforts of candidates who need to retake the exam.

The PE exam is challenging, but analysis of previous pass rates shows that the majority of candidates do pass it the first time. By reviewing appropriate concepts and practicing with exam-style problems, you can be in that majority. Good luck!

ACKNOWLEDGMENTS

I would like to acknowledge the many ideas and suggestions I have received over the years from my electric-power colleagues here in the Klipsch School of Electrical and Computer Engineering at New Mexico State University. These colleagues include Dr. Satish Ranade, Professor William H. Kersting, Dr. Joydeep Mitra, and Dr. Sukumar Brahma. Professor Kersting proposed the instructional approach for fundamental power-flow analysis presented in Chapter 11, "Power Flow Study." I am particularly indebted to Dr. Ranade for his assistance with the power electronics and reliability sections of this textbook.

Lincoln D. Jones provided an excellent foundation for this new edition. Portions of this book are based on work from his *Electrical Engineering PE License Review*, 9th Edition.

The author and publisher are grateful to Steven R. Musial II, PE, for providing extensive accuracy checking and recommendations on the manuscript.

CHAPTER 1

Mathematics Review

OUTLINE

LOGARITHMS 2

CIRCULAR AND HYPERBOLIC TRIG FUNCTIONS
AND THEIR EQUIVALENTS 2

CALCULUS, DIFFERENTIAL EQUATIONS, AND INTEGRALS 4

COMPLEX NUMBERS 5

MATRICES AND SIMULTANEOUS LINEAR EQUATIONS
Matrix Solution by Determinants 6

DETERMINANTS 8
Second Order ■ Third Order

Many calculations in fundamental areas of electrical engineering use rather straightforward mathematics, including arithmetic operations such as addition, subtraction, multiplication, division, square root, and the use of exponents. Problems employ both positive and negative real numbers. We will not review these basic operations here. Beyond these, you may need to refresh your memory on some other math operations, concepts, and techniques. This chapter reviews them and, in addition, presents some useful identities and other equations that may be of help in solving electrical power problems.

Although the PE exam might include some simple number-crunching that you can do mentally, you will likely perform most of the required mathematical operations on a calculator. Required calculations range from simple single-digit manipulations to complex exponentials or trigonometrics, and possibly some matrix algebra and calculus application. You should determine well in advance whether your calculator is allowed in the exam (refer to the Introduction). Further, it is absolutely necessary that you become very familiar with the capabilities of your calculator and how to use it quickly. The exam is not the time to learn and verify new calculator commands!

We have designed this book with numerical examples throughout. You should work each numerical example on your calculator and compare the answers. Be especially careful of modes and units (for example, degrees versus radians, and polar versus rectangular form of complex numbers).

LOGARITHMS

Electric power equations frequently use logarithms, both log (or $\log_{10}$), which is in base 10, and ln, which is in base e (= 2.718). Logarithms were originally developed centuries ago for problems involving navigation, simplifying the processes of multiplication, division, taking roots, and raising to powers. The following are some useful relations:

$$\ln(ab) = \ln a + \ln b \tag{1.1}$$

$$\ln(a/b) = \ln a - \ln b \tag{1.2}$$

$$\ln(a^n) = n \ln a \tag{1.3}$$

Similar equations hold for logarithms to any other base.

Logarithms are widely used in finding the inductive and capacitive properties of circular conductors (such as transmission lines).

CIRCULAR AND HYPERBOLIC TRIG FUNCTIONS AND THEIR EQUIVALENTS

Engineering problems frequently use circular trigonometric operations (most often the sine, cosine, and tangent and their inverses). These operations allow us to relate sides and angles of right triangles. In power engineering, trigonometric functions are used in many situations, including phasors for both single- and three-phase systems and in impedance calculations, for example.

With reference to Figure 1.1, which shows an angle theta (θ), as well as opposite and adjacent sides and the hypotenuse of a right triangle,

$$\sin\theta = o/h \tag{1.4}$$

$$\cos\theta = a/h \tag{1.5}$$

$$\tan\theta = o/a \tag{1.6}$$

$$a^2 + o^2 = h^2 \tag{1.7}$$

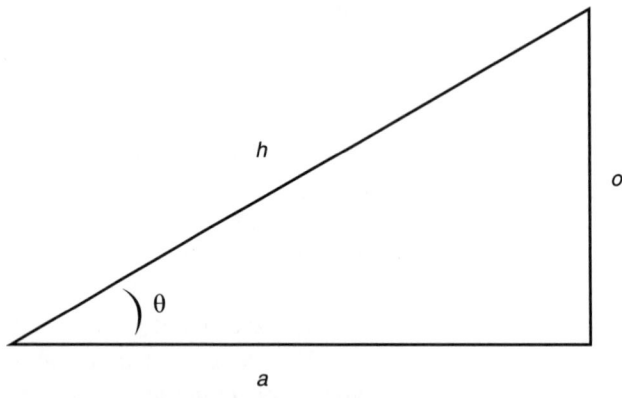

Figure 1.1 Right triangle relationships

It is important to be able to evaluate any of these in any of the four quadrants of a circle. Most modern scientific calculators will handle this issue correctly. It's also worthwhile to note that

$$\sin(-\theta) = -\sin(\theta), \text{ because sine is an odd function where } f(-x) = -f(x) \quad (1.8)$$

and

$$\cos(-\theta) = \cos\theta, \text{ because cosine is an even function where } f(-x) = f(x). \quad (1.9)$$

Note that angles are often denoted by Greek letters, and that the zero-degree reference is often the positive (right) direction along the abscissa (horizontal axis). Also note that counterclockwise angles are usually taken as positive.

Verify the following numerical examples on your calculator:

$$\sin(-35°) = -.5736$$
$$\sin(340°) = -0.3420$$
$$\cos^{-1}(0.5) = 60°$$
$$\tan(-69°) = -2.6051$$
$$\tan(45°) = 1.0000$$
$$\tan^{-1}(0.8) = 38.66°$$

Note that, regardless of the value of θ, $\cos\theta$ and $\sin\theta$ cannot be greater than 1.0 in magnitude, but they could be positive, negative, or zero.

Here are a few more useful trigonometric identities that are sometimes needed in power-system and circuit calculations

$$\sin\theta = \cos(90°-\theta) \quad (1.10)$$

$$\cos\theta = \sin(90°-\theta) \quad (1.11)$$

$$\sin^2\theta + \cos^2\theta = 1. \quad (1.12)$$

Bringing a mathematical handbook to the exam may be wise, but it is unlikely that many problems will require trigonometric identities beyond these.

Although we usually express angles in degrees (°), as in the preceding examples, they are sometimes expressed in **radians** (sometimes abbreviated *rad*, but often a radian value is given without any verbal identification). There are 360° or 2π radians in a complete circle, so to convert radians to degrees, for instance, we multiply the angle by $180/\pi$.

Operations such as sin, cos, and tan are often used in connection with complex quantities, and specifically phasor analysis.

Some aspects of power system analysis, such as transmission line calculations, require the use of **hyperbolic trig functions** (specifically sinh, cosh, and tanh and their inverses). Sometimes, these have complex-number arguments (see the subsection "Complex Numbers" later in this chapter). It is often sufficient to be able to manipulate these using a scientific calculator. Although we will not present a review of hyperbolic trigonometry or hyperbolic trig functions, we will use them as needed. *In each case, ensure that you can solve each of the examples with your own calculator and obtain correct answers.*

Hyperbolic and circular trig functions can be expressed in terms of exponentials. The following equations may be useful, for instance, if your calculator allows complex exponentials but not complex hyperbolic trigonometrics. Here it is assumed that γ is a complex number $\gamma = (\alpha + j\beta)$

$$\cosh(\gamma) = (e^\gamma + e^{-\gamma})/2 \quad (1.13)$$

$$\sinh(\gamma) = (e^\gamma - e^{-\gamma})/2. \quad (1.14)$$

For calculators that can evaluate cosh and sinh with real but not complex arguments, the following equations are useful

$$\cosh(\alpha + j\beta) = \cosh(\alpha)\cos(\beta) + j\sinh(\alpha)\sin(\beta) \quad (1.15)$$

$$\sinh(\alpha + j\beta) = \sinh(\alpha)\cos(\beta) + j\cosh(\alpha)\sin(\beta). \quad (1.16)$$

CALCULUS, DIFFERENTIAL EQUATIONS, AND INTEGRALS

Some understanding of **calculus** (particularly differential equations and integrals) is helpful for the exam, although you will likely find that most problems are solvable by algebraic techniques. It is wise to be able to take the differential or integral of a simple function; review these operations in a calculus book or mathematical handbook. It may be helpful to remember that **differentiation** refers to rate of change, and **integration** is equivalent to finding the area under the graph of a function. In like manner, differentiation is anti-integration, and integration is anti-differentiation.

Here are a few examples of differentiation with respect to a variable x that may be familiar (where a is a constant)

$$da/dx = 0 \quad (1.17)$$

$$d/dx\,(x^2) = 2x\,dx/dx = 2x \quad (1.18)$$

$$d/dx\,(x^3) = 3x^2\,dx/dx = 3x^2 \quad (1.19)$$

$$d/dx\,(\sin\theta) = d\theta/dx\,\cos\theta \quad (1.20)$$

$$d/dx\,(\cos\theta) = -d\theta/dx\,\sin\theta. \quad (1.21)$$

A second or higher derivative involves repeating the differentiation; thus, for example,

$$d^2(x^2)/dx^2 = d(2x)/dx = 2. \quad (1.22)$$

Integrals are sometimes more difficult to evaluate than derivatives, because they require evaluation of limits. A few simple cases are shown below, where x is the variable of integration and a is a constant. Because these are indefinite integrals, there is a constant of integration (not shown) that is to be added to the answer.

$$\int a\,dx = ax \quad (1.23)$$

$$\int dx/x = \log x \quad (1.24)$$

$$\int x\,dx = x^2/2 \quad (1.25)$$

$$\int x^2\,dx = x^3/3 \quad (1.26)$$

$$\int \sin x\,dx = -\cos x \quad (1.27)$$

$$\int \cos x\,dx = \sin x. \quad (1.28)$$

COMPLEX NUMBERS

The use of complex numbers simplifies the solution of many alternating-current problems and is an important part of steady-state sinusoidal ac analysis. A complex number is often used to represent a **vector**, which is a quantity having both magnitude and direction.

Complex numbers arise from a presentation of numerical quantities in a *complex plane* (see Figure 1.2.) in which the horizontal axis represents the "real" part of the number, and the vertical axis represents the "imaginary" part. We define the *imaginary operator j* alternately as the square root of negative one

$$j = \sqrt{-1} \tag{1.29}$$

or as a vector of unity magnitude at an angle of 90° advanced (counterclockwise) from the reference (horizontal) axis

$$j = e^{j90°} \tag{1.30}$$

where e (= 2.718) is the base of the natural logarithm ln.

Note that these two definitions of j are equivalent, since

$$j^2 = -1 = e^{j\,180°}. \tag{1.31}$$

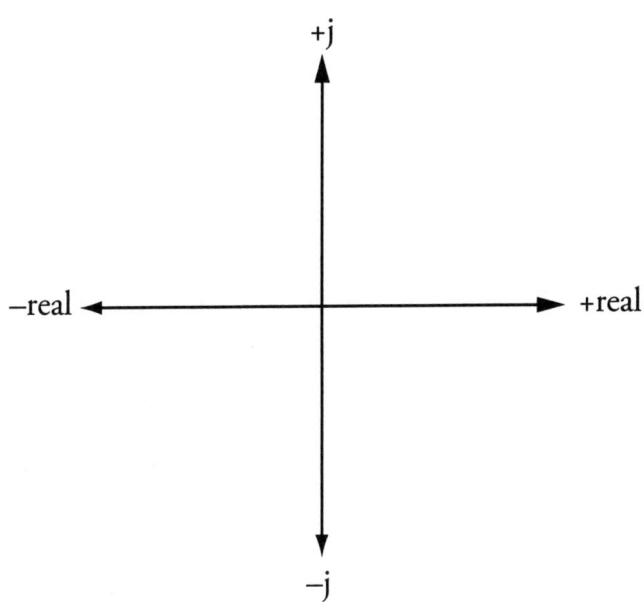

Figure 1.2 The complex plane

The complex quantity **A** (note the boldface, which is often used to denote a vector or complex number) can be defined in rectangular form as

$$\mathbf{A} = a + jb \tag{1.32}$$

or in polar form

$$\mathbf{A} = r\,e^{j\theta} \tag{1.33}$$

where
  r is the magnitude of the vector
  θ is the angle of the vector measured from the reference (real) axis.

The *magnitude* of A is then related to its rectangular components a and b as

$$|\mathbf{A}|^2 = a^2 + b^2 \tag{1.34}$$

and its angle is

$$\theta = \tan^{-1}(b/a). \tag{1.35}$$

A frequently used shorthand way of expressing the vector $\mathbf{A} = r\,e^{j\theta}$

is to write $\mathbf{A} = r\angle\theta$ (read "r angle theta").

Furthermore, we can express the complex number **A** in the *trigonometric form*

$$\mathbf{A} = r(\cos\theta + j\sin\theta) \tag{1.36}$$

where, of course, the term $(\cos\theta + j\sin\theta)$ gives the angle information of the vector. (This is a consequence of what is called *Euler's Theorem*.) It must therefore be true that

$$|(\cos\theta + j\sin\theta)| = 1 + j0 \tag{1.37}$$

which is a useful and important relationship.

Usually, power engineers use a Greek letter to represent an angle, most commonly φ (phi), θ (theta), δ (delta), or sometimes α (alpha), β (beta), or γ (gamma). Often, though, α and β (or a and b) are used to represent the real and imaginary parts of a complex number as we did above.

Alternating-current circuit analysis depends heavily on the use of complex numbers, and thus they will be used extensively in Chapter 4 and, in fact, throughout much of the remainder of this book. For the exam, it is helpful to have a modern calculator that handles complex numbers; you need to be able to use these functions with complete familiarity.

MATRICES AND SIMULTANEOUS LINEAR EQUATIONS

Electric circuit problems are typically amenable to matrix formulation and solution. This is very evident when we consider *linear* circuit problems involving loop, mesh, and nodal solution methods, in which the impedances or admittances are considered constants. Often, we can write a coefficient matrix of impedances or admittances directly from the network diagram (or network data) by inspection. Digital computer (or even calculator) solution is then straightforward if the system is not too large.

For a linear circuit containing n meshes (or loops) or n independent nodes, there will typically be a system of n equations with n unknowns. We must solve for the n unknowns, somehow. A problem on the PE exam might require the solution of such a set of simultaneous equations (probably for no larger than n = 3). Alternatively, the problem might ask you to recognize the correct formulation of the set of equations and not ask you to actually obtain a numerical solution.

For review purposes, we will illustrate a system of three simultaneous equations in three variables, which leads to a 3 × 3 (read "three by three") **coefficient matrix**, as well as a **vector of unknowns** and a **vector of knowns**. Although you don't need to know this in order to understand the math, we are using an impedance-form of solution as occasioned by mesh analysis. Chapter 3, "Basic Concepts of DC and AC Circuit Analysis," discusses mesh analysis and provides an example of how to formulate a set of simultaneous equations from an electric circuit model.

Consider the set of three simultaneous, linear equations, which might represent a circuit problem

$$V_1 = Z_A I_1 + Z_B I_2 + Z_C I_3$$
$$V_2 = Z_D I_1 + Z_E I_2 + Z_F I_3$$
$$V_3 = Z_G I_1 + Z_H I_2 + Z_I I_3.$$

Assume that the Z values (the impedances) and the Vs (the voltages) are all known, and that the Is (the currents) are to be calculated.

This set of equations can be placed in *matrix form* as

$$\begin{bmatrix} V_1 \\ V_2 \\ V_3 \end{bmatrix} = \begin{bmatrix} Z_A & Z_B & Z_C \\ Z_D & Z_E & Z_F \\ Z_G & Z_H & Z_I \end{bmatrix} \begin{bmatrix} I_1 \\ I_2 \\ I_3 \end{bmatrix} \quad (1.38)$$

You should study the relation between the matrix form of the equations and the equivalent set of individual, linear equations. The procedure is straightforward and is the same for a smaller number (two), or a larger number of equations.

We often denote a matrix by square brackets and a vector (that is, a matrix with a single row or column) by curly brackets. Using that notation, Equation (1.38) becomes

$$\{V\} = [Z] \{I\}. \quad (1.39)$$

which is the way we often write it by hand. In printed materials, where boldface is available, the equation is often denoted simply with boldface letters

$$\mathbf{V} = \mathbf{Z}\,\mathbf{I}. \quad (1.40)$$

Note that we use the word **vector** to designate a $1 \times n$, or $n \times 1$ (in this case, 3×1) matrix. We also introduced that term earlier to designate a single complex number with real and imaginary parts (or, equivalently, magnitude and angle). In matrix work, a vector may actually be an *array of complex numbers*. For instance, the voltage vector **V** in Equation (1.40) might represent a vector array of three complex-number voltages. Our multiple use of the term, while confusing, adheres to standard practice. The context should make clear whether "vector" means a single complex number, or an array of real or possibly complex values.

How do we solve for the unknown vector **I** in Equation (1.40)? One way is to use the concept of the **inverse** of a matrix. Rearranging the equation, we can write

$$\mathbf{I} = \mathbf{Z}^{-1}\mathbf{V} \quad (1.41)$$

where $\mathbf{Z}^{-1}$ is the *inverse* of **Z**.

Note that

$$\mathbf{Z}^{-1}\mathbf{Z} = \mathbf{Z}\,\mathbf{Z}^{-1} = \mathbf{U}$$

where **U** is the **unity** or **identity matrix**, a matrix having unity elements along its diagonal and zeros elsewhere.

We will not discuss how to evaluate the inverse here; this is covered in standard math books and many calculators will do it for you. However, we will next show how to solve by hand for the numerical values of the unknown vector using the **determinant**.

Matrix Solution by Determinants

The concept of the determinant provides a straightforward, and thus relatively fool-proof and easy, way of solving for the unknowns in a linear matrix formulation, such as (1.38). Small matrix problems ($n \leq 3$) were routinely solved by determinants in the days before sophisticated calculators. Your calculator might perform the matrix solution for you (including matrix inversion), but it is worth knowing how to obtain answers using the determinant. Following are the determinant solutions to second-order ($n = 2$) and third-order ($n = 3$) sets of equations.

DETERMINANTS

In solving circuit networks, node and loop equations are often written. This results in sets of n equations with n unknowns that are amenable to this kind of solution procedure when the equations are properly ordered. The following are the determinant solutions to second-order ($n = 2$) and third-order ($n = 3$) sets, written here in the notation of a mesh-analysis problem.

Second Order

Circuit equations:

$$V_1 = Z_A I_1 + Z_B I_2$$
$$V_2 = Z_C I_1 + Z_D I_2 \qquad (1.43)$$

Matrix form:

$$\begin{bmatrix} V_1 \\ V_2 \end{bmatrix} = \begin{bmatrix} Z_A & Z_B \\ Z_C & Z_D \end{bmatrix} \begin{bmatrix} I_1 \\ I_2 \end{bmatrix} \qquad (1.44)$$

Determinant solution:

$$I_1 = \frac{\begin{vmatrix} V_1 & Z_B \\ V_2 & Z_D \end{vmatrix}}{\begin{vmatrix} Z_A & Z_B \\ Z_C & Z_D \end{vmatrix}} = \frac{V_1 Z_D - V_2 Z_B}{Z_A Z_D - Z_B Z_C} = \frac{V_1 Z_D - V_2 Z_B}{D}$$

$$I_2 = \frac{\begin{vmatrix} Z_A & V_1 \\ Z_C & V_2 \end{vmatrix}}{D} = \frac{V_2 Z_A - V_1 Z_C}{D} \qquad (1.45)$$

Third Order

Circuit equations:

$$V_1 = Z_A I_1 + Z_B I_2 + Z_C I_3$$
$$V_2 = Z_D I_1 + Z_E I_2 + Z_F I_3 \quad (1.46)$$
$$V_3 = Z_G I_1 + Z_H I_2 + Z_I I_3$$

Matrix form:

$$\begin{bmatrix} V_1 \\ V_2 \\ V \end{bmatrix} = \begin{bmatrix} Z_A & Z_B & Z_C \\ Z_D & Z_E & Z_F \\ Z_G & Z_H & Z_I \end{bmatrix} \begin{bmatrix} I_1 \\ I_2 \\ I_3 \end{bmatrix} \quad (1.47)$$

Determinant solution:

$$I_1 = \frac{\begin{vmatrix} V_1 & Z_B & Z_C \\ V_2 & Z_E & Z_F \\ V_3 & Z_H & Z_I \end{vmatrix}}{D} \quad (1.48)$$

where

$$D = \begin{vmatrix} Z_A & Z_B & Z_C \\ Z_D & Z_E & Z_F \\ Z_G & Z_H & Z_I \end{vmatrix} \quad (1.49)$$

$$= Z_A Z_E Z_1 + Z_B Z_F Z_G + Z_C Z_D Z_H - Z_G Z_E Z_C - Z_H Z_F Z_A - Z_1 Z_D Z_B.$$

Therefore,

$$I_1 = \frac{V_1 Z_E Z_1 + V_3 Z_B Z_F + V_2 Z_H Z_C - V_1 Z_H Z_F - V_2 Z_B Z_1 - V_3 Z_E Z_C}{D}.$$

Similarly,

$$I_2 = \frac{\begin{vmatrix} Z_A & V_1 & Z_C \\ Z_D & V_2 & Z_F \\ Z_G & V_3 & Z_I \end{vmatrix}}{D}, \quad (1.50)$$

$$= \frac{V_1 Z_F Z_G + V_2 Z_A Z_1 + V_3 Z_D Z_C - V_1 Z_D Z_1 - V_2 Z_C Z_G - V_3 Z_F Z_A}{D}$$

$$I_3 = \frac{\begin{vmatrix} Z_A & Z_B & V_1 \\ Z_D & Z_E & V_2 \\ Z_G & Z_H & V_3 \end{vmatrix}}{D} \quad (1.51)$$

$$= \frac{V_1 Z_D Z_H + V_2 Z_B Z_G + V_3 Z_A Z_E - V_1 Z_G Z_E - V_2 Z_H Z_A - V_3 Z_B Z_D}{D}$$

CHAPTER 2

Electric and Magnetic Fields

OUTLINE

RELATION OF ELECTROMAGNETICS TO POWER ENGINEERING 12

ELECTRIC FIELDS 12
Scalar Potential as a Line Integral ■ Charge Density
■ Gradient ■ Divergence ■ Curl

MAGNETIC FIELDS 15

DYNAMIC FIELDS 16

RECOMMENDED REFERENCES 17

This chapter presents some background in **electromagnetics**, which as the name implies, refers to underlying fundamentals of electric and magnetic fields. The general topic is sometimes known as **field theory**.

It is not likely that you will see problems on the PE electric power exam requiring setup or direct solution of electromagnetics equations, such as those that follow in this chapter. Therefore, you may wish to move quickly through this chapter (at least in your initial study) and concentrate on the next several chapters, which are much more directly concerned with what most exam problems will involve. It is worthwhile, though, to return to this chapter after working through much of the rest of this book, because electromagnetic concepts underlie most of electrical engineering, and are of particular importance to several electric power devices. Besides, you might be asked a question concerning how a typical device (such as a transformer) works, in terms of its magnetic field. For example, you might need to know some fundamental magnetic circuit concepts, such as Ampere's circuital law and Faraday's law. Chapter 7, "Transformers," covers them briefly. Some electromagnetic theory is introduced in other relevant sections for simple geometries and situations where particular devices are discussed. The discussion in this chapter is more general, inclusive, spatial, and admittedly mathematical.

This chapter assumes that you are already somewhat familiar with introductory level (that is, static or non–time-dependent) electromagnetic concepts, such as presented in a first electromagnetics class and with some simple time dependencies. However, some background is offered here before presenting dynamic (time-dependent) electromagnetic relationships.

We will also assume a familiarity with mathematical concepts, such as gradient, divergence, cross products, and curl, which are useful in both static and dynamic fields. Most introductory electromagnetics texts discuss these. On the PE power exam, you will probably not have to solve vector problems using these concepts directly. Note, however, that even though these topics are mathematically complicated, especially when dealing with vector quantities, simple problems can often be solved by their direct application.

RELATION OF ELECTROMAGNETICS TO POWER ENGINEERING

Electromagnetics is of fundamental importance to the study of transformers, transmission lines, and rotating machines, and in fact, underlies all electrical circuit and device behavior, as well. Electromagnetics in general describes the continuum, whereas circuit theory typically confines itself to electric circuits consisting of lumped-parameter components, such as sources, resistors, capacitors, and inductors interconnected by ideal conductors (such as wires). In fact, *circuit theory can be considered a subset of electromagnetic field theory.*

In analyzing electrical devices, we suggest that, where possible, you build circuit models rather than work with fields. Circuit models are surprisingly good at replicating and predicting device behavior in a straightforward, understandable sense. In contrast, analysis of fields is usually more complicated in a mathematical and modeling sense, and for all but the simplest geometries, usually requires finite-method, boundary-value, and other complex numerical techniques.

Field theory explains much about the operation of transformers (concerned mostly with magnetic fields), capacitors (concerned with electric fields), transmission lines (where both electric and magnetic fields are of importance) and rotating machines (where the concept of the revolving magnetic field is critical). Furthermore, transmission line calculations are based rather directly on field theory. Even though we usually try to go directly to circuit models for all of these devices, the exam might include a problem involving the relevance of fields concepts to one of these devices.

ELECTRIC FIELDS

In reviewing this topic, we will look at scalar potential, charge density, gradient, divergence, and curl.

Scalar Potential as a Line Integral

The difference in potential between two points in a static electric field due to a single charge source is the same whether the path is the shortest distance between the points or any other path; thus, this potential is single-valued. If the path between two points a and b is located in a uniform electric field of electric field intensity **E** (rather than simply being near a point charge), and if a test charge is moved from point a to point b against the field, the potential difference is

$$V_{ba} = (x_b - x_a)E \cos \theta$$

where θ is the angle between the uniform field direction and the path direction, and x_a, x_b are the coordinates of the points. For an infinitesimal distance dl, the relationship is

$$dV = -E \cos \theta \, dl$$

and the potential difference is

$$V_{ba} = \int_a^b dv = v_b - v_a = -\int_a^b E \cos\theta\, dl = -\int_a^b \mathbf{E} \cdot d\mathbf{l}. \quad (2.1a)$$

It is easily shown that for a point charge, or any number of charges contained in a closed path where the path is in a static electric field, the line integral is given as

$$V = -\int_a^a \mathbf{E} \cdot dl = -\oint \mathbf{E} \cdot dl = 0. \quad (2.1b)$$

Here, the $\oint$ (closed-path integral) symbol implies that one starts and ends at the same point. And, as usual, the direction of $\mathbf{E}$ is the direction that a positive charge would take if released. Also, recall that in a uniform field, the greater the potential difference between the two equipotential lines or surfaces, the greater the field.

Another relationship worth recalling is that of electric flux density $\mathbf{D}$

$$\mathbf{D} = \varepsilon \mathbf{E} = \varepsilon_0 \varepsilon_r \mathbf{E}$$

$$\text{however, } \mathbf{E} = \mathbf{a}_r \left[\frac{Q}{4\pi\varepsilon r^2}\right]$$

$$\text{therefore, } \mathbf{D} = \mathbf{a}_r \left[\frac{Q}{4\pi r^2}\right],$$

where ε_0 is the electric permittivity of free space, ε_r is relative permittivity, Q is a point charge, r is the distance from the point charge, and $\mathbf{a}_r$ is a unit vector in the outward radial direction.

Charge Density

The **charge density** ρ is usually expressed as the ratio of the total charge within a volume to the volume. Similar expressions exist for surface density ρ_s, and line density ρ_L.

This charge density may also be thought of as a charge density *at a point*. Assuming a homogeneous charge throughout a region, consider an amount of charge ΔQ within a volume element Δv. The volume charge density $\rho = \Delta Q/\Delta v$ is the limit as $\Delta v \to 0$.

The potentials for the three kinds of charge densities mentioned above are given as follows. (Please be aware that many texts do not show the double or triple integration sign; instead, they indicate a subscript on the integral, such as S for surface area and V for volume.)

$$V_L = \frac{1}{4\pi\varepsilon} \int \frac{\rho_L}{r} dl = \text{potential due to a linear charge density} \quad (2.2a)$$

$$V_S = \frac{1}{4\pi\varepsilon} \iint \frac{\rho_S}{r} ds = \text{potential due to a surface charge density} \quad (2.2b)$$

$$V_V = \frac{1}{4\pi\varepsilon} \iiint \frac{\rho_V}{r} dv = \text{potential due to a volume charge density.} \quad (2.2c)$$

If these various charge densities are present near a point P, then the total potential at point P is the algebraic sum of the potentials

$$V = V_L + V_S + V_V. \quad (2.2d)$$

Gradient

As previously stated, when equipotential lines or surfaces become closer together, the field becomes stronger; one defines the **gradient** of the potential at a point as the potential rise ΔV along a length Δl (in the direction of the field) divided by Δl. It is easily shown that the gradient of electric potential V (a scalar) is a vector quantity that is the negative of $\mathbf{E}$

$$\text{grad } V = \nabla V = -\mathbf{E}. \quad (2.3a)$$

It will be of interest to define the vector that represents the magnitude and direction for the maximum rate of increase of a scalar (maximum rate will be a normal vector $\mathbf{a}_n$) as the gradient of the scalar; here, the maximum rate will be the shortest distance between two equipotential lines or surfaces

$$\text{grad } V = \nabla V = \mathbf{a}_n (dV/dn). \quad (2.3b)$$

The gradient of V may also be expressed in Cartesian coordinates as

$$\text{grad } V = \nabla V = \left(\mathbf{a}_x \frac{\partial_v}{\partial_x} + \mathbf{a}_y \frac{\partial_v}{\partial_y} \mathbf{a}_z \frac{\partial_v}{\partial_z} \right) \quad (2.3c)$$

where the $\mathbf{a}_x$, $\mathbf{a}_y$, and $\mathbf{a}_z$ represent unit vectors along the respective axes.

Divergence

To understand divergence, it will help to recall the definition of flux lines: lines representing the directions and magnitudes of vector fields. For a two- or three-dimensional field, the number of lines (designated by the density of lines—or sometimes by the length of lines—passing through a unit surface normal to the vector) is a measure of the vector field strength. Consider a specific volume with a closed surface; if inside the volume there is a positive source, than a net positive divergence exists. The divergence is a measure of the strength of the enclosed source.

The divergence for any vector field (for example, a field A) at a particular point is the net outward flux per unit volume as the volume about this point tends toward zero or

$$\text{div } \mathbf{A} = \nabla \cdot \mathbf{A} = \lim_{\Delta v \to 0} \frac{\oint_S \mathbf{A} \cdot d\mathbf{s}}{\Delta v} \quad (2.4a)$$

The divergence operation here is read in words as "del dot A." As with Equation (2.3c), this relationship may be expressed in Cartesian coordinates as

$$\text{div } \mathbf{A} = \left(\frac{\partial A_x}{\partial x} + \frac{\partial A_y}{\partial y} + \frac{\partial A_z}{\partial z} \right) \quad (2.4b)$$

From Equation (2.4a) we see that the volume integral of the divergence of a vector field is the same as the total outgoing flux of the vector through the surface enclosing the volume, where the direction of *ds* is outward and perpendicular to the surface element. This relationship may be expressed as

$$\int_V \nabla \cdot \mathbf{A}\, dv = \oint_S \mathbf{A} \cdot d\mathbf{s} \quad \text{(divergence theorem).} \tag{2.4c}$$

Equation (2.4c) clearly shows a conversion between the volume integral of the divergence and that of a closed surface integral of the vector, suggesting the relationship between the "point form" and "integral form" expressions of field equations.

Curl

The divergence is a measure of the strength of the flow source (that is, a surface integral taken around an infinitesimal surface divided by the volume enclosed). There is another kind of operation similar to this called the **curl**, which is defined in terms of a line integral taken around an infinitesimal path divided by the area enclosed by this path. The curl of a vector field **A** is a vector quantity whose magnitude is the net (maximum) circulation of **A** per unit infinitesimal area; the direction is normal to the surface area when the area orientation is adjusted for maximum circulation. The curl of a function is described in words as "del cross **A**" and is given as

$$\text{curl } \mathbf{A} = \nabla \times \mathbf{A} = \lim_{x \to \infty} \frac{1}{\Delta s} \left[\mathbf{a}_n \int_C \mathbf{A} \cdot d\mathbf{l} \right]_{\max} \tag{2.5a}$$

and for non-time-varying electric fields we can see that

$$\nabla \times \mathbf{E} = 0. \tag{2.5b}$$

The curl of **A** (a vector result) is an expression of the vorticity of the field; it represents the maximum circulation (for a per-unit infinitesimal surface area) and is maximum when the area *ds* is oriented such that it is normal to the direction of curl **A**.

MAGNETIC FIELDS

For static magnetic fields there are no magnetic sources; it is thus clear that the magnetic flux lines always close on themselves. Recall that, for an electric field, the electrostatic force on the charge is proportional to the charge and to the electric field intensity

$$\mathbf{F}_e = q\mathbf{E}.$$

However, the same electrical charge, if in motion in a *magnetic* field, experiences a force $\mathbf{F}_m$ that is proportional to q and whose direction is normal to the velocity vector **u**. This magnetic force leads directly to the definition of magnetic flux density **B**, and is given as

$$\mathbf{F}_m = q\mathbf{u} \times \mathbf{B}, \tag{2.6a}$$

and thus the total force on the charge can be written as

$$\mathbf{F}_{total} = q\mathbf{E} + q\mathbf{u} \times \mathbf{B} = q(\mathbf{E} + \mathbf{u} \times \mathbf{B}) \text{ (Lorentz Force Law)}. \quad \textbf{(2.6b)}$$

Note that a charge has to be in *motion* to experience a force due a *magnetic* field.

For steady magnetic fields in free space, equations (2.6a) and (2.6b) lead to the following postulates

$$\nabla \cdot \mathbf{B} = 0 \quad \textbf{(2.7a)}$$

$$\nabla \times \mathbf{B} = \mu_0 \mathbf{J} \quad \textbf{(2.7b)}$$

$$\nabla \times \mathbf{H} = \mathbf{J} \quad \textbf{(2.7c)}$$

where μ_0 is the **magnetic permeability** of free space.

The integral forms of these equations (again in free space) may be given as

$$\oint_S \mathbf{B} \cdot ds = 0 \text{ (integral form of the law of conservation of magnetic flux)} \quad \textbf{(2.7d)}$$

$$\oint_C \mathbf{B} \cdot dl = \mu_0 I \quad \text{(Ampere's circuital law)}. \quad \textbf{(2.7e)}$$

For Equation (2.7e), one may visualize a circular path around a single current-carrying conductor; the line integral around this closed path is equal to $\mu_0 \mathbf{I}$. The variable **B** is referred to as the **magnetic flux density**, and another quantity, the **magnetic field intensity H**, is defined as being proportional to **B** and in the same direction

$$\mathbf{H} = (1/\mu)\mathbf{B} \quad \textbf{(2.8a)}$$

and thus

$$\int_C \mathbf{H} \cdot d\mathbf{l} = N\mathbf{I} \text{ (generalized form of Ampere's circuital law)}, \quad \textbf{(2.8b)}$$

where N is the number of turns of the conductor and I is the current enclosed by these turns. The units of each side of Equation (2.8b) are frequently referred to as AT (amp-turns).

DYNAMIC FIELDS

For the static electric fields, the electrical quantities **E** and **D** are not related to the static magnetic quantities **B** and **H**, but there is a definite relationship if the fields are changing. For example, when working with static fields, the curl of **E** is zero; see Equation (2.5b). Early in the nineteenth century, however, Michael Faraday discovered that a current was induced in a loop (due to an induced voltage) when the enclosed magnetic field was changing. This discovery led to a number of postulates for time-changing fields including Maxwell's equations:

$$\text{curl } \mathbf{E} = \nabla \times \mathbf{E} = -\frac{\partial \mathbf{B}}{\partial t} \text{ (differential form of the Faraday-Maxwell law)} \quad \textbf{(2.9a)}$$

$$\text{curl } \mathbf{H} = \nabla \times \mathbf{H} = J + \frac{\partial D}{\partial t} = \sigma E + \varepsilon \frac{\partial E}{\partial t} = \sigma E + \varepsilon_0 \varepsilon_r \frac{\partial E}{\partial t} \quad \textbf{(2.9b)}$$

$$\text{div } \mathbf{D} = \nabla \cdot \mathbf{D} = \rho \quad \textbf{(2.9c)}$$

$$\text{div } \mathbf{B} = \nabla \cdot \mathbf{B} = 0. \quad \textbf{(2.9d)}$$

Compare these equations to those of Equations (2.7a) through (2.7e). Their integral forms (again, valid for time-varying as well as static fields) are

$$\oint \mathbf{E} \cdot d\mathbf{l} = -\int_S \frac{\partial \mathbf{B}}{\partial t} \cdot ds = -\iint \frac{\partial \mathbf{B}}{\partial t} \cdot ds \text{ (Faraday's law of electromagnetic induction)} \quad \textbf{(2.10a)}$$

$$\oint \mathbf{H} \cdot d\mathbf{l} = I_c + \int_S \frac{\partial \mathbf{D}}{\partial t} \cdot ds = I + \iint \frac{\partial \mathbf{D}}{\partial t} \cdot ds \quad \textbf{(2.10b)}$$

(where I_c = conduction current)

$$\oint_{SV} \mathbf{D} \cdot ds = Q_f = \int_V \rho dv = \iiint \rho dv \quad \textbf{(2.10c)}$$

(Q_f = free charges)

$$\oint_S \mathbf{B} \cdot ds = \iint \mathbf{B} \cdot ds = 0. \quad \textbf{(2.10d)}$$

Note, for example, that equation (2.10a) does not equal zero in the presence of a time-varying magnetic field, although for non-time-varying fields it does, as shown in Equation (2.1b).

RECOMMENDED REFERENCES

Cheng. *Field and Wave Electromagnetics*. 2nd ed. (1989), 4th printing, Addison-Wesley, 1991.

Claycomb. *Applied Electromagnetics*. Jones & Bartlett, 2010.

Wentworth. *Fundamentals of Electromagnetics with Engineering Applications*. John Wiley, 2005.

CHAPTER 3

Basic Circuit Concepts and DC Circuit Analysis

OUTLINE

NOTATION 20
Single- and Double-Subscript Notation for Voltages and Currents ■ Units

DIAGRAMS AND SYMBOLS 21

BASIC CIRCUIT LAWS 21
Ohm's Law ■ Kirchhoff's Laws

BASIC CIRCUIT-ANALYSIS TECHNIQUES 22
Mesh Analysis ■ Nodal Analysis ■ Thevenin's Theorem ■ Thevenin and Norton Sources

POWER 30

DEPENDENT SOURCES AND MAXIMUM POWER TRANSFER 30

DELTA-WYE TRANSFORMATION 33

SUMMARY 34

RECOMMENDED REFERENCES 35

You probably have a good understanding of basic electrical theory, as well as electric circuit models and their solutions; most electrical engineers use such knowledge on a regular basis. Power engineers become especially familiar with steady-state ac circuits, because they use such circuits to model many aspects of electric power systems. This chapter provides some review of basic electric circuit concepts, with a focus on dc circuits.

We begin with a review of notation. Ohm's law, the Kirchhoff laws, and loop and mesh analyses are covered briefly, including some solved examples. Mastery of these fundamental topics is critical to passing the exam, so be sure you understand the solution process for each example and practice solving exam-style questions, such as those in the companion text, *PE Power Electrical Engineering Sample Exam*.

NOTATION

We have tried to make the notation of this text consistent with that of the PE exam and with standard practice. However, different texts often use slightly different notations. By noting the context, it usually is easy to follow these differences; for example, the sometimes interchangeable use of *e* (for electromotive force) and *v* to denote voltage, or the use of script *F* or mmf to denote magnetomotive force. Similarly, in ac circuit work, we often find VA used to denote volt-amperes (in other words, *apparent power*) instead of the more common *S* (we will see in the next chapter that VA is the abbreviation of the *unit* of apparent power, the volt-ampere).

Recognizing that you probably will review and bring to the exam several texts with somewhat different notations, we have, in a few cases, presented or cited, different but usual notations. Hopefully, these explanations will help you to approach the material without confusion over notation. As with other concerns relative to the PE examination, an application of common sense will often make the meaning clear.

Probably the most usual engineering practice is to denote direct-current (dc) quantities with capital letters, and time-varying quantities with lower-case letters (or functional time-domain notation, as in v(t)). Most texts express phasors or other complex numbers in **boldface** (refer to Chapter 4), while in hand-writing them, we usually use a bar over the capital letter, and a few publications do this also. Phasor magnitudes, which are usually root-mean-square (rms) values, are typically expressed in capital, non-boldface letters, or in *italics*.

PE exam publications seem to show a preference for using boldface for complex numbers and non-boldface regular or italic text for complex-number magnitudes and other real numbers. Because not all references are consistent in this regard, you should always examine the context. In this book, we will typically use lower-case letters for time functions, capital or capital italicized letters for rms magnitudes, and boldface capitals for phasor and other vector quantities. Note that, in this text as well as in exam problems, it is not always possible to achieve complete notational consistency between the text, equations imbedded in the text, and (in particular) illustrations.

Single- and Double-Subscript Notation for Voltages and Currents

Currents are sometimes designated by some unique subscript that tells the loop or element through which they pass. For example, I_a might be the current in loop a, the current through device a, or the current in phase a of a three-phase device. Note that the *direction* of current needs to be identified on a diagram to make it completely unambiguous. This is usually done with an arrow.

Alternatively, a double-subscript notation is often used; for example, I_{12} is the current leaving node 1 and heading toward node 2, while I_{ab} is the current in an element from phase a to phase b in a three-phase situation.

Similarly, voltages should be uniquely defined in terms of polarity. Single-subscript notation is often employed when a specific reference node for the problem is defined, as it usually is in power-systems work. By definition, a voltage noted with double subscripts i and j is related to the voltages at buses i and j as follows:

$$V_{ij} = V_i - V_j, \qquad (3.1)$$

where V_i and V_j are the voltages at nodes i and j with respect to reference or ground. This double-subscript notation can be used for time functions, such as v_{ij} or phasors, such as V_{ij}.

On a circuit diagram, it is important to place a polarity mark to identify each voltage; on scratch paper, this may help you avoid making a serious mistake in the exam. For example, we would usually place a positive (+) sign at buses i and j to specify the polarity of V_i and V_j, and thus of V_{ij}.

In some problems, ground is not the reference, so you should carefully note what the reference is. For example, in unbalanced three-phase ac problems solved by the method of symmetrical components (see Chapter 12), the reference buses for the three sequence networks are in some cases all at different potentials.

Identifying directions for currents and polarities for voltages is particularly important when different voltages and currents are related to each other in a problem, such as a dc circuit, an autotransformer connection, or a three-phase device.

Units

Units for the quantities we employ are usually abbreviated and typically obvious— V for volts, A for amperes, Ω for ohms, S for siemens, s for seconds, and so on. However, it is important to *be careful not to confuse the name of a quantity with its unit*. For example, you might sometimes see an expression such as

$$V = 5 \text{ V},$$

which is read "V equals 5 volts". This ambiguity could have been avoided here by naming the quantity with a subscript, for example

$$V_x = 5 \text{ V}.$$

Unfortunately, the former notation is occasionally encountered.

DIAGRAMS AND SYMBOLS

The symbols for circuit elements are interconnected to form circuit diagrams. Most of the symbols used in these diagrams will be familiar to you, having been used somewhat consistently in recent decades. Unfortunately, one symbol may be used to represent different things; for example, a circle with two collinear lines into it might mean a relay or contactor coil, an ac source, or perhaps a lamp or other device.

Many symbols are *pictorial*, and that is a great help; for example, a battery is denoted by parallel lines of two different lengths representing plates of two dissimilar materials (as in a lead-acid car battery). A capacitor is also designated by parallel lines representing parallel plates, although the lines are the same length. Sometimes, one of the lines in the capacitor symbol is curved, probably to help distinguish the device from a normally-open contact in a motor contactor or relay. Again, careful attention to the context should remove any ambiguity.

BASIC CIRCUIT LAWS

The equations we will use here are primarily *empirical* equations; that is, *equations deduced through observation and experiment and verified through use*.

Ohm's Law

Let's start with the basic building blocks of dc circuit analysis. Probably the most fundamental is Ohm's law. Ohm's law expresses the relation between electrical potential difference V (usually called voltage), current I (the symbol comes from the French term for *intensity*), and resistance R as

$$V = R\,I. \tag{3.2}$$

We usually assume R to be constant wherever possible (and especially not a function of V or I). This yields a *linear* problem which, of course, is usually much easier to mathematically describe and solve than a nonlinear one.

Kirchhoff's Laws

Kirchhoff's voltage law (KVL) states that the algebraic sum of the voltage drops around a closed conducting path is zero. Another way of stating this is that the sum of the voltage rises equals the sum of the voltage drops.

Kirchhoff's current law (KCL) states that the algebraic sum of the currents directed into (or out of) a node is zero. Another way of stating this is that the sum of the incoming currents equals the sum of the outgoing currents with respect to a node.

Actually, we need to qualify KVL by adding "…in the absence of time-changing magnetic fields" and KCL by adding "…in the absence of charge accumulation at the node." These issues can be resolved by adding terms as appropriate, such as by including "induced voltage" in a machine model separate from the IR voltage drops.

Ohm's and Kirchhoff's laws underlie the behavior of dc circuits. Additional considerations will be noted in the chapters dealing with ac circuits, necessitated by the fact that we will then be dealing with time-varying quantities.

BASIC CIRCUIT-ANALYSIS TECHNIQUES

This section focuses on analysis methods in the context of dc circuits to make a simpler presentation. However, they are equally applicable to the ac case. We will examine only basic forms, which are often special cases in their own right despite the fact that they have wide application. For example, we will consider mesh analysis rather than the more general loop analysis, noting that a mesh is a loop that doesn't contain other loops. For most problems you will encounter, these simplifications will be completely adequate.

Mesh Analysis

The circuit shown in Figure 3.1 contains a single dc source (battery) and several resistors. The resistances, and the voltage of the source, are known. Because this circuit has two "open windows" (as they are sometimes called), we can solve it by writing two mesh equations. In general, an n-window planar circuit can be solved by writing n mesh equations.

A mesh current is an artificial concept; currents of course actually flow through circuit elements or branches (here, the resistors and the source). Nonetheless, it is a very useful concept. Like most circuits we encounter, the circuit of Figure 3.1 is a planar network, meaning that it can be drawn without any line crossing another. Note that some planar networks can be drawn so as to appear to be non-planar.

Example 3.1

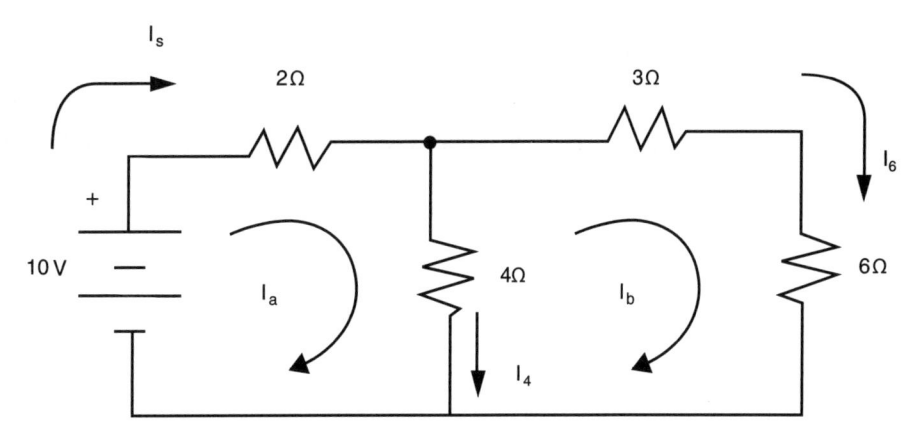

Figure 3.1 Example circuit solvable by two meshes

Solve the circuit of Figure 3.1 for I_6 and I_4, the currents through the 6Ω and 4Ω resistors respectively, by mesh analysis. Also solve for the source current I_s out of the positive terminal of the battery.

Solution
Two mesh currents (I_a and I_b) are first defined as shown in the figure.

For mesh a (the mesh through which mesh current I_a flows), we can write a voltage-drop expression around the mesh as

$$(2 + 4) I_a - 4 I_b = 10.$$

Similarly, for mesh b,

$$-4 I_a + (4 + 3 + 6) I_b = 0.$$

(The zero on the right side of this equation results since there is no voltage source in mesh b)

In standard form, these equations become

$$6 I_a - 4 I_b = 10$$
$$-4 I_a + 13 I_b = 0$$

or in matrix form

$$\begin{bmatrix} 6 & -4 \\ -4 & 13 \end{bmatrix} \begin{bmatrix} I_a \\ I_b \end{bmatrix} = \begin{bmatrix} 10 \\ 0 \end{bmatrix}.$$

Note that each diagonal term of the square resistance matrix is simply the sum of the resistances around that particular mesh, and the off-diagonal term is just the negative of the resistance traversed by the particular mesh currents in question (negative since the meshes are in opposite directions in the element). This will always be true for mesh analysis unless there is coupling between the elements, so we could actually write the matrix by inspection.

Solution for the mesh currents yields

$$I_a = 2.0968 \text{ A}$$
$$I_b = 0.6452 \text{ A}.$$

For this small 2-equation problem, you can simply solve the first equation for I_a, substitute this into the second to calculate I_b, and then return to the first equation to evaluate I_a. The procedure becomes more involved if more than two equations are present. (If you need review in solving for these unknowns, see the section "Matrices and Simultaneous Linear Equations" in Chapter 1.)

To finish the problem, note that mesh a is the only mesh passing through the battery, and mesh b is the only mesh passing through the 6Ω resistor. Thus

$$I_s = I_a = 2.0968 \text{ A}$$
$$I_6 = I_b = 0.6452 \text{ A}$$

while the current in the 4Ω resistor is

$$I_4 = I_a - I_b = 1.4516 \text{ A}.$$

Alternative Solutions

Actually, in this example, the current out of the battery could be more easily solved by finding the equivalent resistance "seen" by the battery, which in terms of the "product over sum" equation for two parallel resistances is

$$R_{eq} = 2 + [(3+6)(4) / ((3+6) + 4)] = 4.7692 \Omega.$$

From which

$$I_s = 10 / 4.7692 = 2.0968 \text{ A}$$

whereas, if we wanted only the current in the 6Ω resistor, it could be easily found by Thevenin's theorem, which we will discuss later in the chapter.

In the alternative solution of Example 3.1, suppose that there had been *three* or more resistances in parallel. Resistances in series simply add, but resistances in parallel can be found by adding the equivalent conductances (designated by the letter G). A conductance is just the reciprocal of the corresponding resistance. For three resistances in parallel, for example,

$$G_{eq} = G_1 + G_2 + G_3 \tag{3.3}$$

and the equivalent resistance of the parallel group is

$$R_{eq} = 1 / G_{eq}. \tag{3.4}$$

Chapter 4 provides more information about series and parallel connections.

Nodal Analysis

Most circuits involving electric power calculations involve nodal (or bus) analysis. In the nodal problem formulation, KCL equations are written explicitly at each independent node, while KVL expressions are accounted for implicitly. For a problem of n nodes (including reference, if identified as such), a total of $(n-1)$ equations must be solved simultaneously to find uniquely the bus voltage differences.

Because most nodal-frame power problems involve calculation of voltages with respect to a reference node, the problem formulation is made easier in such cases by writing a nodal equation at each node other than the reference. An example will illustrate.

Example 3.2

Write the nodal equations for the circuit of Figure 3.2, which contains two nodes (plus reference), two ideal current sources (shown as rectangles), and several resistors. Then, solve to obtain nodal voltages with respect to reference.

Solution

Nodal analysis is more straightforward when the sources are *current* sources (as shown in Figure 3.2) and when the resistances are expressed as *conductances*, where the conductance G corresponding to any resistance R is

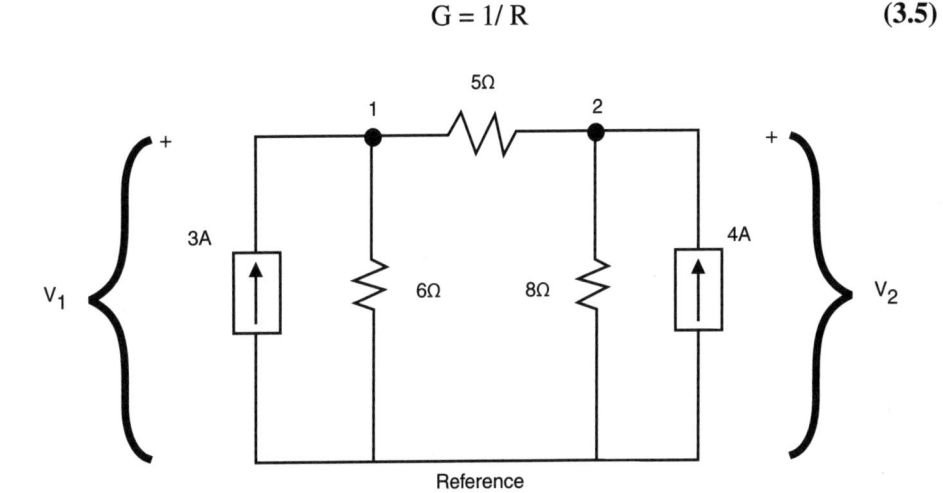

$$G = 1/R \qquad (3.5)$$

Figure 3.2 Example circuit solvable by two nodal equations

The unit of conductance is the siemen. (In the next chapter, in dealing with ac circuits, we will introduce *admittances* for the same task.) An older term for the unit of conductance is mho (ohm spelled backwards) and you may see this terminology on the exam.

At node 1 of Figure 3.2, the sum of the currents directed away from the node is

$$(1/6) V_1 + (1/5)(V_1 - V_2) = 3$$

and at node 2,

$$(1/8) V_2 + (1/5)(V_2 - V_1) = 4.$$

Collecting terms, we obtain the pair of equations,

$$0.3667 V_1 - 0.2 V_2 = 3$$
$$-0.2 V_1 + 0.3250 V_2 = 4$$

which can then be solved to find

$$V_1 = 22.42 \text{ V}$$
$$V_2 = 26.11 \text{ V}.$$

Because matrix formulation of such equations may be important on the exam, let us express these two equations in matrix form as

$$\begin{bmatrix} 0.3667 & -0.2 \\ -0.2 & 0.3250 \end{bmatrix} \begin{bmatrix} V_1 \\ V_2 \end{bmatrix} = \begin{bmatrix} 3 \\ 4 \end{bmatrix}$$

or

$$G_{node} V = I. \tag{3.6}$$

For which the solution is

$$\begin{bmatrix} V_1 \\ V_2 \end{bmatrix} = \begin{bmatrix} 22.42 \\ 26.10 \end{bmatrix} \text{ V}$$

As a check, you might want to obtain the inverse of the matrix G_{node}, and then find the solution by matrix multiplication:

$$\begin{bmatrix} V_1 \\ V_2 \end{bmatrix} = \begin{bmatrix} 4.1047 & 2.5260 \\ 2.5260 & 4.6314 \end{bmatrix} \begin{bmatrix} 3 \\ 4 \end{bmatrix}$$

You should obtain the same solution.

Note that the diagonal terms of the square conductance matrix in Example 3.2 can be written directly from the circuit by summing the conductances at the nodes, while the off-diagonal terms are just the conductances bridging those buses. Thus, it is easy, as with the mesh formulation discussed earlier, to write the coefficient matrix without actually writing the equations (or, equivalently, construct a code to let the computer do this). Recall that a similar observation was made for the resistance matrix for mesh analysis.

For example, the upper left diagonal term of the matrix

$$G_{11} = 1/5 + 1/6 = 0.3667.$$

One reason we use nodal equations to solve a circuit such as Figure 3.2 is that the node voltages (to reference) are the unknowns, while the known source quantities are current injections into the nodes. The ideal current sources are each in parallel with resistances (yielding Norton sources). which can each be converted to Thevinin sources, a conversion that can often be useful to simplify a problem. We will solve this same example using these ideas in the next section.

Thevenin's Theorem

Thevenin's theorem is useful in solving many kinds of electrical network problems. It often allows complex problems to be greatly simplified, thus facilitating their solution. It is especially useful when we are concerned only with what happens in a small part of a larger network, or in cases where we need to solve repeatedly for a voltage or current in terms of changes in one element. For example, we might want to know how the voltage at an electric outlet changes as we change the load connected to the outlet. Knowing the Thevenin model of the "upstream" network obviates the need for using a detailed upstream model for each calculation, and this makes the problem very simple.

Basically, Thevenin's theorem states that a network containing only linear resistances and independent or dependent sources may be replaced, at any given terminal pair, by an equivalent network containing only a single source in series with a single resistance. Figure 3.3 illustrates a general network and its Thevenin equivalent. The Thevenin voltage V_{th} is found as the open-circuit voltage between the pair of terminals in question. The Thevenin resistance R_{th} is the resistance that we see when looking back into the network at the terminal pair with all sources reduced to zero. That is, ideal independent voltage sources are replaced with short circuits and ideal independent current sources are replaced with open circuits (*dependent* sources, if present, must be left in the network). Example 3.3 illustrates the application of this theorem.

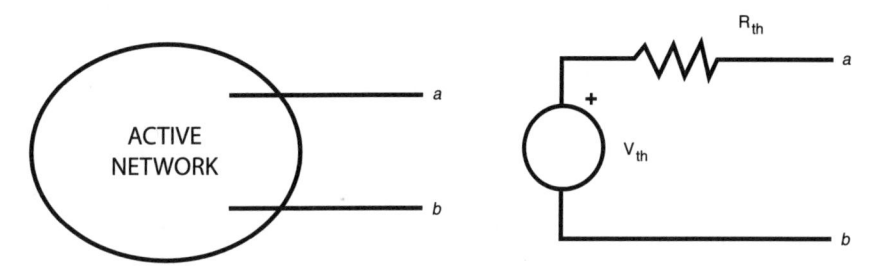

Figure 3.3 A general network and its Thevenin equivalent

Example 3.3

Consider the network shown in Figure 3.4. We wish to find the current I_x in the resistance R_x. This resistance will be considered to be external to what we are calling the network.

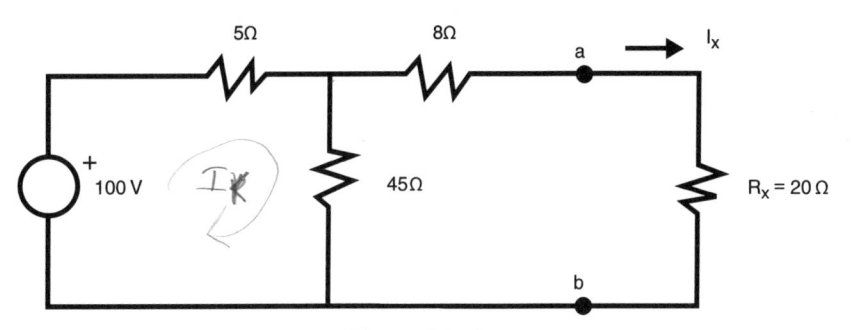

Figure 3.4 Example of use of Thevenin's theorem

Solution
We solve this problem using Thevenin's theorem if we know the Thevenin voltage V_{th}, and the Thevenin resistance R_{th}, seen when looking into the network at terminal pair ab. The Thevenin voltage is simply the voltage V_{ab} at the terminal pair ab without the element R_x connected, and the Thevenin resistance is the resistance seen when looking back into the network at terminal pair ab, again with the external resistance not present.

From basic circuit analysis, we find the Thevenin voltage and resistance to be

$$V_{th} = 90 \text{ V}$$
$$R_{th} = 12.5 \Omega.$$

Knowing these, it is then a simple matter to find I_x as

$$I_x = 90 / (12.5 + 20) = 2.7692 \text{ A}.$$

Thevenin and Norton Sources

In Example 3.3, the combination of the Thevenin voltage V_{th} in series with the Thevenin resistance R_{th} is known as the **Thevenin source**. In general, it is the simplest circuit with which we can replace a complex circuit when we desire to look back from a particular pair of terminals into the network. The only way the Thevenin model could be simpler is if either the Thevenin source, or the Thevenin resistance, were zero. The latter case would of course correspond to an **ideal source**, which is exactly what the source in the Thevenin model is.

The independent Thevenin equivalent can always be replaced with an ideal independent current source in parallel with a resistance. That parallel pair of elements is called a **Norton equivalent**. The resistance is usually treated as a conductance, known as the **Norton conductance** G_{eq}, while the source current in the model is known as the **Norton current** I_n. The Thevenin and Norton circuit models are shown in Figure 3.5. Using a **source transformation**, we can find the elements of the Norton (parallel) circuit that is equivalent to the Thevenin (series) circuit at its terminals, or vice versa. The parameters of the Norton circuit, for example, can be found from the parameters of the Thevenin circuit as

$$I_n = V_{th} / R_{th} = V_{th} G_{eq}. \tag{3.7}$$

This is shown for the two practical sources (the Thevenin source and the Norton source) in Figure 3.5, where the two circuits are equivalent at the terminal pair ab.

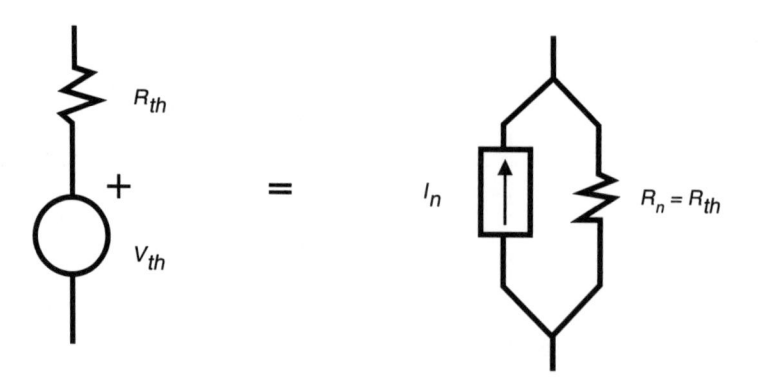

Figure 3.5 Equivalent Thevenin and Norton sources

Example 3.4

Let's use the idea of Thevenin and Norton sources to solve the circuit of Figure 3.2, which we solved previously by nodal analysis.

Solution

Converting what appears as a Norton source at node 1, and similarly, the one at node 2 to equivalent Thevenin sources, yields the simplified equivalent network of Figure 3.6, which contains only a single loop. From this equivalent, we can easily solve for the current I_5 in the loop (which is of course exactly the current through the 5Ω resistor) as

$$I_5 = (18 - 32) / (6 + 5 + 8) = -0.7368 \text{ A}$$

and so, for example,

$$V_a = 18 - 6 I_5 = 22.42 \text{ V}$$

which agrees with what was obtained by nodal analysis.

Note that the fact I_5 is negative simply means that it flows in a direction *opposite* to what we assumed in Figure 3.6.

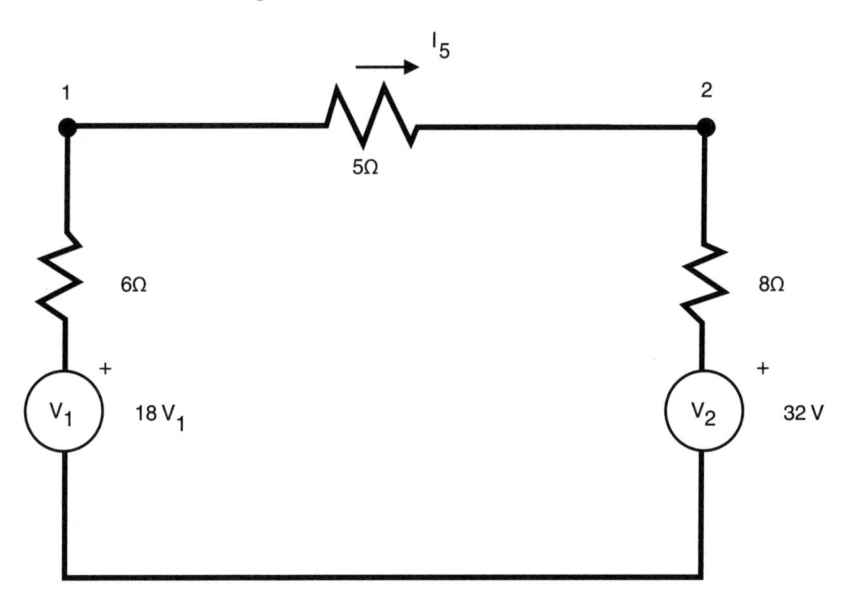

Figure 3.6. Circuit of Example 3.2 with Norton sources replaced with Thevenin sources

To summarize, Thevenin's theorem is valid regardless of the size of the network, and for complex impedances (in the ac case), in general, as well as for resistive elements as shown here. Knowing the Thevenin model at a given pair of nodes, we can immediately determine the effect of an external circuit upon that network at the specified terminal pair, or vice versa.

POWER

Power is the time rate of transfer of energy. For the dc case, power associated with a device is found as the product of the voltage across the device and the current through it

$$P = V I. \qquad (3.8)$$

This applies to sources as well as resistors. From this equation and Ohm's law, we could find the power dissipated in a resistor as

$$P = I^2 R \qquad (3.9)$$

or

$$P = V^2 / R. \qquad (3.10)$$

These equations are so familiar to electrical engineers that they need no further clarification. Experience has taught us, though, that it is easy to make a mistake using $P = V^2 / R$. You need to be careful that V is indeed the voltage across R and not some source or other voltage.

The calculation of power in ac circuits is more complex, and is addressed in the next chapter. Also, in that chapter (and in Chapter 11) we will examine the relation between the polarity of voltage, direction of current, and direction of power flow.

DEPENDENT SOURCES AND MAXIMUM POWER TRANSFER

All the sources we introduced so far are *independent* sources: theoretically, they provide their value of V or I to the circuit regardless of what else happens in the circuit. A class of sources, known as dependent sources, is sometimes used (especially for electronics and signal problems). A dependent source is one whose value is dependent upon some other signal in the network.

A typical circuit problem is to find the maximum power that can be delivered to a circuit element and the resistance of the element that results in this maximum power delivery. This can usually be solved by finding a Thevenin or Norton equivalent circuit such that an unknown load may be matched to the Thevenin impedance. *Power delivery to a load is maximized when the load impedance is equal to the complex conjugate of the Thevenin impedance upstream of it.*

For a dc circuit, this reduces to setting the load resistance equal to the Thevenin resistance. Example 3.5 illustrates these ideas.

Example 3.5

Use Thevenin's theorem to find the maximum power transferable to a resistance in the circuit of Figure 3.7, which contains one independent source and one dependent source (shown as a diamond, where the source current is dependent upon I_1).

Solution

In Figure 3.7a, we wish to find the value of R_L that will absorb maximum power, and the maximum power absorbed by that R_L. The circuit may easily be solved by finding the Thevenin equivalent for the portion of network to the left of the terminal pair x–x. However, if one or more *dependent* sources exist in the circuit (as here), some caution is needed when finding the short-circuit current (this is why we chose an example with a dependent source).

(a) Original circuit **(b)** Thevenin equivalent

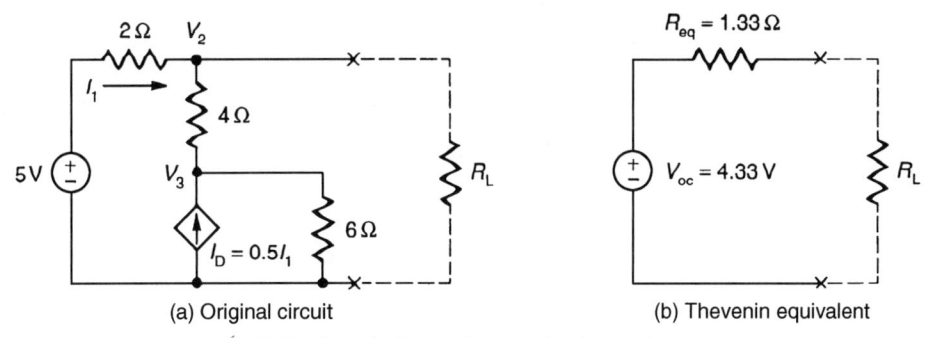

Figure 3.7 A circuit with both an independent and a dependent source

The Thevenin voltage (equal to the open-circuit voltage V_{oc}) is easily found. Let's solve for it by using Kirchhoff's current law (KCL) to sum the currents at each of the voltage nodes V_2 (equal to V_{oc}) and V_3:

$\sum I$'s at node 2:

$$\frac{5-V_2}{2} + \frac{V_3 - V_2}{4} = 0$$

$\sum I$'s at node 3:

$$\frac{V_3 - V_2}{4} + \frac{V_3}{6} - I_D = \frac{V_3 - V_2}{4} + \frac{V_3}{6} - 0.5\left(\frac{5-V_2}{2}\right) = 0.$$

Note that we can solve the second KCL equation above directly for $V_3 = 3$ V. (It should not surprise us that we can do this since, once the load resistance is removed, the circuit, in effect, has only one independent node). This value of V_3 can then be substituted into the first KCL equation to find $V_2 = 4.33$ V, which is, of course, the open circuit Thevenin voltage V_{oc}. (You might wish to check at this point by solving for $I_1 = (5-4.333)/2 = 0.333$ A, which can be combined with the dependent-source current to verify the values of V_2 and V_3).

To find the Thevenin resistance R_{eq}, you could first determine the short-circuit current at the Thevenin terminal pair x–x and find R_{eq} as the ratio of the Thevenin voltage to the short-circuit current. This is the long way, but we will do it here for practice. With the terminal pair x–x short-circuited, we have a different circuit: $V_2 = 0$, and so at voltage point V_3 we could sum the currents as

$$V_3/4 + V_3/6 - 0.5I_1 = 0$$

where, since $V_2 = 0$, I_1 is determined only by the 5V source and the 2Ω resistor: $I_1 = 5/2 = 2.5$ A. The 4Ω and 6Ω resistors are now in parallel, with an equivalent resistance of 2.4Ω (You may want to redraw the circuit at this point to reflect these observations.) The voltage V_3, which is now the voltage across this parallel resistance, must be

$$V_3 = (5/2)(0.5)(2.4) = 3.0 \text{ V}$$

because all of the current out of the dependent source must flow into this parallel resistance.

Here is where it is very easy to make an error. Even though the 4Ω *and* 6Ω resistors are in parallel for these calculations, we need to be careful to note that the current I_{sc} is just the sum of the current through the 4Ω and 6Ω resistors. The current directed upward in the 4Ω resistor is, by the *current divider* rule, 1.25(6)/(6 + 4) = 0.75 A.

Then, summing the currents at the point marked V_2, I_{sc} = 5/2 + 0.75 = 3.25 A

where I_{sc} is the short-circuit current out of the point marked V_2.

Thus, the Thevenin resistance is 4.33/3.25 = 1.33 2Ω, as shown in Figure 3.7b.

Because all independent sources have been temporarily removed from the circuit of our example (in order to find the Thevenin voltage), you must provisionally insert a fictitious voltage or current into the circuit as appropriate. This is done by applying the temporary voltage (in this case) at the terminals x–x (see Figure 3.8). A good value to choose, of course, is the already-calculated Thevenin voltage.

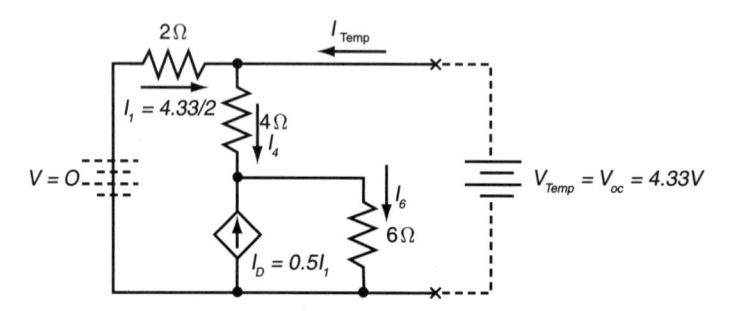

Figure 3.8 Independent source(s) removed; fictitious source inserted

Noting in Figure 3.8 that the original internal source voltage is still zero (shown dashed) when making these Thevenin calculations, we then solve for the current I_{Temp} out of the temporary voltage source as a result of this insertion.

The current I_1 may be found easily since *the inserted source is directly across the 2Ω resistance*: I_1 = –4.33/2 = –2.17 A. This yields a dependent current-source value of I_D = –1.08 A, which allows you to set up a nodal and loop relationship to calculate I_{Temp}. To do this note that, in terms of the currents marked on the diagram,

$$I_{Temp} + (-2.17) = I_4$$

$$I_4 + (-1.08) = I_6$$

and

$$4I_4 + 6I_6 = 4.33.$$

Solving yields

$$I_{Temp} = 3.25 \text{ A}$$

and, thus,

$$R_{eq} = V_{oc}/I_{Temp} = 4.33/3.25 = 1.33 \text{ 2Ω as before.}$$

Interestingly, we would find I_6 = 0 if we calculated it here.

Returning to the original question regarding maximum power transfer, we recall that this occurs when

$$R_L = R_{eq} = 1.33 \text{ 2Ω}.$$

Then the load current

$$I_L = 4.33/(1.33 + 1.33) = 1.6278 \text{ A}$$

and the power supplied to the load is

$$P_L = I_L^2 R_L = (1.6278)^2 \times 1.33 = 3.52 \text{ W}.$$

DELTA-WYE TRANSFORMATION

In various examples in this chapter, we briefly examined the series and parallel connections of resistances. Consider a circuit containing one source and a number of resistors. For such a circuit, no matter how large it is, a succession of series-parallel combining can be applied to reduce the resistors to one single resistance as seen by the source. There are some notable instances, however, where just knowing how to combine elements in series and parallel is not sufficient; additional techniques are needed to find the equivalent resistance. One such example is the bridge circuit (or Wheatstone bridge), a circuit widely used for making accurate measurements in instrumentation contexts.

A basic bridge circuit with five resistors and a battery is shown in Figure 3.9. It is evident that no resistance is in series or parallel with any other resistance in this circuit.

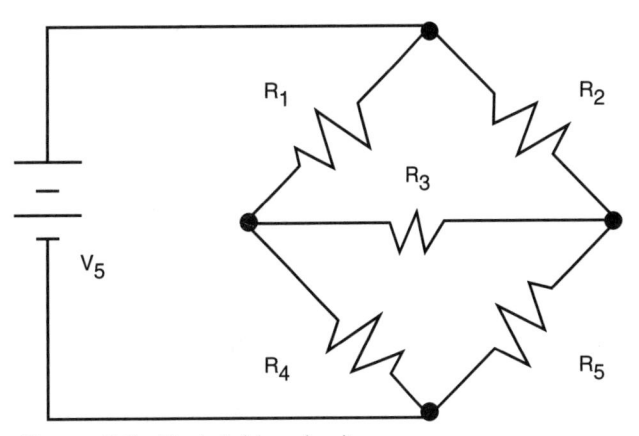

Figure 3.9 Basic bridge circuit

Any such problem should typically be solvable by a circuit solution technique like one of those we covered earlier, such as mesh or node analysis. If you encounter problems that don't fit nicely into the series-parallel category, you can always formulate loop or nodal equations to solve the circuit, or go back to simple branch circuit equations (introduced in elementary circuit theory books).

However, a useful simplification can be made in the case of the bridge circuit. Consider the two connections of three elements each as shown in Figure 3.10. The left one (which contains an internal node) is called a wye (or Y) and the right one (which contains a mesh) a delta (or Δ). The names were chosen long ago for obvious reasons.

Any delta can always be converted to a wye. Similarly, any wye can always be converted to a delta. The equations for making these conversions are given in Figure 3.10. Note that in Figure 3.10 the resistances of the wye are given in lower case, and those of the delta are in upper case, to avoid confusion.

Conversion in one direction or the other can often be used to simplify a complicated network, such as a bridge network, where series-parallel combination techniques do not apply. In the circuit of Figure 3.9, for example, the delta consisting of R_1, R_2, and R_3 can be converted to an equivalent wye. Replacing the delta by the equivalent wye leaves the new r_2 in series with R_4 and r_1 in series with R_5. (Sketch this equivalent to ensure that you see this simplification.) Series-parallel combinations can then be made quickly to find the equivalent resistance seen by the source.

The delta-wye method works for ac circuits in exactly the same way. A particularly important use of both delta and wye connections (of sources and loads) occurs in three-phase circuits, which are covered in Chapter 5. You will want to know how to find and use the equations given in Figure 3.10 readily, since problems using them have appeared on past exams.

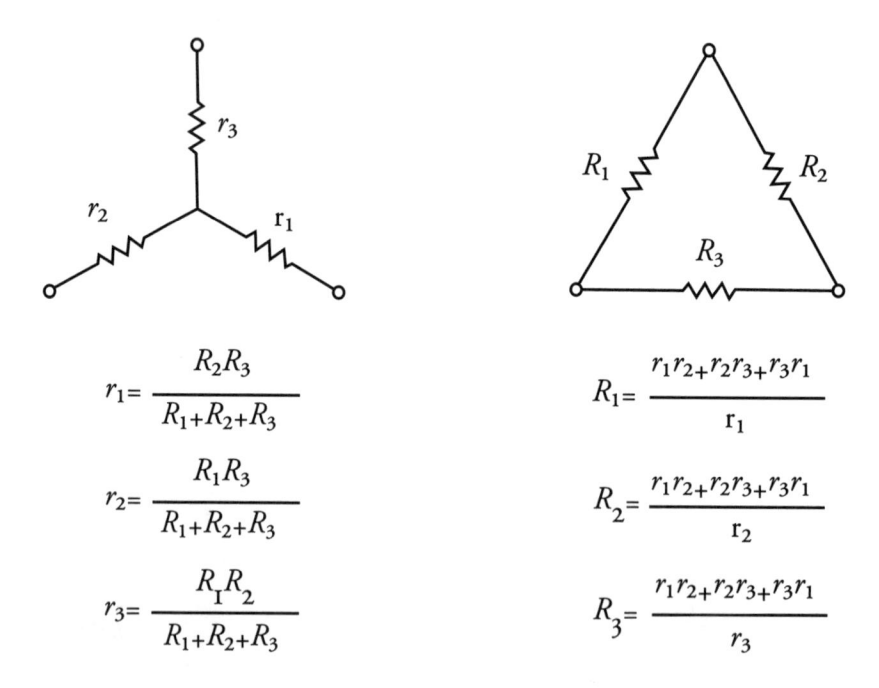

$$r_1 = \frac{R_2 R_3}{R_1 + R_2 + R_3}$$

$$r_2 = \frac{R_1 R_3}{R_1 + R_2 + R_3}$$

$$r_3 = \frac{R_1 R_2}{R_1 + R_2 + R_3}$$

$$R_1 = \frac{r_1 r_2 + r_2 r_3 + r_3 r_1}{r_1}$$

$$R_2 = \frac{r_1 r_2 + r_2 r_3 + r_3 r_1}{r_2}$$

$$R_3 = \frac{r_1 r_2 + r_2 r_3 + r_3 r_1}{r_3}$$

Figure 3.10 Wye-delta transformation for simplifying networks

The Wheatstone Bridge as a Measuring Instrument

As mentioned above, the Wheatstone bridge circuit of Figure 3.9 has been used extensively for making accurate measurements. One specific use has been for measuring resistances. We do not know if problems involving this use of the bridge will appear on the PE power exam, of course. However, traditional use of the bridge for resistance measurement has been mentioned in exam review materials, so we present the following for the sake of completeness. You may wish to refer to Chapter 15 if you are not certain of some of the measurement terminology below.

Suppose the resistance R3 of the center leg of the bridge in Figure 3.9 is replaced by a sensitive high-resistance voltmeter (usually a center-scale microampere D'Arsonval meter movement that can deflect in either direction, in series with a high multiplier resistance). Furthermore, suppose that the values of resistances R1, R2, and R4 are accurately known. R4 is a variable resistance that can be adjusted. When R1 and R2 are properly selected and R4 is adjusted so that the voltage across R3 (now the resistance of the meter and its multiplier) is for all practical purposes zero, the bridge is said to be balanced. It can be shown that this happens when

$$R5 / R4 = R2 / R1$$

from which we can calculate the unknown resistance R5 as

$$R5 = R4\, R2 / R1.$$

For this application, the value of the battery voltage Vs is not critical, which is a great advantage as battery condition is not too relevant. The accuracy of the process is related to the accuracy of the three resistors R1, R2 and R4 and the sensitivity of the meter for null (zero-voltage) detection. One advantage of this circuit for measuring resistances is that very low values of resistance (such as used as ammeter shunts) can be determined very accurately. Such circuits are, in fact, sometimes used in automated instrumentation processes.

SUMMARY

This chapter reviewed some basic strategies for solving dc circuits. Taken together, the methods should allow you to solve nearly any dc circuit you may encounter on the exam. Before starting to write a nodal or loop solution for an exam problem, however, it is wise to look at the problem carefully for a moment and ask yourself if you can see any simpler ways of solving it. Often the requested question can be answered in a very simple way.

Experience in examining and solving circuits, which you have no doubt acquired during your career (and in the study of this book), should help you answer some questions with a minimum of effort. We emphasize again the importance of carefully noting the connections of the elements, the polarity of the voltages, and the directions of the currents.

It is worth stating again that all the analysis methods presented in this chapter, such as mesh and node analysis, the Thevenin/Norton simplifications, and delta-wye conversions, apply to steady-state ac analysis as well, and thus could have been presented in the following chapter. The math is a little more difficult in the ac case, because of the need to work with complex numbers. *Most of the problems on the PE power exam will probably be ac problems.* In the following chapters, you will find many opportunities to check your skills in solving ac circuits.

RECOMMENDED REFERENCES

Nearly any undergraduate circuits text should be appropriate. As an example, we recommend the following reference:

Nilsson and Riedel. *Electric Circuits.* 8th ed. Pearson/Prentice Hall, 2008.

CHAPTER 4

Single-Phase AC Circuits

OUTLINE

THE SINUSOID   38

THE IMPORTANCE OF CIRCUIT MODELS AND SSAC ANALYSIS   39

LINEAR, BILATERAL, PASSIVE CIRCUIT ELEMENTS   39

EFFECTIVE (RMS) VALUES   41

THE PHASOR   43

IMPEDANCE AND ADMITTANCE   45

RESONANCE   47

POWER TO A SINGLE-PHASE LOAD   47
Reactive, Complex, and Apparent Power ■ Units Associated with Power-Type Quantities ■ Complex Power in Terms of Impedance

SERIES AND PARALLEL EQUIVALENTS   51

POWER FACTOR CORRECTION   53

THE NODAL PROBLEM AND THE BUS IMPEDANCE AND ADMITTANCE MATRICES   59
The Bus Admittance Matrix ■ The Bus Impedance Matrix ■ Physical, Global Meanings of the Terms of Y_{bus} and of Z_{bus}

RECOMMENDED REFERENCES   62

The objective of this chapter is to provide a review of some fundamental concepts in steady-state ac (SSAC) circuit analysis. Concepts including the sinusoid, rms values, impedance and admittance, phasor analysis, and types of power are introduced. Solved examples provide a review of analysis principles as well as an introduction to concepts such as power factor correction.

Circuit-analysis techniques including mesh and nodal analysis and Thevenin's Theorem were introduced for the simpler dc case in the previous chapter. It should be emphasized that all of those techniques hold for the SSAC case, with slight adjustments. Because of their unique importance in power-systems work, some of these analysis techniques (particularly nodal analysis) are revisited here in more detail.

This chapter addresses the single-phase case only. Three-phase concepts are introduced in Chapter 5. You may wish to revisit the "Notation" section at the beginning of Chapter 3 before continuing.

THE SINUSOID

Steady-state ac circuit theory begins with the **sinusoid**. This unique smooth curve is not restricted to electrical engineering, but can in fact be developed (in the sense of being unrolled) as the time plot of the projection or height of a line of constant length revolving at constant speed (see Figure 4.1). In fact, that analogy is quite useful when we consider the operational aspects of a synchronous generator in Chapter 10.

The sinusoid is particularly important to electric power studies for at least the following reasons:

- Sinusoidally-varying voltage is naturally developed in a filamentary conductor at constant rotational speed with respect to a steady magnetic field in an electric machine.

- Sinusoidal waveforms are smooth and provide the most uniform steady-state ac transmission of electric energy without discontinuities.

- The basic sinusoidal waveshape is not distorted by any of our linear, passive, bilateral elements (these are described in the section "Linear, Bilateral, Passive Circuit Elements" later in this chapter).

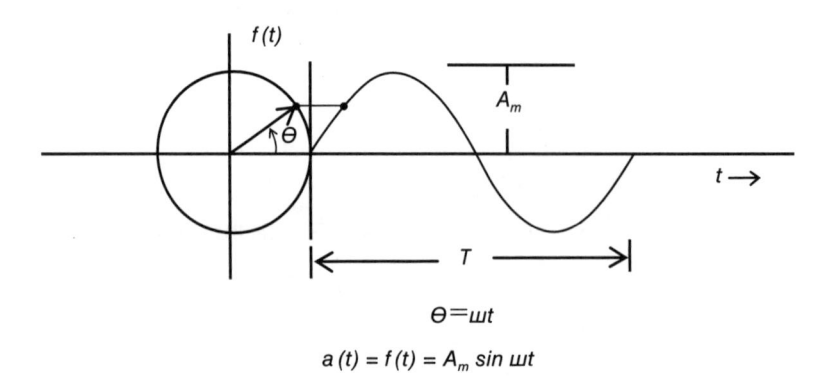

$$\theta = \omega t$$
$$a(t) = f(t) = A_m \sin \omega t$$

Figure 4.1 Development of the sinusoid

Some terms of importance regarding the sinusoid are illustrated in Figure 4.1. They include

amplitude, A_m of the sinusoidal time-function a(t)

period, T, expressed in seconds (s).

Furthermore, we define

frequency, $f = 1/T$, expressed in Hertz (Hz)   (**4.1**)

radian frequency, $\omega = 2\pi f$, expressed in radians per second (rad/s or simply s^{-1})
(**4.2**)

ω is also called the **angular velocity** of the sinusoidal waveform.

Note that **direct current** (dc) is the special case of ac where the frequency is zero.

Being a special case, dc analysis of circuits is generally simpler and is thus presented first in most circuit texts. Since steady-state sinusoidal ac analysis is so important to electric power engineers, we will emphasize SSAC concepts for much of the remainder of the book.

In a circuit where several sinusoidal voltages or currents are of interest, it is awkward to represent all of these currents and voltages as sinusoids on the same plot. The **phasor** method is an analysis technique which overcomes this difficulty and we review it later in this chapter.

THE IMPORTANCE OF CIRCUIT MODELS AND SSAC ANALYSIS

Electrical engineers are often quick to construct circuit models for situations ranging from simple devices to extremely large and complex systems. One reason is that circuit models may be easily analyzed and their results transferred back to the original problem. In fact in the solution of many electric power problems it is standard procedure to:

1. Begin with an actual device or system
2. Develop (and perhaps test and verify) an appropriate circuit model
3. Solve the circuit model for the desired quantities and
4. Relate the results back to the original problem

Circuit models are very helpful in analyzing and predicting the behavior of transformers, power lines, synchronous machines, induction machines, and in fact complete power systems. To use them effectively within the time constraints of the PE exam, you need to be very familiar with their fundamental "building blocks". For the dc case of the previous chapter, these amounted to resistances and both current and voltage sources. For the SSAC case, they will usually consist of the resistor R, inductor L, capacitor C, and both sinusoidal voltage and current sources. (Later, in Chapter 14, we will look at circuits containing diodes and other "electronic" devices.)

LINEAR, BILATERAL, PASSIVE CIRCUIT ELEMENTS

We will treat each of the circuit elements R, L, C (resistors, inductors, and capacitors), and each source, as a two-terminal device. For the passive components R, L, and C, the relations between voltage and current will be identified as in Figure 4.2. Note that in this figure, the current is directed into the positive terminal of the device. This is the conventional configuration, sometimes called **load convention**. When we talk about power quantities, the load convention indicates that those quantities will be directed *into* the device. A negative sign, of course, would imply that they are actually moving oppositely, just as a negative calculated value of current indicates that the actual current is in an opposite direction to the arrow.

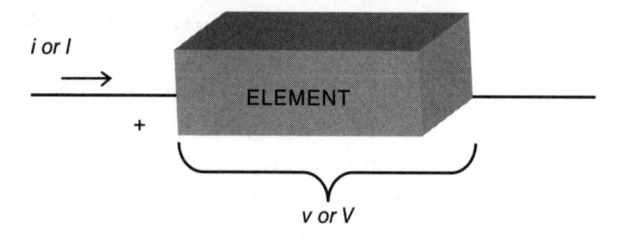

Figure 4.2 Passive element convention for voltage and current

Circuit Elements and Representation

It is usually assumed that the passive devices are linear and bilateral (these terms are defined in the next section). Figure 4.3 shows these idealized devices and the performance equation for each in terms of instantaneous voltage v and current i. Note that the v-i relations for two of these devices are differential equations, while one simply has an algebraic equation (Ohm's Law) describing its behavior.

resistor $\quad v = Ri$

inductor $\quad v = L\dfrac{di}{dt}$

capacitor $\quad i = C\dfrac{dv}{dt}$

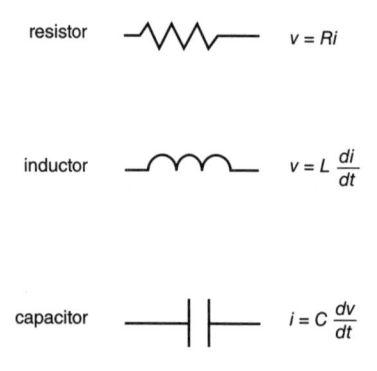

Figure 4.3 Linear, bilateral, passive circuit elements

Terms and Definitions

Engineers often use several simple circuit elements to build a **lumped-parameter model** of a device. This resulting model can then be used to analyze the performance of the device under expected or anticipated operating conditions. Two basic types of circuit elements are used: *active* (sources), and *passive* (R, L, and C).

For simplicity, we often **idealize** the elements of our circuit models. Thus, an ideal resistor is assumed to be a "package of resistance" without any inductance or capacitance; an ideal inductor is assumed to have no resistance, etc. Some of these assumptions break down at higher frequencies (for example, turn-to-turn capacitance begins to become prominent in inductors, leading to resonances), but the assumptions are very useful at the lower frequencies of greatest interest to us. Furthermore, more detailed and comprehensive models can be made whenever necessary. Our approach will be to use the simplest models or, at least, to simplify a more detailed model to the extent that it provides solutions that replicate real-world behavior.

In many cases, a **linear** R, L, or C is assumed. Recall that a linear function is one having a straight-line graph. For a resistor, linearity means that voltage across the device and current through it are related by a fixed proportion (R in Figure 4.3). For an inductor or capacitor, linearity can be defined as the rms value of the voltage across the device, and rms current through it being related by a fixed proportion. (The meaning of rms value is covered in the next section.)

A **bilateral** device is one that behaves the same way for current in either direction. Another way of stating this is that you can flip the device end-for-end and the circuit will behave the same. The R, L, and C elements used to represent devices in power systems are assumed bilateral in most cases. The diode, and other semiconductor devices, are notable exceptions.

Fundamental Power Concepts

Power is defined as the time rate of transfer of energy. The unit of energy is the **joule** (**J**), and the unit of power is the **watt** (**W**). Electrical energy is what is metered into a home or business, and it is usually expressed in the more convenient units of **kilowatt-hours** (**KWH**). One kilowatt is equal to 1,000 watts.

For any of the devices of Figure 4.3 or combinations thereof, **instantaneous power** can be expressed as

$$p = v\,i. \tag{4.3}$$

Note that we use lower-case letters here by convention because v, i, and p are in general time functions.

In an ac circuit, other measures of power become important, as will be shown later. For example, **average power** (also called **real power**) is the average of the instantaneous power over one cycle, which we show as

$$P = P_{av} = av\{p\}, \tag{4.4}$$

where the operator av represents the average of what follows it.

These terms are introduced because we require them in the next section; the other power-related measures will be introduced later.

We will state without proof the following facts:

- *Resistors consume or absorb power* (that is, convert electrical energy into heat or represent other energy-transfer processes in models).

- *Ideal inductors and capacitors do not consume energy but store it.* In sinusoidally-excited circuits with linear devices, this energy is "swapped" back and forth between the device and the rest of the circuit during alternate quarter-cycles of the wave.

EFFECTIVE (RMS) VALUES

The concept of the rms or effective value of a periodic function is inherent in nearly every aspect of steady-state electric power circuit analysis. The problem statement leading to this important concept may be posed as follows:

Consider two identical resistors, each of value R. One is energized by a direct voltage V_{dc} (from a battery, for instance) and the other by a sinusoidal voltage $v = V_m \cos \omega t$, as shown in Figure 4.4. What maximum value of sinusoidal voltage V_m (in terms of V_{dc}) is required to transfer the same average energy to the resistor as the dc source does?

Obviously, the correct answer is not the average of the sinusoid; that is zero. Nor is the correct answer found in setting V_m equal to V_{dc}, because it is obvious that in such a case, the dc circuit would handle more energy because its voltage would not dip from that level.

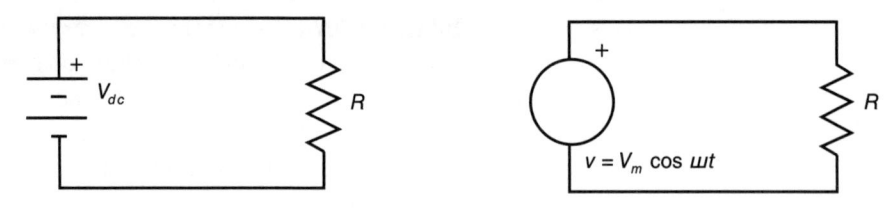

Figure 4.4 Identical resistors receiving same average power

For the dc problem, we know that

$$P_{dc} = V_{dc}^2 / R. \tag{4.5}$$

Assuming the thermal time-constant of the resistor to be much greater than the period of the sinusoid, we express the **average power** for the ac problem as

$$P_{av} = av\{v^2/R\}. \tag{4.6}$$

Setting $P_{dc} = P_{av}$,

$$V_{dc}^2/R = av\{v^2/R\}$$

or

$$V_{dc}^2 = av\{v^2\}$$

from which

$$V_{dc} = \sqrt{av\{v^2\}} \tag{4.7}$$

We see that *the square root of the average of the square* of the sinusoidal function must be set equal to the dc voltage to provide the same heating in both resistors.

We call this value of voltage the **root-mean-square** (**rms**) value. It is also called the **effective** value, because it provides the same effective value of heating as a numerically identical dc value.

The rms value is extremely important in SSAC work, to the extent that *a sinusoidal signal is usually described in terms of its rms value unless specifically noted.* For example, the nominal voltage of electric outlets in the US is, by standard, 120 V (= 120 V rms). We will follow convention and omit the letters *rms* and assume that each sinusoidal V or I is an rms value.

Note that this could have been illustrated using *current*, beginning with the relationships

$$P_{dc} = I_{dc}^2 R$$

and

$$P_{av} = av\{i^2 R\}.$$

Although the idea of rms is of greatest interest to us in connection with the sinusoid, it can be found for a general periodic function of time f(t) as

$$F_{rms} = \sqrt{(1/T)\int_0^T f^2(t)dt}. \tag{4.8}$$

(The integration must be made over a full period, but may start and end at any convenient point; here it is shown starting at the beginning of the wave and ending a period later.)

Very importantly, for a sinusoid, such as the voltage $v = V_m \cos(\omega t)$, we find by calculus that the corresponding rms value V is

$$V = V_m / \sqrt{2} = 0.7071\, V_m. \qquad (4.9)$$

For example, the nominal 120 V rms outlet voltage above would have a maximum value V_m of

$$V_m = 120\sqrt{2} = 169.71\text{ V}.$$

This holds for voltages, currents, magnetic fluxes, etc. To emphasize again, *because rms values are so commonly used, we usually write them as simply V, I, etc., and we will assume that voltages and currents denoted by capital letters are all rms quantities unless otherwise noted.*

THE PHASOR

Typical SSAC problems may require the identification and manipulation of many different sinusoidal quantities. Such operations are facilitated by use of the phasor concept. Furthermore, this concept allows a simple representation of voltages, currents, and fluxes in diagrammatic form. Showing more than a few sinusoidal functions on a time plot is awkward. This difficulty can be overcome by expressing the sinusoidal quantities as phasors, or, as we say, in the frequency domain.

Essentially, a **phasor** is a complex number (often referred to as a **vector** since it can be characterized by magnitude and direction) representing a sinusoidal quantity, in which the magnitude of the vector is the rms value of the quantity and the angle of the vector is the phase angle of the quantity measured with respect to some arbitrary reference. The word *phasor* implies the idea of a phase relationship. Most texts use boldface or an overstrike to denote a vector or phasor quantity.

We sometimes hear the term **rotating phasor**, which implies a line of constant amplitude rotating at constant angular velocity in the complex plane (reference Figure 4.1). This idea is especially worthwhile in envisioning magnetic fields in a synchronous or induction machine, which do in fact rotate in a physical sense.

An important distinction should be noted here: *While quantities such as voltages, currents, and magnetic fluxes can be phasors, and thus vectors, the quantities impedance and admittance (defined in the next section) are vectors but are not phasors.*

Power engineers usually restrict their attention to **rms phasors**, that is, complex-number quantities whose magnitudes are rms values.

Consider the sinusoidal voltage

$$v = V_m \cos(\omega t + \varphi). \qquad (4.10)$$

Here, the angle φ represents the **phase angle**, which is the angular displacement of the quantity from the angular reference. Since φ can take on any real value (positive or negative or zero), the expression for voltage v is completely general.

The rms phasor equivalent of the voltage v above may be expressed, through use of the **phasor transform**, as

$$\mathbf{V} = \left(V_m / \sqrt{2}\right) e^{j\varphi} \qquad (4.11)$$

or, in shorthand notation,

$$\mathbf{V} = (V_m / \sqrt{2}) \angle \varphi. \tag{4.12}$$

The **inverse phasor transform** is

$$v(t) = \sqrt{2} \, \text{re} \left\{ \mathbf{V} \, e^{j\omega t} \right\}. \tag{4.13}$$

To get some practice, you should verify the inverse transformation using the phasor **V** given above. To do this, you might want to use **Euler's Theorem**, which allows us to represent an exponential with an imaginary argument in complex rectangular form:

$$e^{j\varphi} = \cos \varphi + j \sin \varphi. \tag{4.14}$$

Again, note that the $\sqrt{2}$ is necessary because we want all phasor magnitudes to be rms, or effective, values.

For any of the elements shown in Figure 4.3 (or any two-terminal combination of them), we can write the voltage-current pair in the sufficiently-general form

$$i = I_m \cos \omega t \tag{4.15}$$

$$v = V_m \cos(\omega t + \varphi) \tag{4.16}$$

where the angle φ, usually defined as the angle by which phase current lags phase voltage, is called the **phase angle**. A negative angle implies leading current.

Applying the phasor transformation, we can express the phasor-domain voltage-current relationship of each of the three passive devices (R, L, C) as shown in Figure 4.5 (the $j\omega$ enters because of differentiation, as indicated in Figure 4.3). You should study this figure carefully and note the equations and resulting phasor diagrams.

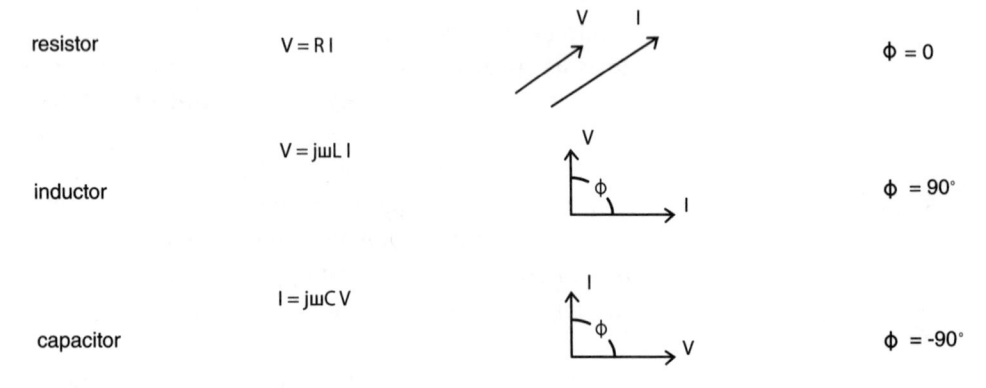

resistor	V = R I	(V and I aligned)	φ = 0
inductor	V = jωL I	(V leads I by 90°)	φ = 90°
capacitor	I = jωC V	(I leads V by 90°)	φ = -90°

Figure 4.5 Voltage-current and phase relationships for ideal R, L, and C under sinusoidal excitation

Some noteworthy observations:

- For a series or parallel connection of an R and L, phase angle is in the range

$$0 < \varphi < 90°. \tag{4.17}$$

- For a series or parallel connection of an R and C, the phase angle φ is in the range

$$-90° < \varphi < 0. \tag{4.18}$$

Also, the following important observation may be made:

If, in any circuit containing linear, passive, bilateral elements, one current or voltage is purely sinusoidal, then all currents and voltages are sinusoidal and of the same frequency.

IMPEDANCE AND ADMITTANCE

For any two-terminal element in Figure 4.3 (or any composite of them), we define **impedance** as the ratio of phasor voltage across an element to phasor current through that element:

$$Z = V/I, \tag{4.19}$$

which is sometimes called **Ohm's Law in phasor form**.

Similarly, **admittance** is defined as

$$Y = 1/Z = I/V. \tag{4.20}$$

If we use the sufficiently-general pair of phasors **V** and **I**

$$\mathbf{I} = I\, e^{j\theta} \tag{4.21}$$

and

$$\mathbf{V} = V\, e^{j(\theta + \varphi)} \tag{4.22}$$

then

$$\begin{aligned} \mathbf{Z} &= V\, e^{j(\theta + \varphi)} / I\, e^{j\theta} \\ &= (V/I)\, e^{j\varphi} \\ &= Z\, e^{j\varphi} \\ &= Z\, /\varphi \end{aligned} \tag{4.23}$$

where φ can be called the **impedance angle**. Note that this angle is identical to the phase angle discussed earlier.

Impedance and Admittance of Series and Parallel RLC Combinations

Two-terminal elements are usually connected in either series or parallel (a notable exception is the bridge circuit described in Chapter 3). For each of these connections, there is a new term or two to review. In this section we will review the quantities *reactance* and *susceptance* as they relate to series and parallel connections, respectively.

For the series connection of Figure 4.6, the total impedance may be written as the sum of the impedance of the resistor, the inductor, and the capacitor

$$\begin{aligned} \mathbf{Z} &= \mathbf{Z}_R + \mathbf{Z}_L + \mathbf{Z}_C \\ &= R + j\omega L + 1/(j\omega C) \end{aligned} \tag{4.24}$$

but because we can show that

$$1/j = -j \tag{4.25}$$

the total impedance becomes

$$Z = R + j(\omega L - 1/(\omega C)) \quad (4.26)$$
$$= R + j(X_L + X_C)$$
$$= R + jX \quad (4.27)$$

where
X_L = inductive reactance = ωL
X_C = capacitive reactance = $-1/(\omega C)$
X = total reactance.

Note that *reactances, like resistances, are real numbers*. Inductive reactance is a positive real number, while capacitive reactance is a negative real number. (We should mention that X_C is sometimes written as a positive number and then subtracted from X_L to obtain total reactance. However, virtually all modern circuits books treat X_C as negative. That is consistent with, for example, Equation 4.24, and this text follows that convention.)

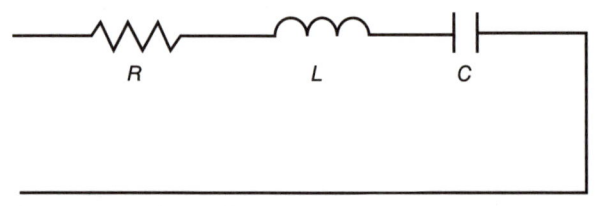

Figure 4.6 Series Combination of RLC

For the *parallel* circuit of Figure 4.7, we can similarly write an expression for the total *admittance* as

$$Y = Y_R + Y_C + Y_L$$
$$= G + j\omega C + 1/(j\omega L)$$
$$= G + j(\omega C - 1/(\omega L)) \quad (4.28)$$
$$= G + j(B_C + B_L)$$
$$= G + jB \quad (4.29)$$

where
B_C = capacitive susceptance = ωC
B_L = inductive susceptance = $-1/(\omega L)$
B = total net susceptance.

Note that *susceptances, like conductances, are real numbers*. Inductive susceptance is a negative real number, while capacitive susceptance is a positive real number. Thus, with regard to sign susceptances behave inversely to reactances.

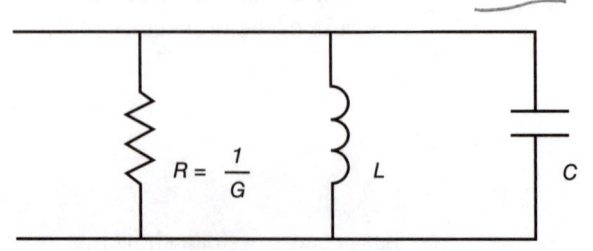

Figure 4.7 Parallel Combination of RLC

RESONANCE

The concept of resonance is probably more important in communication than power circuits. Nonetheless, it is useful when selecting, for instance, filter designs for harmonic problems in power systems. Later in the chapter, when we define power factor, you will see that **resonance** *is just the case of unity power factor.*

In the series RLC circuit of Figure 4.6, suppose that the magnitude of the impedance of the inductor (which is inductive reactance X_L) is equal to that of the capacitor (X_C). Then the reactive components of impedance cancel and the total impedance of the circuit would be resistive:

$$Z = R + j0$$

If, additionally, the resistance were zero, the impedance of the circuit would be zero. This condition is called **series resonance**. Series resonance occurs when

$$|X_L| = \omega L = |X_C| = 1/(\omega C)$$

at a **resonant frequency** of

$$\omega = 1/\sqrt{LC} \text{ rad/s, where } \omega = 2\pi f. \tag{4.30}$$

A series resonant circuit containing low resistance looks very nearly like a short circuit.

Similarly, in the parallel RLC circuit, suppose that the magnitude of the impedance of the inductor X_L is equal to that of the capacitor X_C (or, equivalently, the magnitudes of their susceptances are equal). Then

$$Y = G = 1/R.$$

If, additionally, the resistance in this parallel circuit did not exist—that is, if R = ∞ or equivalently if G = 0—the admittance would be zero and the impedance would be infinite. This condition is called **parallel resonance**. Just like for the series circuit, parallel resonance occurs when

$$|X_L| = \omega L = |X_C| = 1/(\omega C)$$

at a resonant frequency of

$$\omega = 1/\sqrt{LC} \text{ rad/s.}$$

A parallel resonant circuit containing low conductance looks very nearly like an open circuit.

POWER TO A SINGLE-PHASE LOAD

Consider the sufficiently-general sinusoidal voltage-current pair

$$i = I_m \cos \omega t \tag{4.32}$$

$$v = V_m \cos(\omega t + \varphi) \tag{4.33}$$

applied to the load shown in Figure 4.8. Recall that the polarity relation between v and i follows the load convention; that is, current is defined as entering the positive terminal of the device or circuit. If the device consumes power, it will be positive in the direction of the double arrow shown.

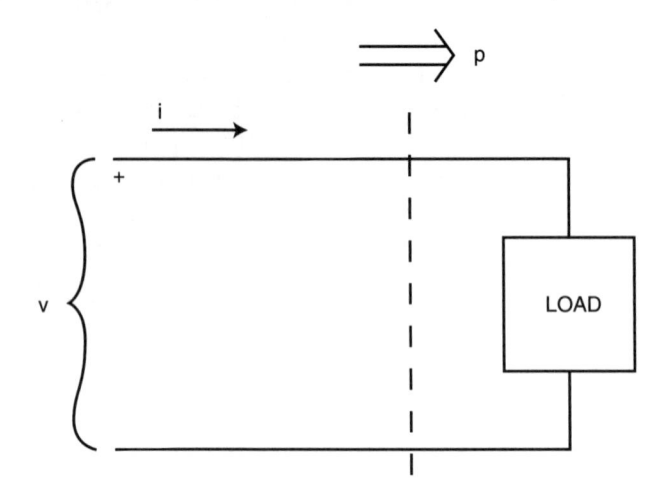

Figure 4.8 Load with instantaneous quantities shown

Instantaneous power, as we know, is given by

$$p = vi$$
$$= V_m I_m \cos(\omega t + \varphi)\cos \omega t.$$

Using the identity

$$\cos x \cos y = [\cos(x - y) + \cos(x + y)]/2,$$

$$p = \frac{1}{2} V_m I_m [\cos \varphi + \cos(2\omega t + \varphi)]. \qquad (4.34)$$

Note that this expression for instantaneous power has two parts: a constant (corresponding to $\cos \varphi$) and a time-varying part of frequency 2ω (corresponding to the $\cos(2\omega t + \varphi)$ part). Therefore, *instantaneous power to a linear single-phase load in the presence of sinusoidal excitation is characterized by a constant plus a double-frequency term.*

Parenthetically, one implication for single-phase loads, including motors, is that there is a pulsating torque component at double frequency for each device excited from a single-phase source. This is one of the advantages that three-phase rotating machines have over their single-phase counterparts.

Figure 4.9 shows the effect of phase angle upon instantaneous power. Note that for a phase angle of zero, the curve is never negative (a resistor cannot supply power back into the source) and for a phase angle of 90°, the average of the curve is zero. This has physical implications for the average power, of course.

The average of the instantaneous power is called **average power** or **real power** (or sometimes simply *power*) and is given as follows:

$$P_{av} = \frac{V_m I_m}{2} \cos \varphi = (V_m/\sqrt{2})(I_m/\sqrt{2}) \cos \varphi = V_{rms} I_{rms} \cos \varphi,$$

which is usually written

$$P = VI \cos \varphi \qquad (4.35)$$

where φ, which we called the impedance angle above, is now called the **power factor angle** and cos φ is the **power factor.**

Note that, in this ac case, *the power factor is the factor by which we must multiply the rms VI product in order to find real or average power.*

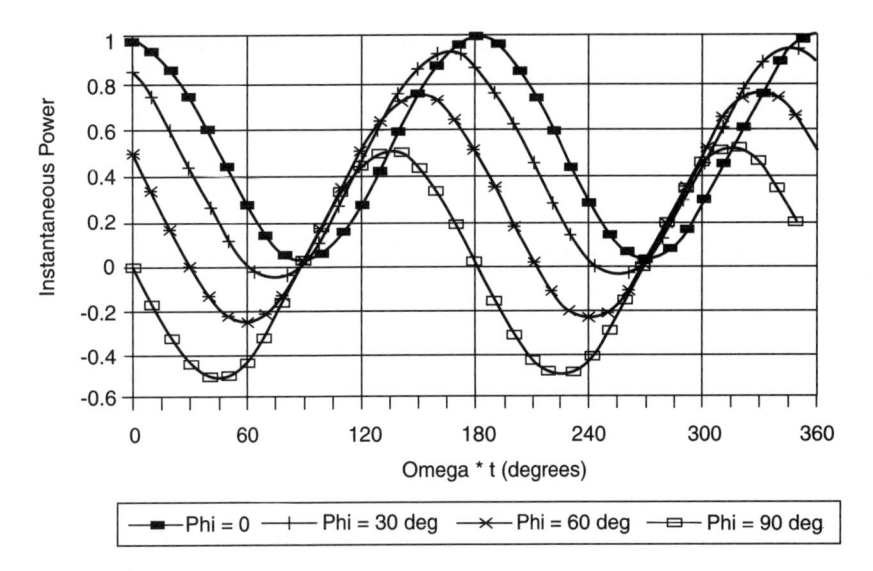

Figure 4.9 Instantaneous power in terms of phase angle

Note that

- for a pure resistance, P = VI cos(0) = VI (4.36)
- for a pure inductance, P = 0 (4.37)
- for a pure capacitance, P = 0. (4.38)

Thus, we see that a resistor consumes real (average) power but ideal inductors and capacitors do not. Furthermore, we identify power factor more concisely by saying that an RL circuit has a **lagging** power factor and an RC circuit has a **leading** power factor.

Reactive, Complex, and Apparent Power

In comparison with Equation 4.35, we now define **reactive** (or **quadrature**) **power** as

$$Q = V I \sin \varphi. \quad (4.39)$$

Note that

- for a resistor, Q = VI sin(0) = 0 (4.40)
- for an inductor, Q = VI (4.41)
- for a capacitor, Q = –VI. (4.42)

Therefore, in accordance with the load convention discussed earlier, a *pure inductance absorbs reactive power and a pure capacitance supplies reactive power.* There is no reactive power associated with a resistor.

Complex power is defined as the complex number

$$S = P + jQ \qquad (4.43)$$

(we remind you that P and Q are scalars).

Therefore, using Equations 4.35 and 4.39,

$$\begin{aligned} S &= V I \cos \varphi + j V I \sin \varphi \\ &= V I (\cos \varphi + j \sin \varphi) \\ &= VI\, e^{j\varphi} \\ &= VI\, \angle \varphi. \end{aligned} \qquad (4.44)$$

Suppose we again let our phasor voltage and current be

$$\mathbf{V} = V\, e^{j(\theta + \varphi)} \qquad (4.45)$$

and

$$\mathbf{I} = I\, e^{j\theta}. \qquad (4.46)$$

Recalling that power should be related to the VI product, examine the phasor product **VI**. This does not equal anything meaningful. However, the product

$$\begin{aligned} \mathbf{V I^*} &= V\, e^{j(\theta+\varphi)} \quad I\, e^{-j\theta} \\ &= VI\, e^{j\varphi} \end{aligned}$$

where I^* (read "I conjugate") = the complex conjugate of I.

This leads to the very useful equation

$$P + jQ = \mathbf{V I^*}. \qquad (4.47)$$

The magnitude of complex power S is called **apparent power**.

In any SSAC circuit, complex power is *conserved*; that is, as much is produced as is consumed. That means that both P and Q are conserved.

Units Associated with Power-Type Quantities

Although it appears that Q and S should have the same units as P, in order to avoid confusion we define units for them as follows:

p, P	watts W
Q	VARs (volt-amperes reactive)
S	volt-amperes

Complex Power in Terms of Impedance

First, we will note that the product of a vector **A** and its conjugate **A*** is equal to the square of the magnitude of the vector:

$$\mathbf{A A^*} = A^2 \qquad (4.48)$$

From the derived relation $P + jQ = \mathbf{V}\mathbf{I}^*$ and the definition of impedance, we can write

$$\mathbf{S} = \mathbf{V}(\mathbf{V}/\mathbf{Z})^*$$
$$= V^2/\mathbf{Z}^*$$
$$= V^2/(R - jX). \tag{4.49}$$

Similarly,

$$\mathbf{S} = (\mathbf{Z}\mathbf{I})\mathbf{I}^*$$
$$= I^2\mathbf{Z}$$
$$= I^2(R + jX)$$
$$= I^2 R + jI^2 X. \tag{4.50}$$

Recalling that

$$P = I^2 R \tag{4.51}$$

we see by comparison that

$$Q = I^2 X. \tag{4.52}$$

This further gives the justification that an inductor absorbs Q, because X is positive for an inductor. The reverse is true for a capacitor.

Caution: Care should be taken when using the expression

$$\mathbf{S} = V^2/\mathbf{Z}^*$$

to ensure that V is the voltage across the impedance Z and not across simply R or X. (This is a very common mistake!)

SERIES AND PARALLEL EQUIVALENTS

When modeling and analyzing power systems, it is often necessary to develop equivalent circuits. In addition to illustrating the foregoing material, Example 4.1 introduces the concept of the equivalent circuit. It must be emphasized that when modeling ac circuits in this manner, the equivalents are valid only at the frequency for which the equivalent is developed.

Example 4.1

We wish to replace the series circuit shown in Figure 4.10 by a terminally-equivalent parallel combination of two elements at a given frequency. What must these elements be? What are their values?

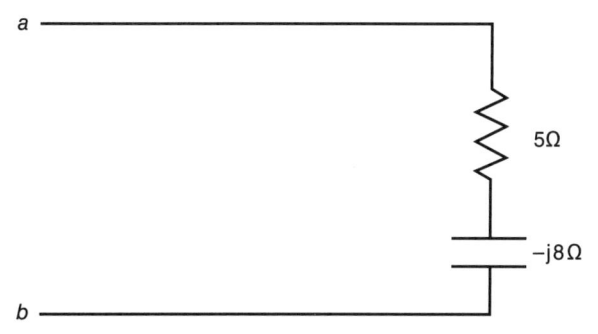

Figure 4.10 Series RC circuit

Solution

It is evident that one of the parallel elements must be a resistance (because the equivalent must also consume P) and that its value will be different from R.

The impedance of the load in the series circuit is

$$\mathbf{Z} = 5 - j8$$
$$= 9.434 \angle{-58.0°} \; \Omega$$

and thus, the admittance of the parallel equivalent must be

$$\mathbf{Y} = 1/\mathbf{Z}$$
$$= 0.1060 \angle{58.0°}$$
$$= 0.0562 + j0.0899 \; \text{S}$$
$$= G + jB.$$

Because B is positive, the reactive element must be a **capacitor**. This can also be shown through energy balance: the original load supplies Q, and thus the equivalent parallel load must supply the same Q.

The parallel equivalent is shown in Figure 4.11, where

$$R_p = 1/G$$
$$X_p = -1/B \quad \text{(note the negative sign here).}$$

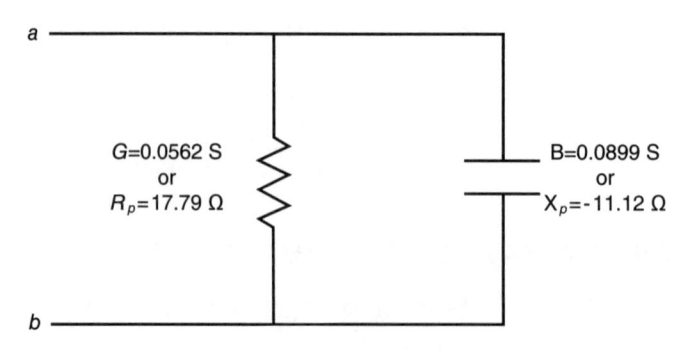

Figure 4.11 Parallel equivalent of Figure 4.10

Example 4.2

Verify that Figures 4.10 and 4.11 are terminally equivalent by applying 120V to each circuit and calculating the resulting P and Q.

Solution
For Figure 4.10,

$$\mathbf{I} = \mathbf{V}/\mathbf{Z} = 120\angle{0°} \; / \; (5 - j8) = 12.72\angle{58°} \; \text{A}.$$

Note that $\mathbf{I}$ leads $\mathbf{V}$, as expected. For practice, you might want to calculate the power factor of this circuit as

$$\cos 58° = 0.5299.$$

Knowing $\mathbf{I}$, we can find

$$P = I^2 R = 12.72^2 \; (5) = 808.99 \; \text{W}$$
$$Q = I^2 X = 12.72^2 \; (-8) = -1294.39 \; \text{VAr}.$$

Note that this load consumes P but supplies Q.

Alternately, we could use

$$P + jQ = \mathbf{V} \mathbf{I}^* = (120\angle 0°)(12.72\angle -58°) = 808.99 - j1294.39 \text{ VA}$$

or we could use

$$P = VI \cos \varphi$$

and

$$Q = VI \sin \varphi$$

to obtain the same result.

Turning now to the parallel equivalent of Figure 4.10, which is shown in Figure 4.11,

$$P = V^2 / R_p = V^2 G = 120^2 (0.0562) = 808.99 \text{ W}$$
$$Q = V^2 / X_p = V^2 (-B) = 120^2 (-0.0899) = -1294.39 \text{ VAr}$$

which verifies the parallel equivalent.

As an exercise to solve on your own, calculate the impedance Z_{eq} seen by the source and the real and reactive power P_L and Q_L supplied to the load of Figure 4.12. Also find the currents I_1, I_2, and I_3. You should obtain the following answers:

$$Z_{eq} = 15.81\angle 18.43° \text{ }\Omega$$
$$P_L = 864 \text{ W}$$
$$Q_L = 288 \text{ VAr}$$
$$I_1 = 7.590\angle -18.43° \text{ A}$$
$$I_2 = 2.4\angle 0° \text{ A}$$
$$I_3 = 5.367\angle -26.57° \text{ A}.$$

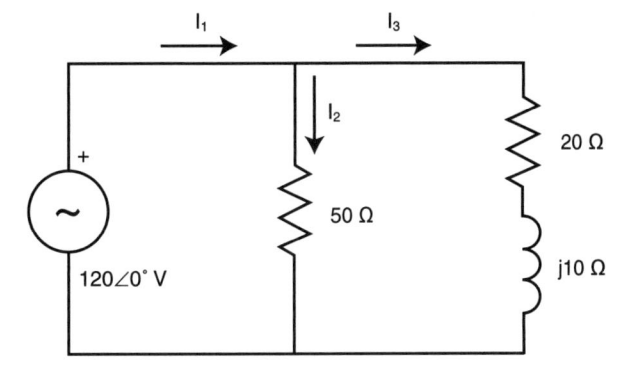

Figure 4.12 Circuit for Example

POWER FACTOR CORRECTION

Suppose a load drawing a fixed amount of real power P is placed across an ideal voltage source. From the equation $P + jQ = \mathbf{V} \mathbf{I}^*$, it is evident that increased values of Q mean an increased value of current. With P constant, this means that as Q drawn by the load is increased, the current increases without any corresponding increase in the usable power P.

Obviously, for this given P and V, the minimum possible value of current occurs when $\cos \varphi = 1.0$, for which $Q = 0$. This would represent an ideal situation as "seen" by the upstream utility, because (in a real world situation) the investment in conductor size and the resulting voltage drop along the line could be reduced.

We noted that inductive loads consume Q and that capacitors produce Q. Thus, it is possible to use a capacitor as a "negative inductor" to correct the total power factor seen by the source to a small value, or even to unity. This is done frequently and is known as **power factor correction**. It is very possible that a PE exam problem will require you to do this. Therefore, we introduce a rather complete example in which power factor is corrected to several different levels to meet desired objectives.

We conclude this review of power factor correction with the following extended example.

Example 4.3

A single-phase induction motor draws 27 A at full load from a 220V, 60Hz line. To make the problem simple, we will assume a stiff (or ideal) source at the point of motor connection; that is, the source has no impedance so anything we do at the load will not change the voltage (in fact, this is equivalent to assuming a Thevenin source model with zero Thevenin impedance).

The motor delivers 6.2 mechanical horsepower (HP) and operates at an efficiency of 87 percent. Find

a) the size of the capacitor needed to be placed in parallel with the motor terminals to raise the power factor to 95 percent (0.95), and

b) the size of the capacitor needed to be placed in parallel with the motor terminals to raise the power factor to 100 percent.

Solution

In problems such as these, we typically assume the terminal voltage of the load to be the zero-degree reference. Also, motors are usually rated in horsepower, so in order to solve this problem you need to know that

$$1 \text{ HP} = 746 \text{ W}. \tag{4.53}$$

You also need to know that an induction motor can be modeled at low frequencies as an RL load, an assumption that is consistent with the fact that the motor absorbs both P and Q (see Chapter 10). We will assume a series RL model, although a parallel RL model (with necessarily different values) would result in the same solution. The steady-state model of the motor is shown in Figure 4.13. Note that, prior to the addition of power-factor correction, the total source current is equal to the current consumed by the motor.

Note that part (a) of the problem statement did not tell us whether to raise the power factor of the motor-capacitor combination to 0.95 lagging or to 0.95 leading. In theory, either solution could be found; in fact, we will find and examine both solutions below in order to become more familiar with the way in which numerical values and the phasor diagrams behave. However, *from a practical standpoint, it is important to emphasize that power factor correction for an induction motor is always made to obtain a lagging value of power factor as seen by the line to the motor/capacitor combination.* One reason for this is that a motor and its power-factor correction capacitor may be switched on as a unit during starting, and an excessive amount of VArs available at the motor terminals might result in high motor voltage, especially if the combination is at the end of a highly inductive line. Thus, if asked to solve this problem on the exam, without being told to correct to a lagging or a leading power factor value, you should correct the power factor to 0.95 lagging.

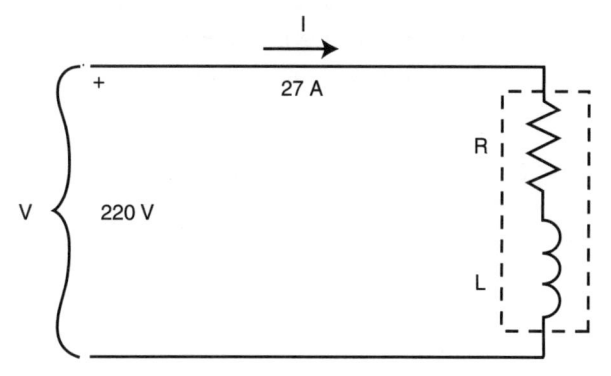

Figure 4.13 Simplest steady-state model of induction motor at constant load

The average electrical power consumed by the motor can be found in terms of the mechanical output and the efficiency as

$$P = (6.2)(746)/0.87 = 5316 \text{ W}.$$

In our example, this value of power drawn from the source will remain constant no matter what size capacitor we place in parallel with the motor, because we assumed an ideal source and an ideal power-factor-correcting capacitor.

The power factor at the motor terminals is

$$PF = \cos \varphi = P / (V\,I) = 5316 / [(220)(27)] = 0.895$$

from which $\varphi = 26.5°$

and so $I = 27 \angle{-26.5°}$ A.

As a reminder, by convention we call this a **lagging power factor load** since I lags V.

The reactive power Q consumed by the motor is

$$Q = V\,I \sin \varphi$$
$$= (220)(27)(0.446) = 2650 \text{ VAr}.$$

The above calculations refer to the motor alone. A capacitor is now added in parallel with the motor as shown in Figure 4.14. Note that *P and Q consumed by the motor* remain the same as before. The total phasor source current now becomes

$$\mathbf{I_T = I_C + I.}$$

We emphasize that these currents must be treated as vectors.

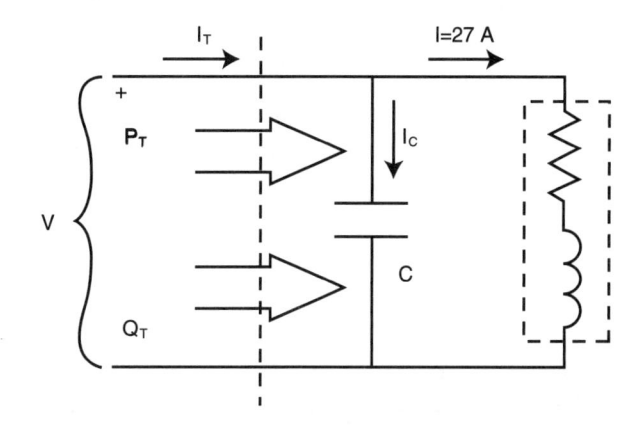

Figure 4.14 Motor with added power-factor correction

Solution (a). 95 percent power factor

In order to have a power factor of 0.95 as seen by the source, the angle between total current I_T and V must be

$$\Psi = \cos^{-1}(0.95) = \pm 18.2°$$

and thus the total Q from the source must be

$$Q_T = V\, I_T \sin(\pm 18.2°) = \pm\, 0.312\, V\, I_T.$$

Note that there are two possible values of angle that will satisfy the theoretical requirement. You can show this by taking the cosine of both the negative and positive angles; each will yield 0.95. The total current I_T is not known. However, the real power is the same as before the addition of the capacitor, and so

$$Q_T / P_T = (V\, I_T \sin 18.2°) / (V\, I_T \cos 18.2°) = \tan 18.2° = \pm 0.329.$$

Therefore,

$$Q_T = \pm\, 0.329\, P_T = \pm\, 0.329\,(5316) = \pm\, 1748\text{ VAr}.$$

Two solutions are possible to achieve a power factor of 0.95 as seen by the source.

Solution (a.1). 95 percent lagging power factor (the typical, practical solution for a motor application)

This corresponds to the source supplying net reactive power ($Q_T = +\,1748$ VAr). The motor-capacitor load combination looks inductive to the source, although the VAr demand from the source has been reduced from that of the original case. The reactive power consumed by the capacitor may be found as

$$Q_C = Q_T - Q = 1748 - 2650 = -902 \text{ VAr.}$$

(The negative sign of course confirms that the capacitor *supplies* reactive power.)

Because

$$Q_C = V^2 / X_C,$$
$$X_C = V^2 / Q_C = 220^2 / (-903) = -53.6\ \Omega.$$

And because

$$X_C = -1/\omega C,$$
$$C = -1/\omega X_C = -1/[(377)(-53.6)] = 49\ \mu\text{F}.$$

Note carefully how the negative signs "fall out" since we have followed the sign conventions diligently. A phasor diagram for this case is shown in Figure 4.15. The dashed lines in the figure show how the phasors add. The current in the capacitor may be found as

$$I_C = j\omega C\, V = j\,(377)\,49 \times 10^{-6})\,(220\angle 0°) = 4.06\angle 90°\text{ A},$$

which, as expected, leads the terminal voltage by 90 degrees.

Recalling that the motor current is

$$27\ \angle{-}26.5°\text{ A}$$

the total phasor current from the source is

$$I_T = I_C + I = 25.5\ \angle{-}18.2°\text{ A.}$$

Note that this current lags the applied voltage by the specified, corrected power-factor angle, and that its magnitude is **less** than that of the motor current alone. In fact, because the line current (in comparison with the current for the motor alone) is reduced by the factor

$$25.5/27 = 0.944.$$

Also observe that the power loss in the line feeding the motor (which we did not take into account, because we assumed an ideal source at the motor terminals) would be reduced by the approximate factor

$$0.944^2 = 0.892,$$

i.e., reduced 10.8 percent below its original value.

This partially explains why utilities often penalize users for low power factor loads. Part of the extra loss occasioned by the higher currents in lower power factor loads must certainly appear in lines owned by the utility and must, therefore, be paid for ultimately by the consumers. In any actual system, the source at the terminals of a load such as a motor is not actually ideal but in fact behaves like a Thevenin source. This means that in an actual situation, as distinguished from the ideal-source case given here, the voltage at the load terminals would be different from (in this case, lower than) the source voltage. Further study would also show that with the power-factor correction given here, the motor terminal voltage would not be as low as for the uncorrected motor; in general, power-factor correction applied to an inductive load tends to boost the load voltage. Power-flow studies (see Chapter 11) can be used to solve for actual load voltages in such cases.

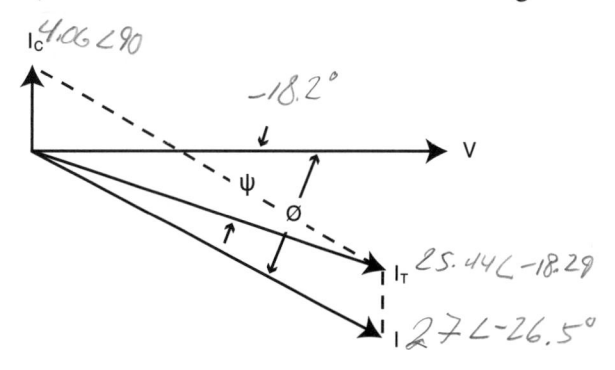

Figure 4.15 Phasor diagram for circuit providing a total power factor of 0.95 lag

Solution (a.2). 95 percent leading power factor.

As mentioned above, although this solution is possible, in practice we would not correct the power factor of a motor to a leading value. We consider the leading solution here only to show, in terms of the numerical results and the phasor diagram, what would happen if it were implemented. For this leading power-factor case, $Q_T = -1747$ VAr and the total circuit looks capacitive to the source, because the capacitor now supplies reactive power to both the motor and the source.

Therefore, the capacitor must now supply

$$Q_C = Q_T - Q = -1747 - 2650 = -4397 \text{ VAr.}$$

(Again, the negative sign confirms that the capacitor *supplies* reactive power, although of course much more in this case).

As before,
$$X_C = V^2 / Q_C = 220^2 / (-4397) = -11.01 \, \Omega$$
and
$$C = -1/ \omega X_C = 241 \, \mu F,$$
which is nearly five times the size of the capacitor for the lagging solution.

A phasor diagram for this case is shown in Figure 4.16. The total current from the source is
$$I_T = I_C + I = 25.5 \angle 18.2° \, A.$$

As you might have expected, this total current has the same magnitude as for the lagging case solved above, and the angle shows that it leads the voltage by the specified power-factor angle. Again, this case is interesting, but not practical from a motor standpoint.

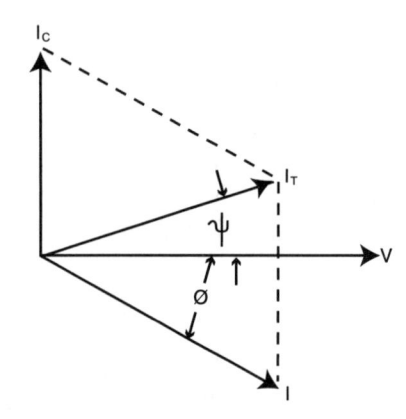

Figure 4.16 Phasor diagram for circuit providing a total power factor of 0.95 lead

Solution (b). Correction to 100 percent power factor

To raise the power factor to 1.0, the capacitor must supply exactly all the reactive power consumed by the motor. This is actually the easiest solution to calculate (although, like the leading solution, it is not usually done in practice).

Here the capacitor must consume
$$Q_C = -2650 \, VAr$$
which, using the same equations above, results in a required capacitance value of
$$C = 145 \, \mu F.$$
The total current supplied by the source is now
$$I_T = I_C + I = 24.2 \angle 0° \, A$$
as you should verify by phasor calculation. This is the minimum current possible at the given voltage to provide the required value of real power, as you can see by multiplying it by the terminal voltage.

The phasor diagram corresponding to this case is shown in Figure 4.17.

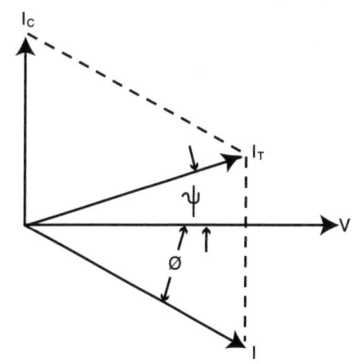

Figure 4.17 Phasor diagram for circuit providing a total power factor of unity

In actual work, capacitors for such applications are often specified in kVAr at nominal voltage, rather than in μF. This is particularly true for motors larger than about one horsepower. Be alert to how the capacitor is specified.

THE NODAL PROBLEM AND THE BUS IMPEDANCE AND ADMITTANCE MATRICES

As discussed in Chapter 3, nodal (bus) analysis is frequently used to solve electric power networks, because bus voltages are often the quantities of interest. Here we will review and expand that material and present the topic in terms of a three-bus power system model with complex number coefficients.

The Bus Admittance Matrix

Consider the single-phase three-bus network shown in Figure 4.18, which contains five admittances and one current-injection source. We wish to write nodal equations enabling us to solve for the voltage from each bus to reference.

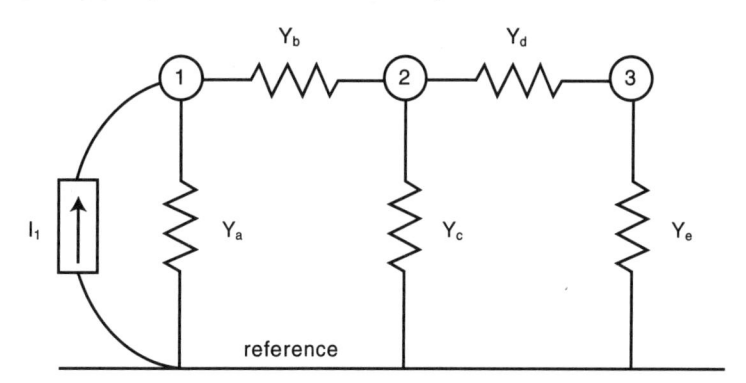

Figure 4.18 Circuit for nodal analysis

In Chapter 3, we wrote the conductance matrix for the nodal problem by inspection. For the ac case, the matrix to be used is the **bus admittance matrix** $\mathbf{Y}_{bus}$. For the network in Figure 4.18, the nodal equation in admittance form becomes

$$\begin{bmatrix} Y_{11} & Y_{12} & Y_{13} \\ Y_{21} & Y_{22} & Y_{23} \\ Y_{31} & Y_{23} & Y_{33} \end{bmatrix} \begin{bmatrix} V_1 \\ V_2 \\ V_3 \end{bmatrix} = \begin{bmatrix} I_1 \\ I_2 \\ I_3 \end{bmatrix}$$

$$\text{or } \mathbf{Y}_{bus} \mathbf{V} = \mathbf{I} \tag{4.54}$$

where

$\mathbf{Y}_{bus}$ is the bus (nodal) admittance matrix.
$\mathbf{I}$ is a vector of nodal injection currents
$\mathbf{V}$ is a vector of node voltages with respect to the reference bus.

In general, all the elements in Equation 4.54 are complex numbers.
In terms of the elements in the circuit of Figure 4.18,

$$\begin{bmatrix} (y_a + y_b) & -y_b & 0 \\ -y_b & (y_b + y_c + y_d) & -y_d \\ 0 & -y_d & (y_d + y_e) \end{bmatrix} \begin{bmatrix} V_1 \\ V_2 \\ V_3 \end{bmatrix} = \begin{bmatrix} I_1 \\ 0 \\ 0 \end{bmatrix}$$

Regardless of the number of buses present in a system, we always note the following pattern for the terms of $\mathbf{Y}_{bus}$:

$$Y_{ii} = \sum \text{ (all admittances incident at bus } i) \tag{4.55}$$

$$Y_{ii} = \text{(negative of admittance bridging buses } i \text{ and } j) \tag{4.56}$$

In building $\mathbf{Y}_{bus}$, any admittances in parallel are first added to yield single equivalent values (this was not an issue in the circuit of Figure 4.18).

Some properties of the bus admittance matrix are as follows:

- Symmetric for a passive network consisting of bilateral elements
- Independent of the injection currents (for a linear problem)
- Can be written from the network by inspection except when mutual coupling is present
- Usually complex
- Very sparse for large networks (that is, a significant percentage of its elements may be zero)
- Diagonally dominant (meaning that the magnitude of each diagonal term is usually greater than the magnitudes of the other terms in that same row and column)

The Bus Impedance Matrix

The bus impedance matrix is the inverse of the bus admittance matrix:

$$\mathbf{Z}_{bus} = \mathbf{Y}_{bus}^{-1} \quad (4.57)$$

and Equation 4.54 can be rearranged to yield

$$\mathbf{V} = \mathbf{Z}_{bus}\,\mathbf{I}. \quad (4.58)$$

We will now investigate some of the physical properties of $\mathbf{Z}_{bus}$.

Consider again the three-bus network of Figure 4.18. Assume that all the network impedances and admittances are known. Therefore, all the values of $\mathbf{Y}_{bus}$ could be found, and thus (assuming a nonsingular $\mathbf{Y}_{bus}$ matrix) the values of $\mathbf{Z}_{bus}$ could theoretically be found by inversion of $\mathbf{Y}_{bus}$. Until we do this inversion or the equivalent, we don't know the values of the terms of $\mathbf{Z}_{bus}$, but we can give them names, such as the position-identifying names Z_{11}, Z_{12}, and so on.

Suppose we inject only one current (I_1 at bus 1) into the network (that is, let $I_2 = 0$, $I_3 = 0$, as shown in Figure 4.18. Then Equation 4.58 becomes

$$\begin{bmatrix} Z_{11} & Z_{12} & Z_{13} \\ Z_{21} & Z_{22} & Z_{23} \\ Z_{31} & Z_{23} & Z_{33} \end{bmatrix} \begin{bmatrix} I_1 \\ 0 \\ 0 \end{bmatrix} = \begin{bmatrix} V_1 \\ V_2 \\ V_3 \end{bmatrix} \quad (4.59)$$

Expanding,

$$Z_{11} I_1 = V_1 \quad \text{or} \quad Z_{11} = V_1 / I_1$$
$$Z_{21} I_1 = V_2 \quad \text{or} \quad Z_{21} = V_2 / I_1$$
$$Z_{31} I_1 = V_3 \quad \text{or} \quad Z_{31} = V_3 / I_1.$$

All of this is of course conditioned on the fact that $I_2 = 0$ and $I_3 = 0$ here.

The equations below the matrix (Equation 4.59) thus show that *by injecting a current into bus 1 we can evaluate all the elements of column 1 of Z_{bus} in a simple and straightforward way*. Moving the source to bus 2 yields column 2 of Z_{bus}, and moving it to bus 3 yields column 3.

This current injection procedure is representative of what we could do in a laboratory with instruments and an actual network at our disposal. In a functioning power system, of course, it is normally not practical to shut the system down and make measurements to determine matrix parameters. However, the current-injection procedure is useful for defining the terms of $\mathbf{Z}_{bus}$, and also in some solutions such as finding short-circuit currents. It also tells us quite a bit of what $\mathbf{Z}_{bus}$ and its elements are, as we see next.

The diagonal terms of Z_{bus} are called **driving point impedances**, while the off-diagonal terms of Z_{bus} are called **transfer impedances**.

Some properties of the bus impedance matrix are as follows:

- Symmetric for a passive network consisting of bilateral elements
- Complex when network contains complex impedances (but real for a network of only resistors and imaginary for a network of only reactors)
- Independent of the injection currents (for a linear problem)
- Full for any connected network (in contrast with the typical sparsity of $\mathbf{Y}_{bus}$)
- Diagonally dominant

Physical, Global Meanings of the Terms of Y_{bus} and of Z_{bus}

From the preceding discussions we draw the following conclusions:

Y_{bus} can be assembled by inspection from a network (or, equivalently, element and connectivity data). In fact, we could alternatively do the opposite and construct the simplest network model corresponding to a given Y_{bus}. This matrix tells us what elements (or at least what simplest equivalents) are connected to what bus, and what their values are. However, each term of Y_{bus} is determined by one or a few network elements and tells us nothing about the rest of the network.

Z_{bus}, by contrast, tells us nothing about the individual element values. Except in the simplest cases, you cannot look at its terms and deduce any information about the values of the passive elements contained within the network (although you might get an idea whether they are all R, all L, or a combination, for instance). What Z_{bus} does do, by contrast, is to tell us what the system looks like electrically as viewed from its buses. In fact, this property is extremely important and simplifies some analysis techniques such as short-circuit study or a determination of the "electrical closeness" of a pair of buses (that is, how significantly a disturbance at one bus will affect what happens at another).

To summarize: While the terms of Y_{bus} are closely related to the values of the elements that make up the system, the terms of Z_{bus} are closely related to system properties as viewed from outside the system. Thus, while the system is properly described by either Z_{bus} or Y_{bus} for nodal analysis, Z_{bus} is more useful when we want to treat the system as an equivalent as viewed at its buses from the outside.

Y_{bus} is widely used in power-flow and stability studies. Z_{bus} is used extensively in short-circuit analysis.

RECOMMENDED REFERENCES

Nearly any undergraduate circuits text should be appropriate for much of the ac review material in this chapter. As an example, we recommend the following:

Nilsson and Riedel. *Electric Circuits*. 8th ed. Pearson/Prentice Hall, 2008.

Brief treatments of the meaning and use of the bus impedance and admittance matrix, with examples, may be found in most undergraduate electric power systems textbooks, for example,

Glover, Sarma, and Overbye. *Power Systems Analysis and Design*. 4th ed. Thomson, 2008.

This book also contains good review sections in which steady-state alternating-current problems are introduced from an ac power systems point of view.

CHAPTER 5

Balanced Three-Phase Circuits

OUTLINE

THE BALANCED THREE-PHASE SOURCE 64
Line-to-Line and Line-to-Neutral Voltages

PHASE SEQUENCES: abc AND acb 65
Reminders About Notation

THE THREE-PHASE SYNCHRONOUS GENERATOR 66

THREE-PHASE CONNECTIONS AND THE PER-PHASE EQUIVALENT 67
Per-Phase Equivalent ■ Wye Connnections and Equations for Total P, Q, and S ■ Delta Connections ■ Combination of Wye and Delta Connections

RECOMMENDED REFERENCE 77

This chapter addresses *balanced* steady-state three-phase operation only. Most three-phase electric power problems will involve balanced networks, because that is typically the desired mode of operation. In contrast, some insight into *unbalanced* networks is provided in Chapters 12 and 13. The presentation in this chapter will illustrate the use of some equations and concepts relevant to balanced three-phase problems as well as provide some review of single-phase, steady-state issues.

As always, our focus is on a practical review of critical topics for the PE exam rather than exhaustive theoretical coverage of these topics. If you feel that you need additional review in this area (or in undergraduate power-systems concepts in general), a good resource is cited in the recommended references at the end of this chapter. Related issues such as symmetrical components, per unit values, unbalances, and other more advanced power-system topics are covered in later chapters of this review, and in much greater detail in the Glover text and other undergraduate power-systems books. Texts on electric circuits and electric machinery often have a chapter or appendix addressing balanced three-phase circuits as well.

THE BALANCED THREE-PHASE SOURCE

Our review of three-phase theory begins with the **typical balanced three-phase source**, which can be viewed as a set of three ideal, synchronized single-phase sources, mutually displaced in phase by 120°. Figure 5.1 shows a schematic diagram of this ideal three-phase source, assumed connected in grounded-wye and with a balanced grounded-wye load attached. The **balanced load**, by definition, is one with identical impedance in each phase. The three individual sinusoidal sources of Figure 5.1 are synchronized to produce the three-phase, abc-sequence waveform shown in Figure 5.2a. The set of resulting line-to-line (phase-to-phase) voltages is shown in Figure 5.2b.

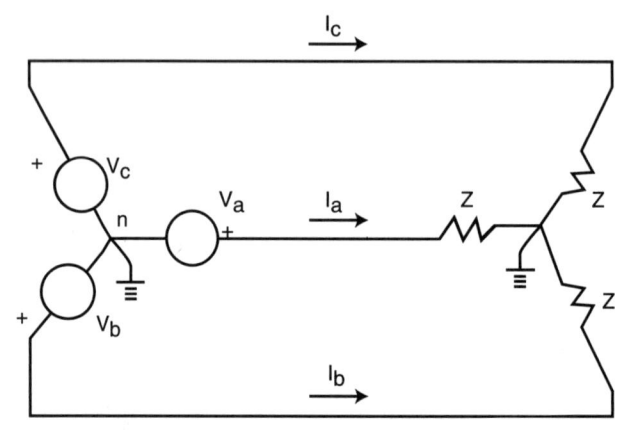

Figure 5.1 Circuit model of balanced three-phase, grounded-wye source and balanced grounded-wye load

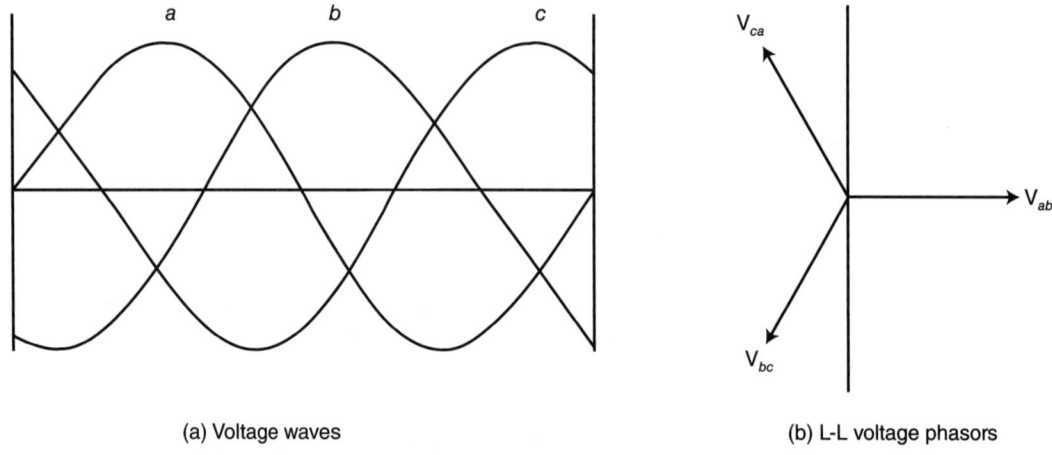

(a) Voltage waves (b) L-L voltage phasors

Figure 5.2 Balanced abc-sequence set of sinusoidal source voltages and line-to-line phasors

In practice, the three-phase source is typically a synchronous machine, and the synchronization of the sources comes about because of the three armature windings being uniformly distributed around the stator (see Chapter 10).

In many cases, this *ideal* three-phase source (having no internal impedance) will be found adequate for solving exam problems, or will be expressly stated in the problem; its use simplifies the problem even more. An actual three-phase generator or *alternator*, of course, will have as its simplest steady-state model an ideal set of synchronized sinusoidal sources each in series with an impedance, which is composed primarily of the *synchronous reactance*.

Line-to-Line and Line-to-Neutral Voltages

Note that, in the circuit model of Figure 5.1, there are two types of voltages: the **line-to-neutral (L-N, or phase-to-neutral)** voltages

$$\mathbf{V}_a (= \mathbf{V}_{an})$$
$$\mathbf{V}_b (= \mathbf{V}_{bn})$$
$$\mathbf{V}_c (= \mathbf{V}_{cn})$$

and the **line-to-line (L-L, or phase-to-phase)** voltages

$$\mathbf{V}_{ab} = \mathbf{V}_a - \mathbf{V}_b$$
$$\mathbf{V}_{bc} = \mathbf{V}_b - \mathbf{V}_c \qquad (5.1)$$
$$\mathbf{V}_{ca} = \mathbf{V}_c - \mathbf{V}_a.$$

The relationship between these balanced sets is shown in Figure 5.3 and some later figures in this chapter. Usually the phase-a voltage to neutral $\mathbf{V}_a$ is taken as the reference, so the phasor diagram orientation of this figure is more usual than that of Figure 5.2b.

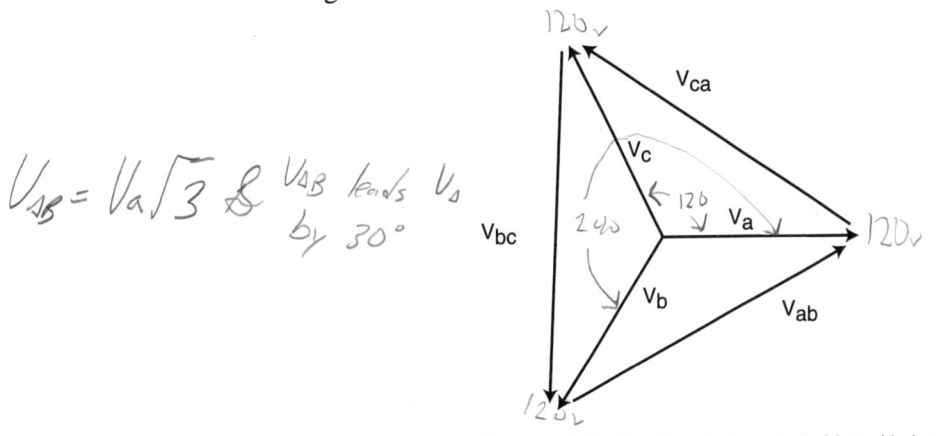

Figure 5.3 Relation between L-N and L-L voltages

Example 5.1

Assuming that $\mathbf{V}_a = 120 \angle 0°$ V in Figure 5.3, calculate $\mathbf{V}_{ab}$.

Solution
Because $\mathbf{V}_a = 120 \angle 0°$ V, we know that $\mathbf{V}_b = 120 \angle -120°$ V (and $\mathbf{V}_c = 120 \angle 120°$ V, although this is not needed in the present example).
Thus,

$$\mathbf{V}_{ab} = \mathbf{V}_a - \mathbf{V}_b = 120 \angle 0° - 120 \angle -120° = 208 \angle 30° \text{ V}, \qquad (5.2)$$

which you can see matches the phasor diagram (more will be said about this later).

PHASE SEQUENCES: abc AND acb

We have made the usual assumption of abc phase sequence so far. Note that there are only two possible sequences of the three voltages in either the L-N or L-L set. In fact, it is evident that once a sequence is identified in terms of L-N values, for instance, the same sequence holds for the corresponding L-L quantities. The sequence abcabc… is called the **abc or positive sequence** and is what we usually assume. It corresponds to the voltage progression $\mathbf{V_a}\,\mathbf{V_b}\,\mathbf{V_c}$ or, equivalently, $\mathbf{V_{ab}}\,\mathbf{V_{bc}}\,\mathbf{V_{ca}}$. The sequence acbacb… is called the **acb or negative sequence** and is important in considering unbalanced three-phase systems; it will be treated in Chapter 12. Note that we can always relabel the phases to achieve an abc sequence.

When a three-phase generator is driven by a prime mover, it produces a particular sequence as defined by rotational direction. We define the abc sequence to coincide with the specified rotational direction and label the terminals to reflect this accordingly. Thus, negative sequence implies phase rotation that is opposite to what the generator produces. One can understand the fact that there are only two possible phasor sequences in a three-phase system as a consequence of there being only two possible rotational directions for a generator shaft.

Reminders About Notation

In using equations concerned with steady-state ac analysis, it is important to know whether only the magnitudes or the actual phasor quantities are needed. As mentioned earlier in this book, complex quantities will usually be shown in **boldface** and their magnitudes in standard print. This may not always be done on exams. Careful attention to the problem statement will usually clear up what is given and what is required.

Just to emphasize the difference in vector and scalar quantities, recall from Chapter 4, in particular, that real or average power P and reactive power Q are scalars and, therefore, not complex numbers (this is a very common mistake). Thus, P and Q may be found from magnitudes (as in $P = V\,I\,\cos\varphi$) or as the real and imaginary parts, respectively, of an equation involving complex numbers ($P + jQ = \mathbf{VI}^*$). Also recall that resistance, conductance, inductance, capacitance, reactance, and susceptance are always real numbers; susceptance and reactance may be positive or negative. Steady-state voltages and currents may be phasors. Complex power S is always a complex number although its magnitude, called *apparent power*, is of course real. All of this is addressed in detail in Chapter 4.

As with single-phase circuits, let's again emphasize that the voltage from any point to neutral is sometimes expressed in single-subscript notation (where a reference is implied) and sometimes in double-subscript notation (which emphasizes, correctly, the point-function nature of an electric potential difference). The two notations carry the same meaning; thus, $\mathbf{V_{an}} = \mathbf{V_a}$ as shown earlier. Both notations will be found in practice.

THE THREE-PHASE SYNCHRONOUS GENERATOR

In order to get a better picture of where a set of three-phase source voltages comes from, we will begin with a brief introduction to an actual three-phase source, the three-phase synchronous generator (alternator). A simplified alternator, with stator windings and controllable dc rotor field winding, is shown in Figure 5.4. A much more detailed treatment of it is given in Chapter 10.

(a) Synchronous generator stator

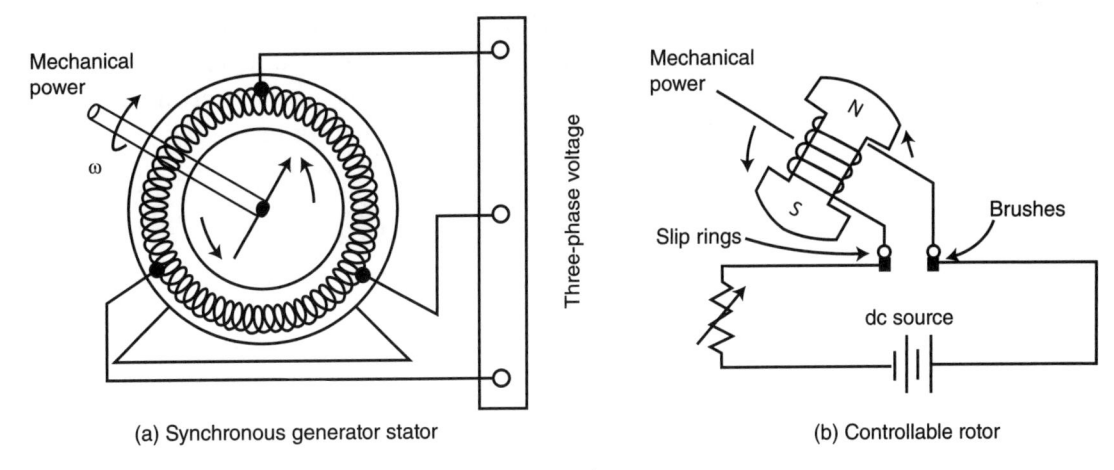

(b) Controllable rotor

Figure 5.4 Synchronous alternator in generator mode

The **stator** (stationary part of the machine), shown in Figure 5.4a, carries the **armature coils** which deliver the three-phase voltages. In an actual machine, these are usually much heavier than the rotor winding. The **rotor** (the rotating part of the generator) is assumed to be driven at a fixed speed ω by some mechanical source. The rotor consists of a ferromagnetic pole set carrying a field winding, as shown in Figure 5.4b. Direct current is conveyed to the field winding by means of slip rings, making the rotor a controllable electromagnet.

Note that the polarity of the rotor magnetic poles in Figure 5.4b conforms to the **right-hand rule** in terms of the current setting up that field, as provided by the battery symbol (whose larger end plate is positive). The right-hand rule states that if the fingers of the right hand are wrapped around a coil carrying current in the direction the fingers point, then the thumb points in the direction of the resulting north magnetic pole, or equivalently the direction of magnetic flux. Positive current is defined as being directed out of the large (positive) plate of the battery symbol, as illustrated in Figure 5.4b.

THREE-PHASE CONNECTIONS AND THE PER-PHASE EQUIVALENT

Balanced three-phase sources or loads may be connected in one of two configurations: **wye (Y)** or **delta (Δ)**. A wye may or not be grounded; for a balanced system it theoretically makes no difference since neutral current is zero in a balanced system operated at fundamental frequency. Note that the generator shown in Figure 5.4 happens to have a delta-connected set of armature coils. Most generators, at least smaller machines, are connected in wye (actually *grounded wye*), primarily to provide a firmly-grounded output. The neutral of the wye may be grounded through a small impedance to reduce the effect of ground fault currents should they occur. For simplicity, a wye source connection is normally assumed if the emphasis is not on the actual generator connection. Three-phase *motors* are often connected in ungrounded wye.

Per-Phase Equivalent

Note that in the balanced circuit of Figure 5.1, if the voltage of one phase to neutral (or equivalently, the voltage between a pair of lines) is known, then the other voltage phasors can be immediately drawn because we know what the sequence is (abc here). Similarly, if current I_a is known, for instance, then currents I_b and I_c are known also, because in a balanced case they are all identical in magnitude and their angles are known because they follow in sequential order.

This leads to the idea of the **per-phase equivalent**. In solving a balanced three-phase problem, a per-phase model for one of the phases (usually phase a) is first drawn and solved. From this we can obtain any or all of the other phase voltages and currents if desired.

The phase angle φ referred to in three-phase relationships is the angle between a given phase voltage and its corresponding current; e.g., the angle by which $\mathbf{V}_a$ leads $\mathbf{I}_a$ in Figure 5.1.

The per-phase equivalent circuit corresponding to the simple case of Figure 5.1, containing one three-phase source and one three-phase load, is shown in Figure 5.5.

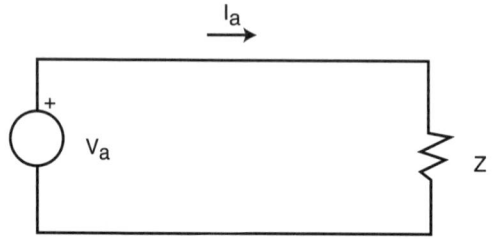

Figure 5.5 Per-phase equivalent of Figure 5.1

Balanced three-phase problems are almost always solved on a per-phase, line-to-neutral basis. This per-phase equivalent lends itself especially well to wye-connected devices, so delta connected transformer windings, loads, etc. are often converted to an equivalent wye, although for some primarily delta-connected circuits, this step is not necessary. It is usually a great solution aid to draw the per-phase equivalent circuit for any given problem. Voltages and currents in this per-phase equivalent are usually taken to be those of phase a. Of course, if we know what is happening in phase a, then the corresponding conditions in phases b and c may be readily evaluated. Furthermore, each phase supplies or consumes one-third of the total real, reactive, or complex power, so any such value calculated for the per-phase equivalent can be multiplied by 3 to obtain the total. In simple problems on the exam, such as those involving one source and one load (with perhaps a connecting line of given impedance), it may not be necessary to draw or solve a per-phase equivalent in order to obtain what is requested. However, sketching the per-phase equivalent is usually helpful, at least to help you avoid errors.

Three-phase devices are usually specified in terms of their line-to-line voltage and total three-phase apparent power. Thus, a load rated 208 V, 5 kVA will imply a line-to-line voltage rating of 208 V and it will consume a total of 5 kVA when this voltage is applied.

The per-phase representation of a three-phase generator is simply a Thevenin equivalent, with an ideal sinusoidal source in series with an impedance. You can expect that some exam problems might specify a generator and some may simply specify an ideal source (perhaps called a constant-voltage source or just a source).

Understanding the need to develop a per-phase equivalent circuit for each balanced circuit problem, let's next examine the two possible balanced connections and develop their per-phase models.

Wye Connections and Equations for Total P, Q, and S

First we consider the wye connection. A simple system consisting of an ideal, balanced three-phase source and resistive load is shown in Figure 5.6, where, to make a point, each of the phases is connected as separate single-phase circuits (we will see shortly how they are actually interconnected in practice). The voltage source in each phase (which gives rise to a *phase voltage*) is connected to each load resistance in the diagram by lines representing zero impedance. The presence of the three separate single-phase circuits in this illustration emphasizes the somewhat independent aspect of each of these phases.

As is typical, we have let phase *a* be the reference by setting its source voltage to zero degrees.

As an example, suppose that each phase voltage in Figure 5.6a is 100 V and that each resistor is of value 5Ω. Then in phase *a* the current is $\mathbf{I}_a = \mathbf{V}_a/R$, or 20 A at an angle identical to that of the voltage $\mathbf{V}_a$, and the power dissipated in the resistor is $P_a = I_a^2 R$ or 2kW. This is clearly seen in the single-phase equivalent of Figure 5.6b, which typifies conditions in phase *a*. In the complete balanced three-phase circuit, of course, the total power dissipated in all of the loads is merely the sum of the individual powers, or 6 kW. Also, because this is a wye connection, the phase currents are the same as the line currents.

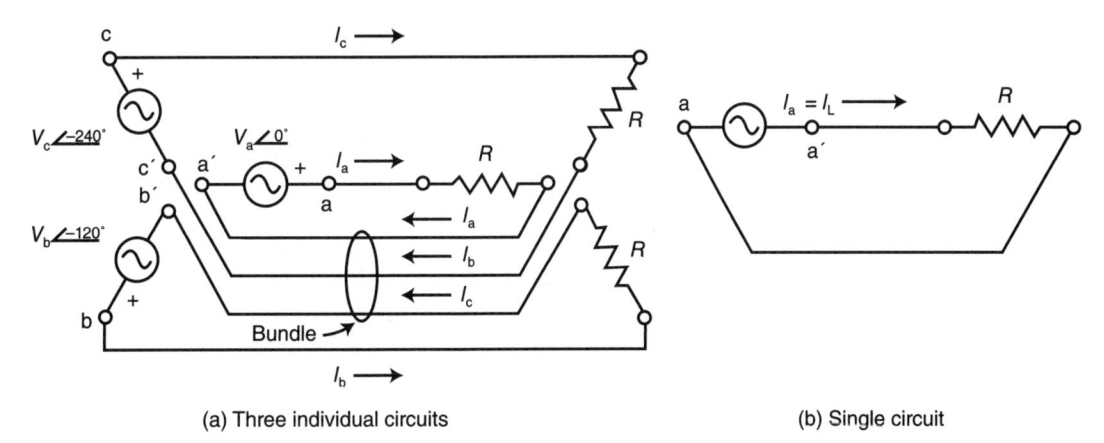

(a) Three individual circuits   **(b)** Single circuit

Figure 5.6 A conceptual "three source" system

All three currents represented by the bundle of return lines are, of course, the same as their corresponding phase currents. Because the return currents are independent, the wires in the return bundle could be connected together as one single conductor, called the **neutral**. This is in fact done in practice, as shown in Figure 5.7a. The current in the neutral wire is, by Kirchhoff's current law,

$$\mathbf{I}_n = \mathbf{I}_a + \mathbf{I}_b + \mathbf{I}_c, \tag{5.3}$$

which for this balanced case, where the currents are all equal in magnitude and mutually displaced by 120 degrees, becomes

$$\mathbf{I}_n = \mathbf{I}_a [(1\angle 0° + 1 \angle -120° + 1 \angle +120°)] = 0. \tag{5.4}$$

Because it carries zero current, there is no theoretical need for the (dashed) neutral conductor in a perfectly-balanced system. Usually the neutral conductor is in fact present; some of its purposes are to carry any unbalanced or harmonic currents, to provide a safety ground, and to keep the potential of the neutral point close to zero.

A phasor diagram showing the voltages and currents in the simple wye-connected circuit is shown in Figure 5.7b. Note that the phase sequence is *abc*. A graphical construction is presented on the phasor diagram to emphasize that

$$\mathbf{I}_a + \mathbf{I}_c = (-\mathbf{I}_b).$$

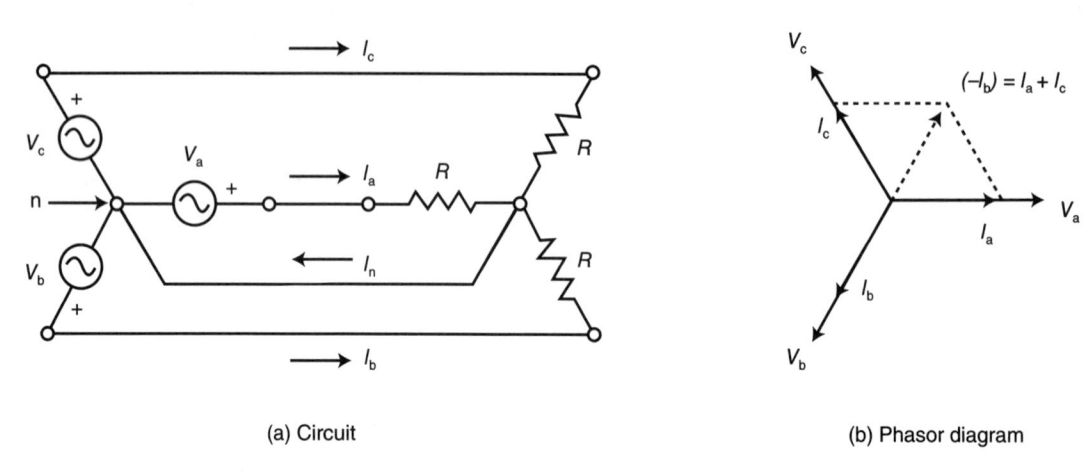

(a) Circuit

(b) Phasor diagram

Figure 5.7 A wye-wye three-phase circuit with ideal source and resistive load

A line-to-line voltage is shown as a measurement in Figure 5.8a. From the phasor diagram of Figure 5.8b, we see, as before, that this line-to-line voltage $\mathbf{V}_{L1}$ or $\mathbf{V}_{ab}$ is $\sqrt{3}$ times as large as $\mathbf{V}_{an}$ in magnitude and leads V_{an} by 30°. Of course the other two line-to-line voltages may be found in the same manner. It should be emphasized that for balanced loads, only the one leg need actually be solved and the required values in the other two are found by similarity.

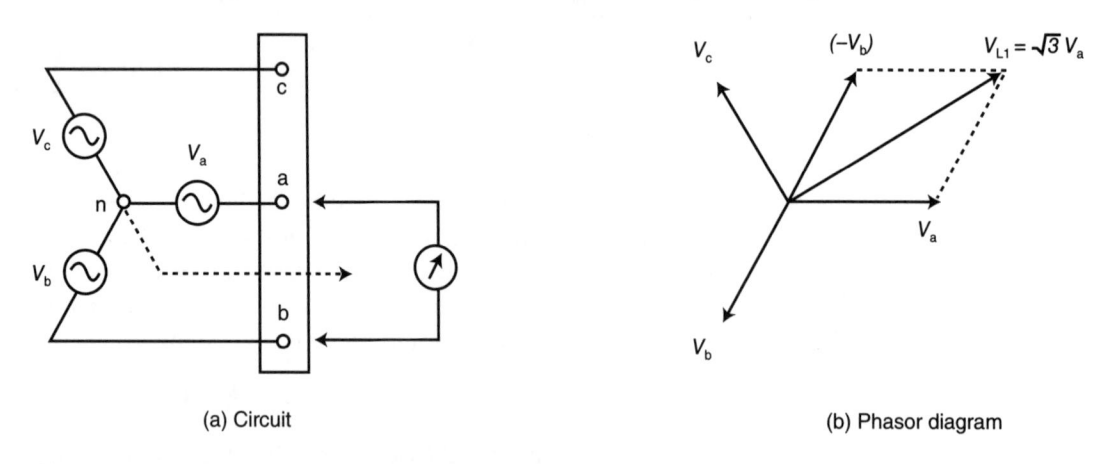

(a) Circuit

(b) Phasor diagram

Figure 5.8 Phase and line voltage relationship for a wye connection

The total power taken from the source or dissipated by the full load (assuming no power loss in the line) is given for balanced conditions by the following easily-derived three-phase formula which is in fact valid for both delta and wye connections:

$$P_{total} = \sqrt{3}\, V_{LL}\, I_L \cos \varphi \qquad (5.5a)$$

where

V_{LL} = the magnitude of a line-to-line voltage
I_L = the magnitude of a line current
φ = the power-factor angle.

Be especially careful to note that φ *is the phase angle between the voltage and current in the reference phase* (that is, the angle between $\mathbf{V}_a$ and $\mathbf{I}_a$); as explained in Chapter 4, it is said that the angle is *lagging* if the current is behind the voltage phasor and *leading* if ahead. The *power factor* (*PF*) is of course cos φ. Note that these definitions follow the single-phase practice of Chapter 4 exactly. Another way of writing Equation 5.5a is given in Equation 5.5b:

$$P_{total} = \sqrt{3}\ V_{LL}\ I_L\ (PF). \qquad (5.5b)$$

Recall that complex power, designated S or VA (for volt-amps), was defined in Chapter 4 as

$$\mathbf{S} = P + jQ$$

where *P* is the real power (in watts) and *Q* is the reactive power (in VArs). This equation is valid for a single-phase device or for any leg of a three-phase circuit if the values are interpreted as per-phase.

For a three-phase balanced load, the magnitude of the total complex power (that is the apparent power) is

$$S_{total} = \sqrt{3}\ V_{LL}\ I_L\ \text{volt-amps}. \qquad (5.5c)$$

The total complex power in this case is

$$\mathbf{S}_{total} = \sqrt{3}\ V_{LL}\ I_L\ (\cos \varphi + j \sin \varphi) \qquad (5.5d)$$

where, again, φ is the power-factor angle associated with one phase (that is, the angle between a line-to-line voltage and its corresponding phase current in a delta connection, or a line-to-neutral voltage and its corresponding line or phase current in a wye connection).

Remember that, just as in single-phase circuits, *real (or average) power P can only be dissipated in a pure resistance, and reactive power Q is produced by capacitors and consumed by inductors.* As a scalar, real power dissipation can always be found by summing all the I^2R values in the circuit; the only caution being that in ac machinery analysis there may be a fictitious resistance that represents mechanical power. Reactive power can be found similarly by summing up all the I^2X terms.

Because the concept is so important, we restate that equations 5.5a through 5.5d actually hold for any balanced load, whether wye or delta. When using them with a delta load, be sure that the power-factor angle is the angle between the voltage across one delta-connected impedance and the current through *that particular impedance*.

Of course, because the wye load above is balanced, and because line and phase currents are the same in the wye load, the total load power is also. Therefore the resistive load discussed above,

$$P_{total} = 3\ I_L^2\ R \quad \text{for per-phase load resistance R.}$$

Similar equations to those for P hold for Q in terms of sin φ, as Example 5.2 illustrates.

Example 5.2

For a balanced three-phase wye circuit, the voltage-current relationship in one leg of the wye, consisting of a resistor in series with an inductance, yields the following rms measurements: current = 5A, the voltage across R is 70.7 V, and the voltage across L is likewise 70.7 V. Determine the magnitude of the reactive power Q taken by the total three-phase configuration.

Solution

Because there are two load elements in series, it makes sense to assume the current as reference. Equal voltages across the series R and L of course imply equal impedance magnitudes, so the phase (line-to-neutral) voltage across the load must be

$$\mathbf{V}_p = 70.7 + j70.7 = 100\angle 45° \text{ V}.$$

Then,

$$V_{LL} = \sqrt{3}\,(100) = 173.2 \text{ V}$$

so

$$Q_{total} = \sqrt{3}\,V_{LL}\,I_L \sin\varphi = \sqrt{3} \times 173.2 \times 5 \times \sin 45° = 1061 \text{ VAr}.$$

Alternatively,

$$Q_{total} = 3\,I^2 X_L = 3 \times 25(70.7/5) = 1061 \text{ VAr}.$$

Delta Connections

In a delta connection, the three load (or source) elements are connected line-to-line as shown for the load of Figure 5.9a, rather than line-to-neutral. It is obvious that there is no neutral return and that the phase voltage across each leg of the delta is equal to the line-to-line voltage. These voltages are, of course, 120° out of phase with each other in the balanced case. One of these line-to-line voltages is usually made the reference in solving delta-connection problems that are not converted into an equivalent wye.

The phase currents are merely the respective line voltages divided by the phase impedances; these three currents, of course, are also 120° out of phase with each other (see Figure 5.9b which, as an example, gives conditions for a unity power-factor load). The subscript p on the currents in the figure emphasizes that these are phase currents; i.e., the currents in the actual phase components (the resistors in this case) as distinct from line currents, which are currents *into* the delta.

Because there are two types of currents associated with a delta load (line and phase currents), we have adopted a different notation for the delta of Figure 5.9 compared with that of the previous (wye) illustration. As shown on the figure, line currents are sometimes designated L_1, L_2, and L_3 even where letter names are used for the phases. (This is worth knowing because many actual devices have line current, and even voltage, designations in terms of letters and numerals.)

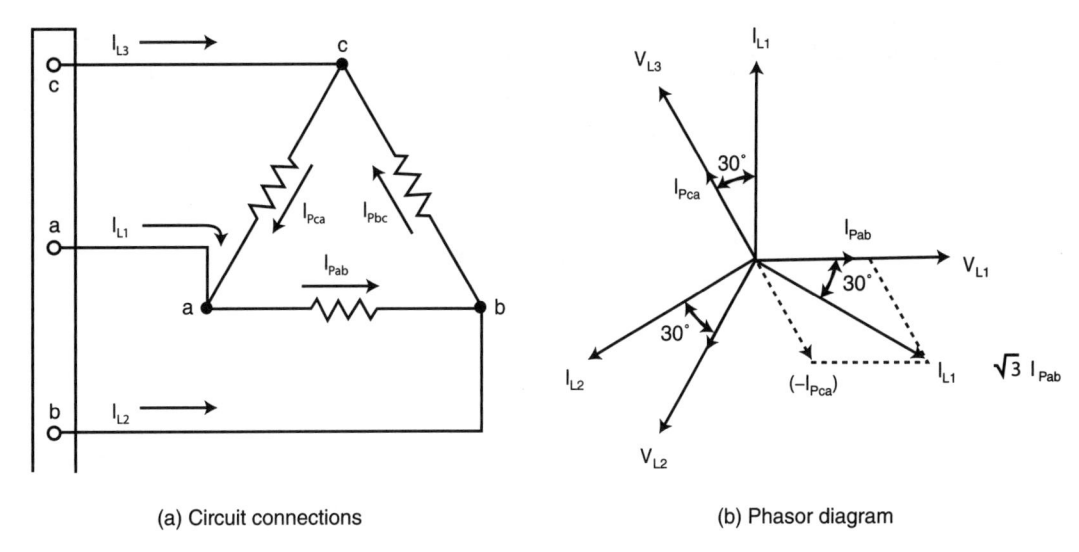

Figure 5.9 The delta relationship

(a) Circuit connections

(b) Phasor diagram

In the balanced delta, the phase current is the same in magnitude for each leg. The line currents are the phasor sums at each junction of the respective phase currents, and it is easily shown that each line current is equal to $\sqrt{3}$ times the phase current in magnitude, with the two sets of currents separated by respective angles of 30°. For a delta circuit, then, we could say that line-to-line voltages appear across the loads and the relationship between the phase and line current magnitudes is

$$I_L = \sqrt{3}\, I_p. \tag{5.6}$$

As an example, with reference to Figure 5.9, the line current associated with phase a can be written in phasor form specifically as

$$\mathbf{I}_a = \mathbf{I}_{L1} = \sqrt{3}\, \mathbf{I}_{pab} \angle{-30°}$$

where $\mathbf{I}_{pab}$ is the phasor phase current in leg ab.

Example 5.3

As an example of a balanced delta connection, consider a load having three phase impedances of 5 + j5 Ω each, connected as a delta, with the applied line (line-to-line) voltage equal to 173.2 V. Determine the power per phase and total power.

Solution

Choose a line-to-line voltage to be the zero-degree reference. Then, the corresponding reference phase current within the delta is

$$\mathbf{I}_P = (\mathbf{V}_P)/\mathbf{Z} = (173.2 \angle 0°)/(5 + j5) = 24.5 \angle{-45°}\,\text{A}.$$

The power per phase is found as

$$P = (I_P)^2\, R = (24.5)^2\, 5 = 3\,\text{kW},$$

and thus the total power is 9 kW.

Again, using the general power Equation 5.5b as a check, the power taken from the source is given by

$$P_{total} = \sqrt{3}\, V_{LL}\, I_L\, PF = \sqrt{3}\,(173.2)(\sqrt{3}\,\, 24.5)(0.707) = 9\,\text{kW}.$$

Example 5.4

A balanced delta load is connected to a three-phase source of 200 V. If each leg of the load has an impedance of $10 + j10\ \Omega$, determine the magnitude of the line currents and the total power dissipated in the load.

Solution

$$I_L = \sqrt{3}\ I_P = \sqrt{3}\ [200 / |(10 + j10)|\ ] = 24.5\ \text{A in each line,}$$

so

$$P_{total} = \sqrt{3}\ V_L I_L \cos \varphi = \sqrt{3} \times 200 \times 24.5 \cos 45° = 6\ \text{kW.}$$

Combination of Wye and Delta Connections

Both delta and wye connections are often found in the same problem. For instance, a wye-connected source might supply a delta-connected load, or loads of two different configurations may be in parallel. Following are some suggestions for solving such problems.

An ideal source can always be viewed as either delta or wye, depending upon what is more convenient to solve the circuit. That is because once the L-L voltages are given, the L-N voltages may be determined immediately, or vice versa (see the phasor diagram of Figure 5.3). It is often convenient to assume a delta source if the load is connected in delta directly across it, because you then know the voltage across each leg of the load.

However, it is usually more convenient to convert all the delta loads to wye, and assume a wye-connected source. As mentioned earlier, this fits the guidelines of the per-phase approach best. If a circuit contains only a source, a line, and a wye load, then no conversion is necessary. If a delta load is given, convert it to wye.

You can always convert a balanced delta to a balanced wye, or the reverse, using

$$Z_{wye} = Z_{delta} / 3. \tag{5.7}$$

(This was cited for the more general, unbalanced case in Chapter 3.)

Note that the magnitudes of the delta and wye impedances are related by the factor 3 and their angles are the same.

Example 5.5

Two balanced three-phase loads are connected to the same voltage source of 173.2 V (line-to-line). Each load has an impedance of $10 + j10\ \Omega$ in each leg, but one load is connected in wye, while the other is connected in delta. What is the magnitude of the line current as measured at the source?

Solution

Although the problem doesn't state it, a parallel connection of loads is logical to assume. For instructional purposes, we will solve this problem in three ways.

First, let us assume that a line-to-line voltage is reference. Then, for the wye, the reference phase voltage is $100\angle-30°$ V.

For the wye-connected load,

$$I_L = I_P = V_P / Z = (100\angle-30°)/(10 + j10) = 7.07\angle-75°\ \text{A.}$$

For the delta load, $V_L = V_P$ and so

$$I_P = V_P/Z = 173.2\angle 0° /(10 + j10) = 12.23\angle -45° \text{ A}$$

and

$$I_L = \sqrt{3}\, I_P \angle -30° = \sqrt{3}\,(12.25\angle -45°)(1\angle -30°) = 21.2\angle -75° \text{ A}.$$

The line current is the phasor sum of the total load current:

$$I_{L\,total} = I_{L\,wye} + I_{L\,delta} = 7.07\angle -75° + 21.2\angle -75° = 28.27\angle -75°$$

and thus the magnitude of the total current is I_L total = 28.3 A.

This method of solution is correct, but requires careful consideration of the reference angle. Thus, it is not the way in which this kind of balanced problem is usually solved.

Perhaps a less risky method of solution is to first find the values of power using the current *magnitudes* in the separate loads. Using the values for the currents above

$$P_{wye} = 3(I_P)^2 R = 3(7.07)^2 10 = 1500 \text{ W}$$
$$P_{delta} = 3(I_P)^2 R = 3(12.25)^2 10 = 4500 \text{ W}.$$

Because real power values are scalars, the total is just the sum of the individual values

$$P_{total} = P_{wye} + P_{delta} = 6000 \text{ W}.$$

We know, because both loads have the same power factor, that the source sees a power-factor angle $\varphi = 45°$

and that $P_{total} = \sqrt{3}\, V_L I_L \cos \varphi = \sqrt{3} \times 173.2 \times I_L \cos 45°$.

Solving for the total line current,

$$I_L = 6000/(1.73 \times 173.2 \times 0.707) = 28.3 \text{ A}.$$

Probably the *simplest and least error-prone way* to solve this kind of problem, however, is to convert the delta load to a wye as suggested above, and build the per-phase equivalent as suggested in the text.

The per-phase equivalent of the delta-connected load is then $(10 + j10)/3 = 3.33 + j3.33\ \Omega$, and paralleling this with the original wye-connected load of $10 + j10\ \Omega$ yields an equivalent impedance by the product-over-sum rule for parallel impedances of

$$Z_{eq} = (10 + j10)(3.33 + j3.33) / (10 + j10 + 3.33 + j3.33) = 3.54 \angle 45°\ \Omega.$$

Letting the phase-to-neutral voltage be the reference this time (which is what is most frequently done in practice),

$$I_{L\,total} = (173.2/\sqrt{3})/3.54\angle 45° = 28.3\angle -45° \text{ A}.$$

Example 5.6

The real power input to a 240 V three-phase motor is 10 kW. The motor's efficiency is 80 percent and the power factor of the device is also 80 percent lagging. Determine the horsepower output of the motor and the magnitude of the input line current.

Solution

Because the efficiency is 80 percent, the power output is equivalent to 8000 W. In horsepower, this is

$$hp_{out} = 8000 \text{ W} / 746 \text{ W/hp} = 10.7 \text{ hp}.$$

The equation for power input is

$$P = \sqrt{3}\ V_{LL}\ I_L\ (PF);$$

thus, $I_L = P/[\sqrt{3}\ V_{LL}\ (PF)] = 10000/[\sqrt{3} \times 240\ (0.8)] = 30.1$ A.

Note that this is the same *magnitude* of current as would be obtained if the power factor had been 80 percent leading. (For a more comprehensive but single-phase example of power-factor correction, see Example 4.3 in Chapter 4.)

Example 5.7

If the motor in Example 5.6 were connected to the terminal board shown in Figure 5.10 and the meter connected between terminals *a* and *b* reads 240 V, determine both the (a) line-to-neutral voltage magnitude and the (b) angle between the phase-a voltage and the phase-a current.

Solution

a) $V_{LN} = V_{LL} / \sqrt{3} = [240/\sqrt{3}] = 138.6$ V

b) Because the power factor is given, we immediately find the phase angle:

$$PF = \cos \varphi = 0.8, \text{ and thus } \varphi = 36.9°$$

Because it is a lagging power factor, $\mathbf{I}_a$ lags $\mathbf{V}_a$. Choosing $\mathbf{V}_a$ to be the reference,

$$\mathbf{V}_a = 138.6 \angle 0° \text{ V}$$

and then

$$\mathbf{I}_a = 30.1 \angle{-36.9°} \text{ A}.$$

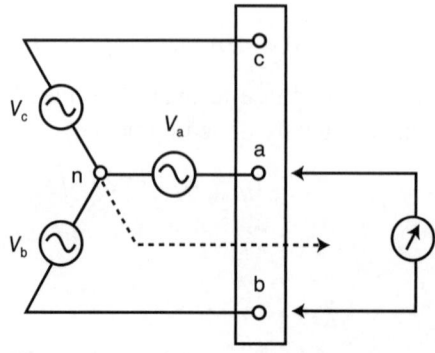

Figure 5.10 A three-phase voltage source connected internally as a wye

RECOMMEND REFERENCE

Glover, Sarma, and Overbye. *Power Systems Analysis and Design*. 4th ed. Thomson, 2008.

CHAPTER 6

Per-Unit Methods and Calculations

OUTLINE

PER-UNIT, PERCENT, AND BASE QUANTITIES 79
Application to Single-Phase AC Circuits ■ Application to Three-Phase Circuits
PRINCIPAL ADVANTAGES OF PER-UNIT REPRESENTATION 83
OTHER NUMERICAL EXAMPLES 83
RECOMMENDED REFERENCE 83

Normalized numerical values, such as per-unit and percent quantities, are widely used in electric power studies and problem solutions. Most numerical problems involving transformers and machines on the PE exam will require at least some knowledge of per-unit methods. Thus, you should be familiar and comfortable with this topic, which is straightforward and easily learned.

PER-UNIT, PERCENT, AND BASE QUANTITIES

It is often convenient to express a quantity as a ratio or percentage of some base quantity. For instance, if a process requires two hours to complete, we say that in one hour it is 50 percent complete. Alternatively, we could say that in one hour, the task was *0.5 per-unit* complete.

Many electric-power problems are formulated or solved using per-unit quantities. This *normalized* representation, which often has significant advantages, is so pervasive that it is essential to become familiar with per-unit (and percent) formulation and solution. For instance, it is nearly universal to find machine impedances expressed per-unit, while transformer impedances are almost always given in percent values (in each case on the base ratings of the device). In power systems work, solving complete problems in per-unit values rather than with actual values is often advantageous. Some of the advantages will be explored and summarized later in this chapter.

A per-unit number is usually written with the abbreviation *pu* following it, while a percent number is usually followed by the percent sign (%).

The **per-unit** value of a quantity is the ratio of an **actual value** of the quantity to the **base value**. Equation 6.1 summarizes this relationship. Note that the actual and base quantities must be the same type of quantity.

$$\text{per-unit value} = \text{actual value} / \text{base value} \tag{6.1}$$

In ac circuits, we must deal with both scalar and vector quantities. Base quantities are always scalars; thus, it is important to note that *the per-unit process scales the magnitudes but not the angles of vector quantities*. For example, per-unit phasor voltage (often written puV or $\mathbf{V}_{pu}$) can be expressed as

$$\mathbf{V}_{pu} = \mathbf{V}_{actual} / V_{base} \tag{6.2}$$

where the boldface $\mathbf{V}_{pu}$ and $\mathbf{V}_{actual}$ are vector quantities.

As another example, per-unit impedance magnitude (often written puZ or $\mathbf{Z}_{pu}$) can be expressed as the ratio of actual vector impedance to base impedance:

$$\mathbf{Z}_{pu} = \mathbf{Z}_{actual} / Z_{base}. \tag{6.3}$$

This, of course, applies to the magnitude and to the real and imaginary parts R and X of Z as well:

$$Z_{pu} = Z_{actual} / Z_{base}$$
$$R_{pu} = R_{actual} / Z_{base}$$
$$X_{pu} = X_{actual} / Z_{base}.$$

Base quantities relate by conventional physical laws just like their actual counterparts in the real world, and per-unit quantities do also. Thus, for instance, base admittance must be the reciprocal of the base impedance:

$$Y_{base} = 1 / Z_{base}$$

and per-unit admittance is just the reciprocal of the corresponding per-unit impedance:

$$\mathbf{Y}_{pu} = 1 / \mathbf{Z}_{pu}. \tag{6.4}$$

Note that a **percent quantity** is just a per-unit quantity multiplied by 100%.

Example 6.1

Find the pu voltage corresponding to the actual phasor voltage $V_a = 116\angle 25°$ V in a system where base voltage is 120 V.

Solution

$$V_{a\,pu} = (116\angle 25°) / 120 = 0.9667 \angle 25°) \text{ pu V}$$

We read pu V as "per-unit volts." Strictly speaking, a pu quantity does not have units; however, because we need to distinguish between a number of different pu quantities, we usually print or state the units corresponding to the actual quantity after the per-unit value as we did in this example.

Application to Single-Phase AC Circuits

Per-unit methods are applicable to dc as well as ac circuits, and for polyphase as well as single-phase problems. Power engineers work most frequently with steady-state single- and three-phase problems, of course. Single-phase circuits will be explored first.

Four base quantities are of importance at any point in an ac circuit:

- V_{base} base voltage
- S_{base} base volt-amperes (identical to base P or Q)

- I_{base} base current
- Z_{base} base impedance (identical to base R or X)

(This is in fact true for three-phase as well as single-phase systems, although the four quantities take on slightly different meanings for the three-phase case.) Because base quantities relate to each other like their actual counterparts, it can be seen that only two of these base quantities above are independent. That is, selection of two of them determines the other two. In electric power work, we normally select V_{base} and S_{base}, or as they are usually expressed in power-system problems, kV_{base} and either MVA_{base} or kVA_{base}. The following relations between base quantities apply:

$$I_{base} = VA_{base} / V_{base} \tag{6.5}$$

$$Z_{base} = V_{base}^2 / VA_{base} \tag{6.6}$$

or, as more usually found in electric power problems,

$$I_{base} = kVA_{base} / kV_{base} \tag{6.7}$$

$$Z_{base} = (kV_{base}^2 * 1000) / kVA_{base} = kV_{base}^2 / MVA_{base}. \tag{6.8}$$

Base quantities carry the units of their actual counterparts. Thus, base current is usually expressed in A or kA, base voltage is expressed in V or kV, etc.

Per-Unit Calculations and Transformers

The per-unit system poses particular problems, as well as very important advantages, for transformers. This is because a transformer "transforms" voltage, current, and impedance, as discussed in the next chapter. Chapter 7 provides a detailed section on the use of per-unit calculations with transformers. Be sure to examine this section, because transformer per-unit problems are certainly possible on the exam.

Change of Base

Once we select two of the base quantities (such as kV_{base} and kVA_{base}) in one part of a circuit, all the other base quantities in this part of the circuit *and elsewhere in the circuit* are automatically fixed. It is very likely, therefore, that a device introduced into the circuit will have a rating or ratings different from the corresponding system base(s) in that region. How do we handle such a situation?

Placing a device into a region of the network having different bases from the device ratings requires modification of the per-unit impedance of the device. The equation for doing this, known as the **change-of-base equation**, may be stated as

$$Z_{pu\,sys} = Z_{pu\,device} * (V_{base\,device} / V_{base\,sys})^2 * (VA_{base\,sys} / VA_{base\,device}) \tag{6.9}$$

where

$Z_{pu\,device}$ = the per-unit impedance on the device base (nameplate value)

$Z_{pu\,sys}$ = the corresponding per-unit impedance on the new (system) base

$V_{base\,device}$ = the base (rated) voltage of the device

$V_{base\,sys}$ = the system voltage base at the point of connection of the device

$VA_{base\,sys}$ = the system base VA at the point of connection of the device

$VA_{base\,device}$ = the base (rated) VA of the device.

Because the rated kVA of a device is not likely to be the same as the base kVA of the circuit, this equation becomes very important and must be applied frequently. A word of caution: Note carefully the multipliers (k, M, etc.) in base and actual quantities in your problem.

Application to Three-Phase Circuits

Application of per-unit methods to three-phase circuits involves a straightforward extension of the single-phase case. Recall that for three-phase balanced circuits the specified kVA is understood to be the total for all three phases and the kV is understood to be the line-to-line (or phase-to-phase) kV. Therefore, for a three-phase system we usually specify kVA_{base}, meaning total three-phase base kVA, and kV_{base}, that is, line-to-line base kV. From these, the remaining base quantities can be found as

$$I_{base} = kVA_{base} / (\sqrt{3}\, kV_{base}) \tag{6.10}$$

$$Z_{base} = (kV_{base}^2 * 1000) / kVA_{base} \tag{6.11}$$

$$= (kV_{base}^2 *) / MVA_{base}. \tag{6.12}$$

From an examination of these relationships, the two calculated base quantities may be interpreted as follows:

- I_{base} is the line current which would flow in any of the three phase conductors with the device operating at 1.0 per-unit kV and 1.0 per-unit kVA.

- Z_{base} is the equivalent impedance of one phase of a wye-connected load, operating at 1.0 per-unit kV, which produces 1.0 per-unit kVA.

Example 6.2

In a certain part of a three-phase circuit, the following base quantities have been selected:

$$kV_{base} = 345\ kV$$
$$MVA_{base} = 100\ MVA$$

A certain transformer leakage reactance is given as 0.1 pu on this base. Calculate the actual value of this reactance, in ohms.

Solution
This example is typical of a problem that might be on the exam: one in which you don't really need to know much about the device in question in order to solve for what is required.

As is typical, we interpret kV_{base} to be line-to-line and MVA_{base} to be total three-phase. Using Equation 6.12,

$$Z_{base} = 345^2 / 100 = 1190.25\ \Omega.$$

Then, from Equation 6.3,

$$X_{T\ actual} = (0.1)(1190.25) = 119.025\ \Omega.$$

We will look at transformer issues further in Chapter 7.

PRINCIPAL ADVANTAGES OF PER-UNIT REPRESENTATION

Here is a summary of advantages of per-unit representation that may simplify your approach to solving several types of problems on the PE exam:

- Per-unit values of voltage, current, and impedance are automatically referred to the proper voltage level when going across a transformer. Therefore, in circuits where an actual transformer is present, the ideal transformer can be omitted from the model (see Chapter 7).

- Many per-unit quantities lie within a narrow range despite significant differences in device rating. This is especially true of impedances of transformers and rotating machines. Thus, if the value of one of the parameters is unknown, it is often possible to make a reasonable estimate in per-unit.

- Per-unit values are often easier to compare than actual values. This is especially helpful in a power-flow or other system study where several different nominal voltage levels are present. For instance, a power-flow study might include buses whose nominal voltages are 345 kV, 115 kV, 34.5 kV, etc. Comparing the voltage magnitudes at buses on the two sides of a transformer is much easier when these values are expressed in per-unit than if they were given in kV. In fact, power-flow outputs usually show bus voltages (but not line P and Q flows) in per-unit (see Chapter 11).

- Manufacturers often specify data, especially machine and transformer impedances, in per-unit or percent. (These are understood to be expressed in terms of the device ratings as base.) This practice is so widespread that it would be unlikely to find a machine reactance or transformer leakage impedance given in actual ohms.

- In many numerical studies, and especially in more involved power-system studies such as dynamic machine modeling in stability evaluation, normalized quantities offer some computational advantages. This is partially because the per-unit values fall into narrower ranges than actual values, making numerical computation (especially in iterative and related processes) more stable. Numerical analysts have long exploited these advantages.

Although they are similar (differing only by a scaling factor of 100), per-unit quantities have an additional advantage over percent quantities. *The product, sum, quotient, square, etc. of a per-unit quantity or quantities is another per-unit quantity. This is not true for most operations on percent quantities except for addition and subtraction.*

OTHER NUMERICAL EXAMPLES

Numerical examples of the use of the per-unit method are found throughout this book in the relevant topics. For example, Chapter 13 provides an extended numerical example of the use of the per-unit method in three-phase short circuit study. Chapter 7 treats in further detail the use of per-unit methods in connection with transformers.

RECOMMENDED REFERENCE

Glover, Sarma, and Overbye. *Power Systems Analysis and Design.*
    4th ed. Thomson, 2008

CHAPTER 7

Transformers

OUTLINE

ELECTROMAGNETIC FUNDAMENTALS OF TRANSFORMERS 85
The Basic Two-Winding Ideal Transformer and Polarity Considerations ■ Polarity Conventions and Winding Names ■ Impedance Referral and Power Relations ■ Transformer Ratings

PER-UNIT CALCULATIONS AND BASE QUANTITIES 89
Voltage Regulation

MULTI-WINDING, BUCK-BOOST, AND AUTOTRANSFORMER CONNECTIONS 92
Buck-Boost Transformers ■ Autotransformer Connections

REAL TRANSFORMERS: LOSSES, IMPEDANCES, AND MODELS 96
Efficiency ■ Voltage Regulation Revisited ■ Importance of the Phasor Diagram

TRANSFORMER TESTS TO DETERMINE MODEL PARAMETERS 100
Short-Circuit Test ■ Open-Circuit Test ■ Application of Test Results ■ Effect of Frequency

BALANCED THREE-PHASE TRANSFORMER CONNECTIONS 102

RECOMMENDED REFERENCES 104

This chapter reviews transformer fundamentals, including basic principles of operation, equivalent circuits, multiple windings, loss mechanisms, instrumentation and testing to determine equivalent-circuit parameters, and autotransformers. The application of per-unit techniques to the transformer is discussed. Some mention is made of three-phase transformers to the extent that you can understand the basic connections and solve some simple problems if they appear on the exam.

ELECTROMAGNETIC FUNDAMENTALS OF TRANSFORMERS

Because transformers transfer energy through a time-changing magnetic field, they are often considered to be in the machinery category even though they are not designed to involve relative mechanical motion. In fact, a transformer is very much like the special case of an induction machine (see Chapter 10) in which there is no relative motion between primary and secondary.

Steady-state energy transfer in magnetic systems functions only with time-changing signals (see Chapter 2). Thus, a transformer will block dc but will pass ac signals (which need not necessarily be sinusoidal or even periodic). We will restrict our attention to transformers operating in the sinusoidal steady state. Again, we will designate rms voltage and current magnitudes by capital letters and phasors by boldface capitals.

Transformers are also considered with electric machines because they have been, for more than a century, important components of electric power systems. Furthermore, the electromagnetic fundamentals underlying transformer analysis are the same as for machines. Thus, a review of transformers should help you understand the chapters on rotating machinery and assist you in working problems in the machines area.

The Basic Two-Winding Ideal Transformer and Polarity Considerations

Consider the basic transformer whose core and windings are shown in Figure 7.1a. Here we have a single magnetic path with two windings. Technically, this is called the **core form** of transformer, in contrast to the **shell form**, in which both windings are on the same, common center path, as shown in Figure 7.1b. In most single-phase applications, the shell form has better coupling between windings and is thus used more frequently than the core form. Because the core form is probably easier to understand, however, we will concentrate on it here.

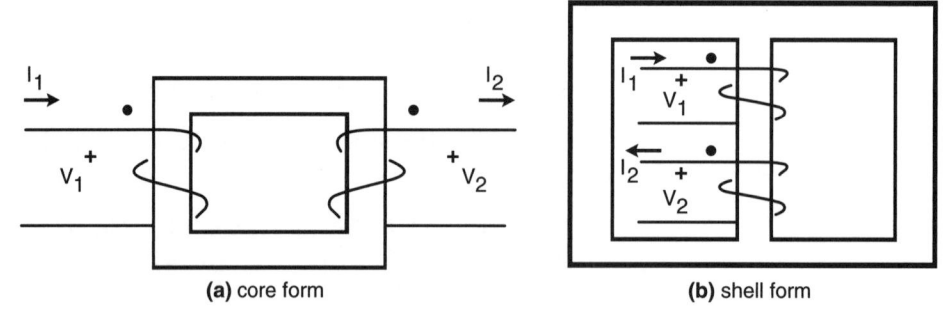

Figure 7.1 Pictorial representation of basic two-winding transformer structures

From a modeling and calculation standpoint, and for an easier understanding of the device, it is very useful to assume an **ideal transformer** whenever possible. In an ideal transformer, magnetic flux is assumed to be totally confined to the core and thus the same flux links both windings. No **leakage flux** (flux which links one winding but not the other) is present. An ideal transformer would also be lossless and its presence would not be detectable from the source side unless load was present on the other winding.

To summarize, an ideal transformer possesses the following theoretical characteristics:

- Zero winding resistance, and thus no "winding loss" (power loss in the windings)

- Infinite **core permeability** μ, and thus zero reluctance

- No **leakage flux**; i.e., all magnetic flux in the core links both windings

- No core loss (that is, no power loss in the ferromagnetic core)

- 100 percent efficiency

Of course, we cannot achieve any of these objectives completely. However, well-made power transformers have relatively small winding resistance, very low core reluctance (causing the flux to be nearly confined to the core, meaning that there is very small leakage flux), and relatively low core loss, for reasons to be discussed shortly. Thus, idealization can be approached fairly closely in practical manufacture by using quality core and winding materials and computer-aided design and assembly. In fact, very significant improvements in transformers (and in rotating machines as well) have continually appeared during the history of the devices.

The fundamental mathematical relations between voltage, current, complex power, and impedance are easy to establish for the ideal transformer in the sinusoidal steady state. For the ideal transformers shown in the pictorial view of Figure 7.1 (or in the schematic diagram of Figure 7.2, which is discussed below), a relation for the rms phasor primary and secondary voltages $\mathbf{V}_1$ and $\mathbf{V}_2$ may be found. To do this, we employ *Faraday's Law* (often expressed by the differential equation e = N dφ/dt) in terms of the mutual flux φ and the number of turns N_1, N_2 of each winding as

$$\mathbf{V}_1 / \mathbf{V}_2 = N_1 / N_2 = a. \tag{7.1}$$

Similarly, using *Ampere's circuital law* the currents $\mathbf{I}_1$ and $\mathbf{I}_2$ are found to be related to the turns ratio inversely:

$$\mathbf{I}_1 / \mathbf{I}_2 = N_2 / N_1 = 1/a \tag{7.2}$$

where a is the turns ratio.

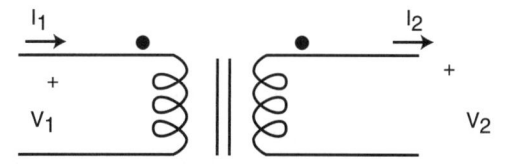

Figure 7.2 Schematic diagram of two-winding transformer with polarity identification

Polarity Conventions and Winding Names

In the pictorial device drawings of Figure 7.1, there is no ambiguity of polarity of voltages or direction of currents. Faraday's law mandates that the polarity of V_2 be as shown, given the polarity of V_1 (for phasors, this means that $\mathbf{V}_1$ and $\mathbf{V}_2$ must be in phase with each other as shown). Similarly, because of Lenz's Law, the directions of the currents $\mathbf{I}_1$ and $\mathbf{I}_2$ must be related as shown, because they produce fluxes that oppose each other. However, we only know these proper polarity relations if we can see the way in which the actual windings are placed. For situations where the windings are hidden in a case or for schematic models such as shown in Figure 7.2, we need some kind of polarity marking.

To accomplish this, we use a polarity convention often called the **dot convention**, because it is usually delineated by dots (or similar marks) placed on the windings. Two such large dots have been placed on each transformer in Figure 7.1 so as to follow the convention. Speaking in terms of phasors, note that $\mathbf{V}_1$ and $\mathbf{V}_2$ are in phase with each other, and $\mathbf{I}_1$ and $\mathbf{I}_2$ are in phase with each other when they follow the dot convention shown. This dot convention is also used in the *schematic* model of Figure 7.2 where the polarity relation would be ambiguous without it.

Winding polarity is of particular interest when transformer windings are interconnected with each other, as in the autotransformer configuration described below, or where multiple transformers or multiple windings are used, as in three-phase applications.

Windings are often designated by numerical subscripts, as done above. Another frequent winding designation method, for a two-winding transformer, is to call the winding near the source the **primary** and the winding near the load the **secondary**. We will also find occasion to talk about the **high voltage** versus the **low voltage** winding, which of course are more unambiguous designations (at least in the usual case where the windings have different numbers of turns). Depending on how it is used, the primary winding might be the low-voltage or the high-voltage winding.

Impedance Referral and Power Relations

From the voltage-current relations of Equations 7.1 and 7.2, it is easy to show that a complex load impedance Z_L on the secondary side is "seen" when looking into the primary side as

$$\mathbf{Z}_{pri} = a^2 \mathbf{Z}_L \tag{7.3}$$

where the **turns ratio** a is

$$a = N_1/N_2 = |V_p/V_s| = |V_1/V_2| = |I_s/I_p| = |I_2/I_1|. \tag{7.4}$$

Thus, we see that *the ideal transformer refers impedances according to the square of the turns ratio*. Here the subscripts p and s (signifying primary and secondary) are introduced, because, as mentioned, transformer problems are sometimes expressed in those terms.

Equation 7.3 is useful for **impedance matching**, such as is often done in audio or high-frequency communication equipment. This topic is not discussed frequently in connection with electric power circuits, but the principles still hold. It is of interest, however, in establishing the per-unit behavior of transformers.

More important to us is the volt-ampere relationship for the ideal transformer, which can be obtained from the Faraday and Ampere relations (Equations 7.1 and 7.2) in complex form as

$$\mathbf{V}_1 \mathbf{I}_1 = \mathbf{V}_2 \mathbf{I}_2.$$

Because it is then true that

$$\mathbf{V}_1 \mathbf{I}_1^* = \mathbf{V}_2 \mathbf{I}_2^* \tag{7.5}$$

we can see that the complex power on each side of the ideal transformer must be the same, i.e.,

$$\mathbf{S}_1 = \mathbf{S}_2 \tag{7.6}$$

and thus the real and reactive power on each side (which are scalars) are related as

$$P_1 = P_2 \tag{7.6a}$$

$$Q_1 = Q_2. \tag{7.6b}$$

This confirms that the ideal transformer does not consume *real* or *reactive* power. As mentioned earlier, many *actual* transformers (especially large ones) approach this ideal fairly closely, but do not attain it in practice. This difference is addressed later.

Transformer Ratings

Large transformers are typically designated by their winding voltage ratings, the total volt-amp capability, and one or more impedance parameters (often only the leakage impedance which will be discussed later). Voltages are given in V or kV, and the volt-amp rating is usually given in kVA or MVA. Impedances are typically given in per-unit with the device ratings as base.

A frequent way used particularly to designate small transformers (when we are not concerned with their internal impedances) is simply by rated voltages and rated VA. For instance, a two-winding, 30 VA door-chime transformer with rated winding voltages of 120 V and 24 V might be designated

$$120/24 \text{ V}, 30 \text{ VA}$$

or

$$120{:}24 \text{ V}, 30 \text{ VA}.$$

Although it is assumed that transformers for most power systems in the United States are designed to operate at 60 Hz, the rated frequency is usually identified on the device as well. Large transformers, like large machines, will typically have a **nameplate** giving such information as well as much other data.

PER-UNIT CALCULATIONS AND BASE QUANTITIES

The per-unit system that we discussed in Chapter 6 poses particular problems and provides particular advantages for transformers. That is because a transformer "transforms" voltage, current, and impedance, as discussed above.

For a device such as a transmission line, all base quantities remain unchanged as we go from one end of the line to the other. However, this is not true for a transformer with other than a 1:1 turns ratio. Some base quantities (V, I, Z, Y) change as we go across a transformer, while others (S, P, Q) are unchanged.

Because they relate like actual quantities, *base voltage, current, and impedance refer (change) from one side of a transformer to the other just like their counterpart quantities in the ideal transformer.* Voltages refer as the turns ratio, currents refer inverse to the turns ratio, and impedances refer as the turns ratio squared. Also, from ideal-transformer theory, real power P, reactive power Q, and apparent power S entering the transformer primary are equal to the corresponding values leaving the secondary, and thus are each the same on both sides of an ideal transformer, as we have just shown above; and these items when used as base quantities do not change in an actual transformer.

Expressed mathematically, where the subscripts 1 and 2 represent the two windings of the ideal single-phase transformer and N represents the number of turns on the corresponding winding, we restate the ideal-transformer relations for emphasis as

$$\mathbf{V}_1 / \mathbf{V}_2 = N_1 / N_2 \tag{7.7}$$

$$\mathbf{I}_1 / \mathbf{I}_2 = N_2 / N_1 \tag{7.8}$$

$$\mathbf{Z}_1 / \mathbf{Z}_2 = (N_1 / N_2)^2 \tag{7.9}$$

$$\mathbf{S}_1 = \mathbf{S}_2. \tag{7.10}$$

The same relationships of course hold for scalar V, I, S, and Z as for the corresponding vectors, *and they hold for corresponding base values.*

From these considerations, in terms of per-unit analysis we can see that base voltages refer as the turns ratio, base currents refer inversely to the turns ratio, and base impedances refer as the square of the turns ratio as we go from one side of a transformer to the other. Furthermore, S_{base} (and thus P_{base} and Q_{base}) are the same for all parts of a given circuit. An important result of this is that *the per-unit impedance of a device is the same regardless of whether we look at it through a transformer or not.* This means that *no ideal transformers are needed when the problem is formulated in the per-unit system* (except for unity-ratio ideal transformers to model isolation if necessary). More fundamentally stated, *the turns ratio, when expressed in per-unit, is always unity.* This provides a significant advantage for the use of the per-unit system of calculation where transformers are present.

To facilitate identifying base quantities, a circuit is usually broken into **regions** or **zones** by base voltage (i.e., at transformer boundaries), and base quantities are calculated in each region. These are sometimes shown in the form of a table, called a **table of bases**, especially if there is more than one transformer (and thus more than two zones) in the circuit. It often helps to build such a table of bases prior to problem solution. This helps keep straight what the base quantities must be. Furthermore, it encourages easy calculation of all the needed unknown base values.

Example 7.1

A single-phase load having an impedance $Z_L = 20 + j10\ \Omega$ is connected to the low-voltage side of a single-phase step-down transformer rated 10 kVA, 4160 / 480 V, having a leakage impedance of 8 percent. A voltage of 4200 V is applied to the high-voltage side. Assuming a base of 4400 V and 50 kVA on the source side, calculate the per-unit and actual current in both the load and the source. Also find the per-unit and actual voltage at the load.

Solution

We have given a different system base from the transformer rating and a different value of source voltage from both of these in this example to provide more practice in the use of the per-unit method. Furthermore, we have given the leakage reactance of the transformer in percent as is typical, which is always assumed to be given on device base. We have also assumed that this is the only transformer parameter that will have a significant effect upon the desired calculations.

The equivalent circuit model is shown in Figure 7.3, which gives some per-unit values (which we will calculate below). A goal is to have all quantities on our diagram expressed in per unit.

$Z_T = j0.3576$ pu

$V_S = 0.955 \angle 0°$ pu

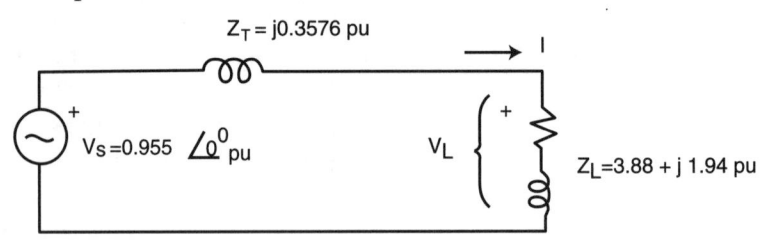

V_L

$Z_L = 3.88 + j 1.94$ pu

Figure 7.3 Circuit model for Example 7.1

Because we will need them shortly, we find the base currents on the source (high voltage) and load (low voltage) sides as

$I_{baseHV} = 50000 / 4400 = 11.36$ A

$I_{baseLV} = 11.36\ (4160 / 480) = 98.48$ A.

Let's look at the load first. Because the load impedance is given in actual ohms on the LV side, we need to find base impedance on that side in order to calculate the per-unit impedance of the load. We will also need to know the base voltage of that side for voltage calculations a little later. On the LV side,

$$V_{baseLV} = 4400\,(480/4160) = 507.69 \text{ V}$$

and

$$Z_{baseLV} = (507.69)^2 / 50000 = 5.155\ \Omega.$$

In terms of this base, the per-unit value of the load impedance is

$$Z_L = (20 + j10) / 5.155 = 3.88 + j\,1.94 \text{ pu } \Omega.$$

Examine the transformer next. Because both its VA and V ratings are different from system bases, its impedance needs to be corrected for both. From the change-of-base equation,

$$X_T = 0.08\,[(4160/4400)^2\,(50/10)] = (0.08)(0.8939)(5) = 0.3576 \text{ pu}.$$

Note how much the *pu* value of X_T has changed by placing it in the circuit where the bases are different! Also observe that the pu load impedance is much greater than the *pu* transformer impedance (on the same base), a desired condition which means that the voltage drop in the transformer will be much smaller than the load voltage. Use of the per-unit system makes this immediately obvious.

Finally, the source voltage in *pu* is $V_S = 4200 / 4400 = 0.955$ pu and we will assume it to be the zero-degree reference:

$$V_S = 0.955\,\angle 0°\text{ pu}.$$

We can now calculate the *pu* current as

$$I = 0.955\,\angle 0° / (3.88 + j\,1.94 + j\,0.3576) = 0.212\angle{-30.6°} \text{ A pu}.$$

(This *pu* current is the same for the whole circuit because there is only one path in the *equivalent* circuit. Remember that the per-unit current on both sides of a transformer is the same. Again this points out an advantage of per-unit calculations.)

The actual current in the source is then

$$I_S = 11.36\,(0.212\angle{-30.6°}) = 2.41\,\angle{-30.6°} \text{ A}$$

while in the load it is

$$I_L = (98.48)(0.212\angle{-30.6°}) = 20.85\,\angle{-30.6°} \text{ A}$$

and the *pu* voltage across the load must be

$$V_L = IZ_L = 0.212\angle{-30.6°}\,(3.88 + j1.94) = 0.9184\,\angle{-4.07°} \text{ V pu}.$$

or

$$V_L = (507.69)(0.9184\,\angle{-4.07°}) = 466.24\,\angle{-4.07°} \text{ V}.$$

As an aside, note that, had the transformer been ideal, the voltage magnitude at the load would have been $4200\,(480/4160) = 484.6$ V.

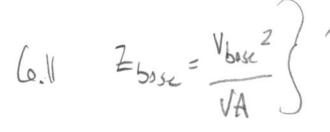

Voltage Regulation

Voltage regulation, sometimes denoted VREG, is an important parameter in the analysis of devices such as transmission lines and transformers. It is usually defined in terms of no-load and full-load voltage at the specified load point as

$$\text{VREG} = (V_{R\,NL} - V_{R\,FL}) / V_{R\,FL} \qquad (7.11)$$

where

$V_{R\,NL}$ = the no-load voltage at the receiving point R and

$V_{R\,FL}$ = the full-load voltage at the receiving point R.

The parameter VREG is a measure of the effect that a load will have when switched on to a system. For instance, if you switch on a space heater and the light in another room of your home dims excessively as a result, we say that this part of the power system has poor voltage regulation (which translates to a high numerical value of VREG). Ideally, we would usually prefer a voltage regulation of zero (in which, for the home example just mentioned, the outlet voltage would not change as load is added). Many power transmission systems tolerate a voltage regulation of up to 10 percent, because values in this range can be compensated fairly easily by transformer taps. Real systems, because they always have some impedance between source and load, will always have a nonzero voltage regulation.

We should point out that in dealing with capacitive loads in an otherwise inductive system, there may be a *rise* of voltage as load is switched on. This voltage rise would translate into a *negative* value of VREG in Equation 7.11. This situation is not often encountered in low-voltage systems, because most loads are RL but could happen as a consequence of excessive power-factor correction (see Chapter 4).

The voltage regulation at the load in Example 7.1 could be found as

$$\text{VREG} = (0.955 - 0.9184) / 0.9184 = 0.04 = 4 \text{ percent,}$$

which is probably an acceptable value.

Because voltage regulation is used frequently in different contexts in power-systems work, you will see it discussed in other chapters of this book.

MULTI-WINDING, BUCK-BOOST, AND AUTOTRANSFORMER CONNECTIONS

Obviously, a transformer can have more than two windings. For instance, cutting the secondary winding at some point and bringing out all four resulting leads, plus the two primary leads, yields a three-winding transformer. For such a device, if assumed ideal, the three voltages are related, with reference to Equation 7.1, as

$$V_1 / N_1 = V_2 / N_2 = V_3 / N_3. \qquad (7.12)$$

In analyzing a multi-winding transformer, we usually assume a single primary and multiple secondaries; this is the way such units are used in practice. The sum of the vector VA loads on the secondaries will equal to that of the primary if the transformer is assumed ideal. Thus, for an ideal three-winding transformer, with loads on windings two and three,

$$S_1 = S_2 + S_3. \qquad (7.13)$$

Many older and some modern consumer-electronic devices use transformers with multiple secondaries. For instance, older televisions with vacuum tubes typically used a large filament transformer which stepped down the applied 120 V to several windings of about 6 V each, and had one or more step-up high-voltage windings as well, which might be center-tapped for rectification. Some three-phase utility power transformers (in kVA and MVA sizes) have a primary, secondary, and a **tertiary** winding; the tertiary (third) winding might be connected in delta in three-phase applications with the other two windings connected in wye.

Buck-Boost Transformers

One frequently-encountered single-phase transformer application in industry is the **buck-boost transformer**. This is simply a transformer which usually has several primary, as well as several secondary, windings which can be interconnected in various ways, usually for the purpose for making small adjustments in voltage.

For instance, a typical buck-boost unit might be of 5 kVA rating and have two 120 V windings and two 12 V windings (see Figure 7.4). Among the connection possibilities of such a transformer are

- connect the two 120 V windings in parallel and the two 12 V windings in parallel to yield a 120/12 V transformer,

- connect the two 120 V windings in series and the two 12 V windings in parallel (this provides a 240/12 V unit),

- connect the 120 V windings in series and the 12 V windings in series (this results in a 240/24 V transformer), or

- connect the 120 V windings in parallel and the 12 V windings in series (this results in a 120/24 V transformer).

In each of these cases, the total VA rating is the same as given for the transformer. Of course, polarity conventions must be followed carefully in each case because the windings must be interconnected. A typical buck-boost transformer will either show polarity marks or will give terminal designations with an indication of how to interconnect these terminals to achieve the stated connections.

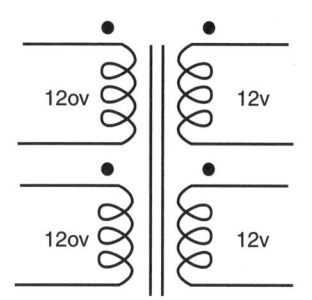

Figure 7.4 Typical four-winding buck-boost transformer

Each of the four connections mentioned above still results, effectively, in a two-winding transformer as far as it appears to the outside world. Primary and secondary are still isolated from each other. However, buck-boost transformers are more frequently used in *autotransformer connections* as described below. Their name derives from the fact that, in this connection, they can be used to adjust voltages up (boost) or down (buck) by small amounts to compensate for voltage drops or other changes in the power system.

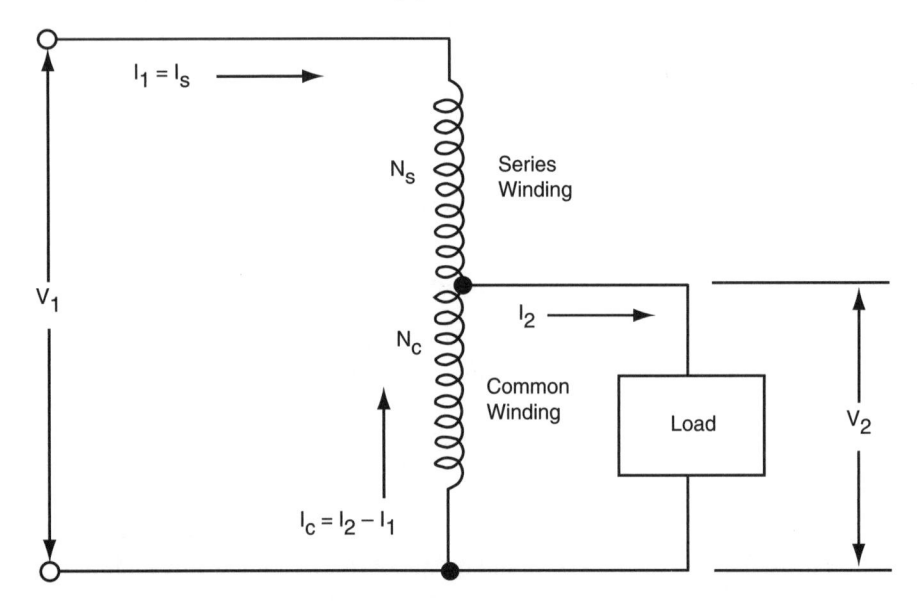

Autotransformer Connections

Transformers can be connected externally in many different ways. There are some advantages in connecting a transformer so that the primary and secondary share a common winding. Such a connection leads to what is called an **autotransformer**.

Consider the connection shown in Figure 7.5, where we have referred to the two windings in terms of their functions in the autotransformer connection: *common* (of number of turns N_c) and *series* (of number of turns N_s). The two windings have been placed in "series aiding", like two cells in a flashlight. A voltage V_1 is supplied across this series combination, which in turn supplies a voltage of V_2 to the load placed across the common winding. Assuming a polarity dot at the top of each winding in the figure yields the usual *step-down* connection. Note that the voltages V_1 and V_2 in this context refer not to winding voltages (although in this case V_2 is in fact the voltage across the common winding), but to voltages on the two sides of the autotransformer *configuration*.

Figure 7.5 An autotransformer connection

Example 7.2

Suppose we have a two-winding transformer rated 120/12 V, 1.0 kVA. We will assume it ideal. We wish to connect this device as an autotransformer to provide an output of 12 V for a 132 V input. Examine the connection, and calculate the rated currents for each side of the autotransformer connection. Also examine the kVA capability of this connection.

Solution

The windings can be connected as in Figure 7.5. In terms of the autotransformer winding designations, the voltage ratings of the windings shown on the figure must be

$$V_s = 120 \text{ V}$$
$$V_c = 12 \text{V}.$$

When connected as in Figure 7.5, note that in order to satisfy the transformer-winding voltage ratings, it must be true that

$$V_1 = 120 + 12 = 132 \text{ V}$$

and, of course,

$$V_2 = 12\text{V}.$$

This means that the transformer, when connected in the autotransformer connection shown, effectively functions like a step-down unit of voltage rating 132/12 V (or equally well as a step-up transformer in the opposite direction), thus satisfying our stated voltage guidelines.

We will now determine the current ratings of the two sides of the connection and look at the resulting kVA ratings of the two sides. These will be the values that we can get through this connection without exceeding the ratings of either transformer winding.

We first calculate the rated current of each winding from the rated voltages and rated VA:

$$I_1 = I_s = VA_{rated} / V_1 = 1000/120 = 8.33 \text{ A}.$$

This is the rated current that the series winding can safely carry, because the current I_1 is the current that passes through the series winding.

For the common winding, with reference to the original transformer, the current rating of that winding must be in terms of the VA rating and the rated voltage *of that winding*, or

$$I_c = 1000 / 12 = 83.33 \text{ A}.$$

However, the connection of Figure 7.5 dictates that

$$I_c = I_2 - I_1$$

or

$$I_2 = I_c + I_1 = 91.67 \text{ A}.$$

The rated VA on the primary side must be

$$S_1 = V_1 I_1 = (132)(8.33) = 1100 \text{ VA}.$$

Is something wrong here? We began with a two-winding transformer rated 1 kVA and just found that when connected as in Figure 7.5 it could carry 1.1 kVA.

To answer this, note that *for the autotransformer connection, not all the load current flows through the secondary (common) winding of the original transformer.* The rated VA for the *transformer itself* is still 1 kVA. But the added metallic path allows additional VA to be transmitted to the load external to the two-winding transformer's magnetic field. Thus, *the autotransformer connection can in fact provide more "volt-amp capability" than the original two-winding unit* without overheating any winding. Alternatively, *using the autotransformer connection, a much smaller transformer can be employed to achieve the same resulting VA capability.*

As a check, let us calculate the rated VA at the load; because we are dealing with an ideal transformer, we would expect it to equal that at the source.

The rated VA on the load side must be

$$S_2 = V_2 I_2 = V_2 (I_c + I_1) = (12)(91.66) = 1100 \text{ VA}.$$

Although we have illustrated this for a small transformer (that one could easily pick up and hold), the same principle is true for larger utility units.

Example 7.3

As an interesting aside, suppose we use the same two-winding transformer as in Example 7.2 but interchange the two windings (again, see Figure 7.5). Calculate the kVA rating of this connection.

Solution

Now the autotransformer connection will have a voltage rating of 132/120 V, and the VA capability of this connection (calculated at the source side) will be an amazing

$$S = (132)(83.33) = 11 \text{ kVA}.$$

This can be checked, as before, by calculating S at the load side. You will get the same answer.

In cases where the windings are arranged so that the voltages on the two sides of the autotransformer are not very different from each other, as in this example, the autotransformer connection can often carry *many times* the VA rating of the basic transformer. Such is often the case with buck-boost transformers, which are primarily designed for making small changes in the voltage.

You should note that, unlike the two-winding connection, the autotransformer connection does *not* provide *isolation* between input and output circuits. This is often a factor in small devices (such as cell phone chargers and other small ac-powered rechargeables where the low-voltage circuit must be isolated from line voltage for safety reasons) but is usually not of concern in utility applications, because in such systems both sides are typically grounded.

It is also important to realize that an autotransformer is not a special device with different properties from those of a two-winding transformer; it is just a two-winding transformer with the windings interconnected in a particular way.

REAL TRANSFORMERS: LOSSES, IMPEDANCES, AND MODELS

The foregoing equations should enable the solution of many simple transfer problems, certainly those in which the transformer is stated to be ideal. It only remains to account for the non-ideal behavior of real transformers. At low frequencies such as 60Hz, we can develop a fairly simple RL model; at higher frequencies, such as in the kHz range, capacitances between turns and between turns and core might become prominent.

Actual (as opposed to ideal) transformers, of course, do consume real and reactive power when energized by a source, whether they are supplying a load or not. This behavior can be modeled in a straightforward way. The most frequently-used, general circuit model that takes into account the non-ideal behavior of the transformer is shown in Figure 7.6a. This model is called the **T-model** because of its resemblance to the letter T. A source and load are shown connected as well. This transformer equivalent circuit includes the resistance and inductive reactance of the primary windings (as R_1 and X_1 respectively) in series with the referred secondary values a^2R_2 and a^2X_2 (all of this together is called the **series branch**, or **leakage branch**, of the model).

The T-model also includes the shunt-branch parameters (which will be discussed below). The choice of referring all quantities to primary (as in Figure 7.6), or all to secondary, is dependent upon the particular problem. Note that the process of referring quantities from one winding to another, as in Figure 7.6a, moves or does away with the ideal transformer, at least as far as turns ratio is concerned (more on this later).

The resistances in the series branch, of course, model the resistances of the respective windings and thus allow calculation of power loss in the windings (sometimes called **winding loss**). The series reactances (**leakage reactances**) model the flux leakage (flux which links one winding but not the other). Of course, the most desirable transformer would have extremely low leakage flux and winding resistance, making these four series parameters very low (zero for the ideal transformer). *The total leakage reactance (such as $X_1 + a^2 X_2$ above) is usually the most important parameter in the analysis of a large transformer*, especially at high values of load current.

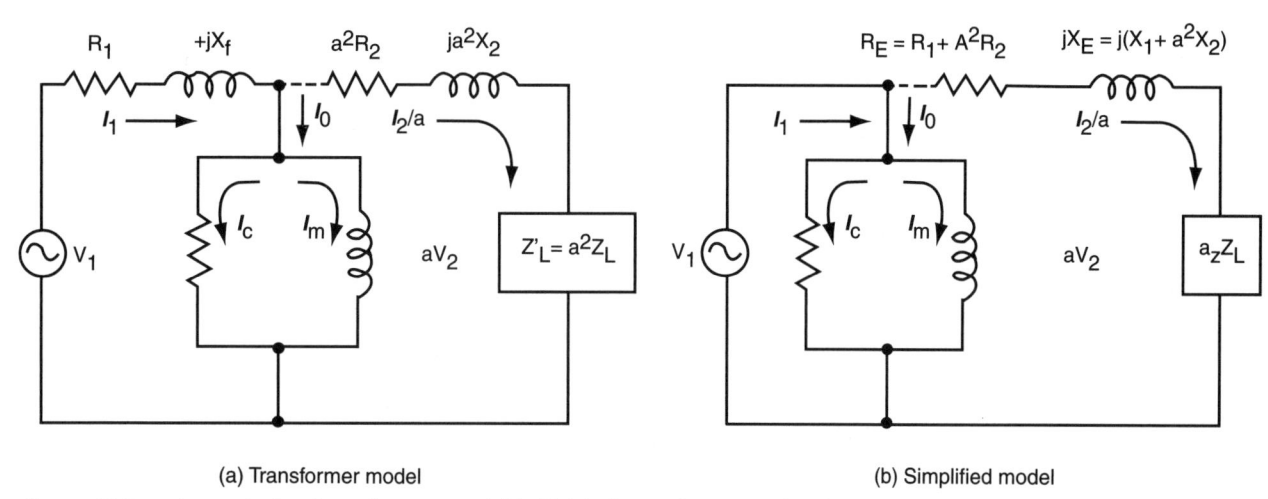

Figure 7.6 A two-winding transformer model (with ideal transformer omitted)

(a) Transformer model   (b) Simplified model

The **shunt branch** in the transformer model of Figure 7.6a consists of a resistance and a reactance. Let's now consider the need for these two elements in the model. The changing magnetic field in the core causes power loss within the transformer core by **eddy-current** and **hysteresis** loss mechanisms; these core losses are represented by the current I_C. The current parameter I_M represents core **magnetization**; that is, I_M is the component of primary current that must flow just to set up the magnetic field in the core. In an ideal transformer, of course, both of these shunt current components would be zero; in a good practical power transformer they are much smaller than (perhaps on the order of a few percent of) rated full-load current. For this reason, the shunt branch is sometimes neglected, especially when dealing with load current near the rated value, although it is necessary for efficiency studies (recall that efficiency = P_{out}/P_{in}, and thus its evaluation is based upon loss).

Core loss can be calculated if the core-loss current and equivalent shunt resistance or conductance are known. Conversely, it can be measured approximately by performing an **open-circuit test**, in which one winding is energized and instrumented for voltage, current, and power, with the other winding open-circuited. In fact, all of the impedance parameters shown in the model of Figure 7.6a may be easily obtained through simple laboratory tests (the series parameters are obtained by means of the *short circuit* test). The significant (several orders of magnitude) difference between the shunt and series impedances of the model allows these two effects to be "decoupled" for most practical purposes.

It is this decoupling that allows the simplification of Figure 7.6b, in which the shunt branch has been moved to the source terminals and the resulting series impedances combined. These series impedances can be added directly in the circuit of Figure 7.6b, because the secondary leakage parameters have been referred to the primary. When you work such a problem, *make sure that all parameters appear on a common base* as has been done here, by referring all quantities to one side of the transformer. A telltale sign of this referral, of course, is that the ideal transformer does not appear between any groups of impedance.

The phasor sum of the two shunt current components $\mathbf{I}_C$ and $\mathbf{I}_M$ is the total shunt current $\mathbf{I}_O$ (this current, sometimes called **exciting current**, is essentially constant at constant applied voltage, whether a load current is present or not). We can view the simplification of Figure 7.6b in terms of this shunt branch to get a different perspective. Because I_1 or I_2 is much greater in magnitude than I_O, it is usually possible to move the shunt branch to one end or the other of the model without incurring extensive error. (If the shunt branch is moved to the left side as shown in Figure 7.6b, the result is said to be an equivalent circuit as referred to the primary side, because the source is on that side.) Given this type of equivalent circuit, one may easily compute the primary current, the power input, the efficiency, the secondary voltage with and without a load (and the resulting voltage regulation, as introduced earlier), and other performance characteristics for a given load.

As an aside, we might expect that *core losses in transformers and other devices with ferromagnetic cores can be reduced* significantly by proper design and materials, and this is in fact the case. Hysteresis loss is reduced by choosing a "soft" magnetic material for the core, one that is easily demagnetized and remagnetized in the opposite direction with each half-cycle of the ac wave (in contrast to the "hard" magnetic material desired for a permanent magnet, which is expected to retain its magnetism for long periods of time). Eddy-current loss is minimized by **laminating** the core; that is, by constructing it from thin sheets of core material. This breaks up the electrically conductive paths in the core, greatly reducing the currents that are inadvertently induced in the core and their respective power loss and heating. Similar procedures are followed for the cores of rotating machines, even for the rotors of dc armature machines. (The armature winding on a dc machine in fact carries ac currents, which must be rectified by the commutator-brush arrangement, as will be shown in Chapter 9.) In the schematic diagram of Figure 7.2, the parallel lines between the windings hint at the presence of a laminated core.

Winding loss can be reduced as well, by making the windings of highly conductive material and with the largest cross section possible, subject to other design constraints.

Efficiency

Once we have an equivalent circuit for the actual transformer, we can easily obtain performance measures for a given application. For example, the efficiency *Eff* may be obtained as

$$\text{Eff} = (P_{out} / P_{in}) = P_{out} / (P_{out} + \text{copper loss} + \text{winding loss})$$
$$= P_{out} / \left[P_{out} + I_1^2 (R_1 + a^2 R_2) + I_C^2 R_C \right]$$

(7.14)

where R_C is the equivalent core-loss resistance in the shunt branch, through which I_C flows.

Example 7.4

A step-down transformer (with turns ratio of 4:1) has an input current I_1 of 10A when the load is a pure resistance of 10 Ω. It is known that the transformer equivalent circuit parameters are $R_1 = X_1 = 1$ Ω, $R_2 = X_2 = 1/16$ Ω. It is also known from tests that the core loss of the transformer is 125 W; but for this problem, you may ignore the exciting current I_O, when considering I_1. Determine the transformer efficiency when operating under the above assumptions.

Solution

We have seen exam-type problems stated this way, so it is wise to examine this kind of problem statement. Because R_1, X_1, R_2, and X_2 are given in this manner, with no implication that any are referred values, it is logical to assume that they are on the original winding bases. The statement concerning exciting current and core loss seems inconsistent, but we will follow the directions in the problem statement.

It is probably easier here to solve this problem by referring secondary quantities to the primary, as in Figure 7.6. Because the current is already given and only the series portion of the circuit need be considered (as the core-loss power is indicated), one only has to add the two I^2R winding losses and the core loss to get the total loss and thence to find the efficiency.

Ignoring I_0 means, of course, that the referred secondary current is equal to the primary current (=10A), and this simplifies the problem greatly. Referring the load resistance to the primary yields a value of referred resistance of $(4)^2 (10) = 160$ Ω, which consumes a power of $(10^2) (160) = 16000$ W. The referred secondary winding resistance is 1 Ω, making a total $R_{eq} = 2$ Ω. Thus, the power loss in the windings will be $(10^2) (2) = 200$W. Then,

$$\text{Eff} = P_O/(P_O + \text{all losses})$$

$$= 16000 / (16000 + 200 + 125) = 0.98 \text{ or } 98 \text{ percent.}$$

Voltage Regulation Revisited

The voltage regulation of a typical power transformer is determined primarily by the leakage impedances. As we did for the simple problem of Example 7.1, in which the shunt branch and both winding resistances were neglected, we find VREG by looking at both the open-circuit and loaded cases, referring all appropriate quantities from one side to the other as appropriate so as to be able to work on a common base.

Consider the phasor diagram of Figure 7.7, in which secondary (winding two) quantities have been referred to the primary (winding one). This typical example assumes a load at rated secondary voltage (which when referred becomes $a\mathbf{V}_2$) with a lagging power factor (and thus a lagging load current of $\mathbf{I}_2/a$). Equivalent leakage impedance $R_E + jX_E$, expressed on the base of the primary side, causes a voltage drop between source and load. Note that it takes a much greater value of $\mathbf{V}_1$ in order to achieve this value of $\mathbf{V}_2$ in a lagging circuit. The magnitudes of these voltages may be calculated and the resulting voltage regulation evaluated. You can rearrange the phasor diagram easily to show that a highly capacitive load (e.g., a large bank of power-factor-correction capacitors) might actually force the sending-end voltage to be lower than that at the receiving end.

Importance of the Phasor Diagram

In problems like these, the phasor diagram can serve as a check on your understanding and your calculations. *It is highly recommended that you sketch appropriate phasor and circuit diagrams when solving problems involving lossy transformers and other lossy devices.* The phasor diagram is easy to sketch quickly and is a very good check against error. You should get into the habit of doing this while working ac problems in your review, which will make the process come naturally during the test. Furthermore, having seen and interpreted many such phasor diagrams by test time, you will be able to verify your solution procedures quickly and spot errors and inconsistencies. For example, in a RL load you know that the current and voltage (assuming that the current is directed into the positive load terminal) will be between 0° and 90° apart, with the voltage leading. If your phasor diagram shows something else, you know that something is wrong.

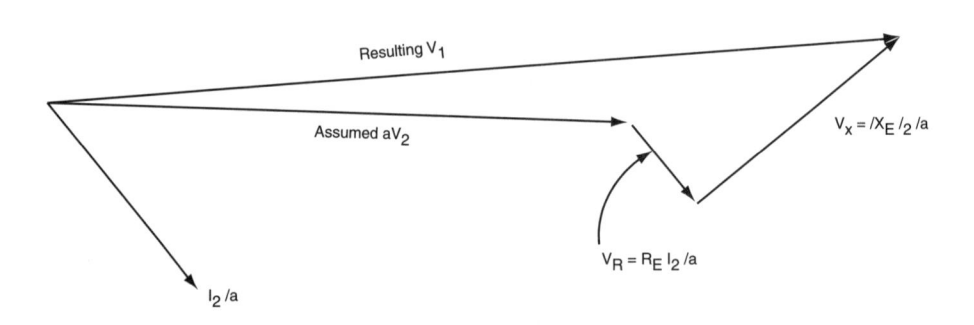

Figure 7.7 Transformer phasor diagram for lagging load

TRANSFORMER TESTS TO DETERMINE MODEL PARAMETERS

The impedances of the lossy transformer model of Figure 7.6 may be obtained for an actual transformer through simple tests. Measurement of voltage, current, and real power (or, alternatively, reactive power) are required.

For a well-designed power transformer, the impedance of the shunt branch (as referred to, for instance, winding one) is typically far larger (by orders of magnitude) than the impedance of the series branch referred to the same winding. This allows the measurements to be decoupled into two separate tests. One test gives the parameters of the series (leakage) branch, and the other gives the parameters of the shunt branch.

It is very important to recognize that *when making either of these tests, the parameters that are calculated from the measurements are automatically referred to the instrumented side.* Thus, if winding one is energized and instrumented, the parameters calculated from these measurements will already be referred to winding one.

The two required tests are known as the *short-circuit test* and the *open-circuit test*.

Short-Circuit Test

For this test, one winding (let us say winding two) is short-circuited, and a *very low* voltage, usually sufficient to give rated current, is applied to the other winding. Instruments are placed at the energized winding to measure V_{SC}, I_{SC}, and P_{SC}.

From this test, the leakage branch parameters ($R_E + j X_E$ in Figure 7.6) can be calculated.

The math involved is simple. From the voltage and current measurements, the magnitude of equivalent impedance is found by

$$Z_E = V_{SC} / I_{SC}$$

and the equivalent resistance R_E is found as

$$R_E = V_{SC}^2 / P_{SC}.$$

Then X_E is found from the Pythagorean relation

$$X_E = (Z_E^2 - R_E^2)^{0.5}.$$

Open-Circuit Test

The procedure is similar. One winding (let us say winding two) is left open-circuited, and a voltage, usually *rated* voltage, is applied to the other winding. Instruments are placed at the energized winding to measure V_{OC}, I_{OC}, and P_{OC}.

From this test, the shunt (core) branch parameters, which we shall call $G_C + j B_M$, can be calculated.

From the voltage and current measurements, the magnitude of equivalent admittance is found by

$$Y_C = I_{OC} / V_{OC}$$

and the equivalent conductance G_C is found as

$$G_C = P_{OC} / V_{OC}^2.$$

Then the magnetizing susceptance can be found from the Pythagorean relation

$$B_M = -(Y_C^2 - G_C^2)^{0.5}.$$

Note that B_M is negative since an inductance is involved. Also observe that we have used admittances in the open-circuit test, because the two shunt-branch elements are in parallel. However, an equivalent shunt branch impedance could of course be found as

$$R_C + X_M = 1 / (G_C + j B_M).$$

which is the representation some authors prefer. Note that, in general, R_C is *not* the reciprocal of G_C and X_M and is *not* the negative reciprocal of B_M.

Application of Test Results

The numerical values obtained from the short-circuit and open-circuit tests can now be used in the T-model. Often, especially in short-circuit study, we simplify the model by neglecting the shunt parameters (because their impedance is relatively great and their effect upon short circuits is miniscule), and sometimes we even neglect the series resistance R_E (because the X_E / R_E ratio is often >10). *For most power studies, X_E is the most important of the four calculated parameters. In* *fact, it is often the only parameter given for a transformer in some kinds of study data.* We used this simple model in Example 7.1.

In some detailed applications, involving the short-circuit and open-circuit tests, it is useful to make a small correction to the results of one test in terms of the other, but the improvement in accuracy that this procedure produces is often unnecessary, especially for high-quality power transformers where the series and shunt parameters are significantly decoupled. We should point out that there are certain kinds of specialized transformers (such as high-voltage transformers for powering neon signs) where nontypical characteristics (such as significant leakage impedance) are desired, but these devices do not fall into the major classification of power transformers.

A further concern for the *open-circuit test* is to select a wattmeter (or transducer) that is capable of reading accurately for very small values of current and power. In the past, special wattmeters were made for this application; modern transducers can eliminate such problems.

Finally, one may ask which winding should be instrumented in each of the tests. The answer is to instrument the one in which it is convenient to make the measurements, using the voltages and currents required. Often, this means (at least for a distribution transformer) energizing the low-voltage winding in the open-circuit test and the high-voltage winding in the short-circuit test.

Effect of Frequency

Transformers can be used somewhat satisfactorily at higher frequencies, but use at lower frequencies will result in higher losses and heating. A simple explanation of this is in terms of inductance: at lower frequencies, inductive reactance is lower and thus current is higher for a given rms voltage. Thus, you might be able to use (at reduced performance level) a transformer rated for 50 Hz applications in a 60Hz situation, but doing the reverse might cause excessive heating and losses.

Similar concerns apply to rotating machines, solenoids, and any other device that uses ferromagnetic cores.

BALANCED THREE-PHASE TRANSFORMER CONNECTIONS

For ease of understanding, the foregoing discussions were restricted to single-phase transformers. Much of this information applies to three-phase transformer connections as well.

Although most three-phase transformers are built with a common three-phase core, we can solve balanced problems with them as if they were built up with banks of single-phase units, which can be done in practice and is sometimes necessary (e.g., in cases where a three-phase transformer cannot be moved into a tight location such as a vault). Furthermore, three-phase distribution "transformers" are typically composed of banks of single-phase transformers. Often there are only two such single-phase transformers present in an unbalanced configuration, and they might be of unequal rating, especially where there is a small three-phase load and a large single-phase load. Such unbalanced configurations are beyond the scope of this book. However, you could easily find voltage relationships for them by using basic transformer principles, especially if you can assume the transformers to be ideal.

Recall that most *balanced* three-phase problems are most easily solved by assuming a per-phase (positive-sequence) model. This simplifies the analysis of three-phase transformers as well. For instance, a three-phase transformer may be connected as an autotransformer, in which each phase has an identical autotransformer connection. Furthermore, this per-phase connection is what we use to analyze the device under balanced voltages and currents.

The following six balanced connections are possible for a two-winding, balanced three-phase application:

- Grounded wye/grounded wye
- Grounded wye/delta
- Delta/delta
- Grounded wye/ungrounded wye
- Ungrounded wye/delta
- Ungrounded wye/ungrounded wye

For this list we have assumed that, for instance, a delta/grounded wye connection is the same type as a grounded wye/delta connection, only with the primary and secondary connections switched.

The first three of these connections are widely used. The last three, because of the uncertain potential at the "floating" ungrounded point(s), are not used often, except that an ungrounded connection might be used in electronic circuits such as rectifiers and inverters, or perhaps in a motor.

By convention, the terminals of three-phase transformers are labeled H_1, H_2, and H_3 on the HV side and X_1, X_2, and X_3 on the LV side. Terminals H_0 and X_0 imply neutral on the HV and LV sides, respectively, where present. In the case of a grounded wye/grounded wye transformer, terminals H_0 and X_0 should be bonded together to avoid ferroresonance. ABC phase labeling is usually attached by the user. Often the a-phase is selected to be the one involving H_1 and X_1.

Any connection of ideal transformers involving a wye on each side (grounded or ungrounded), *or* a delta on each side, can be labeled so that the voltage of phase a on the HV side is in phase with the voltage of phase a on the LV side. When one side has a delta and one side has a wye winding, there is no way that this can be done. American-standard labeling for a delta-wye (or wye-delta) configuration calls for the voltage at H_1 (with respect to its reference) to lead the voltage at X_1 by 30°, assuming an ideal transformer. Knowing that a transformer has been connected to this standard inside its case allows us to make external connections properly and understand the angles and polarities involved.

Three-phase transformers can be paralleled. However, a transformer with a wye connection on one side and a delta connection on the other can never be paralleled with a wye-wye or delta-delta configuration, because there is no way to achieve a phase match, as we have just indicated.

In more complicated three-phase transformer problems, such as one which gives the angle of a certain voltage and then asks for the angle of another voltage, constructing a phasor diagram is particularly useful. In such cases, the following steps can help organize your solution process:

1. Draw a phasor diagram (usually of abc sequence) for the line-to-line and line-to-neutral voltages (or phase and line currents in a delta).

2. Sketch a diagram of the transformer, after carefully noting the connections, and identify voltages (or currents) in the windings with those on the phasor diagram. Apply appropriate polarity marks to the windings, and identify voltage polarities and current directions.

3. Identify the unknown quantity and determine its magnitude and angle.

RECOMMENDED REFERENCES

Fitzgerald, Kingsley, and Umans. *Electrical Machinery*. 6th ed. McGraw-Hill, 2002.

Chapman, Stephen J. *Electric Machinery Fundamentals*. 4th ed. McGraw-Hill, 2003.

CHAPTER 8

Transmission and Distribution Lines

OUTLINE

LUMPED PARAMETER MODEL OF LINE 106

CONDUCTOR TYPES AND CONFIGURATIONS 107

UNITS, MEASUREMENTS, AND CONVERSION FACTORS 108

OBTAINING THE PARAMETER VALUES FOR THE PER-PHASE MODEL 108
Resistance ■ Inductance ■ Shunt Capacitance
■ The Circular Mil and the Conductor Tables ■ Conductor
Bundling ■ Geometric Mean Distance (GMD) ■ Calculation of Inductive
Reactance and Capacitive Susceptance

SHORT AND MEDIUM-LENGTH MODELS 112

ABCD PARAMETERS 112

LONG-LINE EQUATIONS 114

WAVELENGTH 114

THE LOSSLESS MODEL AND SURGE-IMPEDANCE LOADING 115

FULL-LOAD AND NO-LOAD PERFORMANCE CALCULATIONS 115

RECOMMENDED REFERENCES 117

This chapter presents some material that will enable you to calculate the electrical parameters for open-wire transmission and distribution lines. Some performance measures and guidelines for these devices are also given. By the end of the chapter, you should be able to construct a long-line ("exact") model and perform a complete steady-state performance calculation on an open-wire line operating under balanced conditions.

Transmission-line theory is based upon electromagnetic field theory (see Chapter 2). However, exam problems involving lines will probably involve calculating parameters for a lumped-parameter circuit model and using those parameters in calculations of performance measures. Accordingly, we will concentrate here on defining a model, presenting equations for its parameters, and exploring some performance measures such as voltage regulation, voltage ratio, loss, and efficiency, rather than providing derivations from field theory.

LUMPED PARAMETER MODEL OF LINE

At a frequency of 60 Hz, and for lines of realistic length (perhaps 150 miles at the most), we can use a relatively straightforward procedure to concentrate on how the line behaves when viewed from the ends. The calculations are relatively simple but involve complex numbers. When line length exceeds one-fourth wavelength, some interesting reversals in parameter trends appear. Wavelength at 60 Hz is about 3100 miles though, so these effects are only of academic interest in most cases. No matter the length of the line, its parameters can always be accurately found, at least for balanced operating conditions, and a simple equivalent model, as far as terminal behavior is concerned, can always be obtained. Acceptable modeling of longer lines, and, therefore, accurate modeling of most lines, requires evaluation of hyperbolic trigonometric functions with complex arguments or, equivalently, complex exponentials. These are presented later in the chapter.

However, let's start at the beginning. Using common sense, we can postulate the per-phase, lumped-parameter circuit model of Figure 8.1 for a line. The elements of this model are

R = ac resistance of conductor
X_L = inductive reactance of conductor
B_C = shunt susceptance corresponding to capacitance between conductors.

Note that we have broken the shunt capacitance into two equal parts and placed one at each end of the line. This model is called a **pi (π) model**, because it looks like the Greek letter pi. It is a **lumped-parameter** model, because we assume that the parameters of the line can be lumped into those shown in Figure 8.1, at least as far as the line appears electrically to the remainder of the system. We will discuss these ideas, and another lumped-parameter structure (the T-model, such as used in the previous chapter for the transformer) later in this chapter.

We will assume balanced conditions for all of the work in this chapter, and thus the models will provide positive- and negative-sequence parameters only, but not zero-sequence (see Chapter 12 for the meaning of these terms). However, the presentation here will apply to most cases of interest. If desired, zero-sequence impedance of an open-wire line can be estimated as approximately three times the positive-sequence impedance value calculated in this chapter, while negative-sequence impedance is equal to positive-sequence impedance for a line. More detailed treatments of this can be found in the Recommended References.

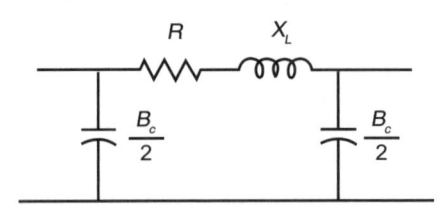

Figure 8.1 Transmission line model

The model of Figure 8.1 can be made to include factors such as the capacitances between phase conductors in the three-phase system as well as capacitances to ground and the inductive couplings between all conductors. Furthermore, although the reference bus in the figure is shown as an equipotential surface, the resistive and inductive effects of the earth can be accounted for as well.

With proper parameter values, the model of Figure 8.1 gives good results for lines up to about 150 miles in length. For longer lines, or in general for better accuracy, the parameters in the model must be modified by the *long-line equations* cited in the section "Long-Line Equations" later in this chapter.

CONDUCTOR TYPES AND CONFIGURATIONS

Most three-phase lines are open-wire, supported by porcelain, fiberglass, or (more frequently in new work) polymer insulators; only about 1% of transmission in the United States is underground. Thus, we concentrate on open-wire structures. Many of the principles here apply to cables as well, however.

A three-phase line should have, of course, three phase-conductors. There may also be one or more **ground wires** (**static wires**) whose purpose is not only to carry return currents (which should be small in balanced, fundamental-frequency situations) but, more importantly, to provide shielding against lightning strikes. Typically, the angle from vertical between the shield or ground wire and the outermost phase conductor should not exceed 30 degrees for good lightning protection. Ground wires may be stranded steel, because they bear mechanical load but are not designed primarily to carry current. (Steel has lower conductivity than copper or aluminum.) There may be more than one conductor per phase. **Bundling** refers to the practice of using more than one conductor per phase, both electrically and physically in parallel. In practice, a bundle of two conductors, usually 1.5 feet apart, is typically used in lines of voltage greater than about 220 kV. This not only increases the current-handling capacity (by doubling the conductor cross-sectional area) but has important advantages such as reducing the inductive reactance and the radio-frequency interference given off by the circuit. Three, and even four, conductors in symmetrical configurations (equilateral triangle or square) are sometimes used in bundles for high-voltage lines. We will examine this below.

The conductors discussed so far are typically not single, solid wires but instead are **composite conductors**; that is, they are each composed of a number of smaller solid wires, spiraled in alternate layers, and are typically bare (see Chapter 16 for related safety concerns of bare wires). Thus, they are somewhat like large versions of the stranded conductors in extension cords. The largest conventionally-used composite conductors may be nearly two inches in diameter.

Although copper wire, in various forms, was traditionally used in early power lines, most modern power conductors are made of aluminum or one of its alloys, which is lighter in weight and less expensive than copper. Another advantage of aluminum is that a larger-diameter conductor than a copper equivalent is necessary to obtain the same current-carrying capacity (**ampacity**). The electrical advantage of a larger conductor will become evident in the equations below, and one could show that the larger-diameter conductor leads to less radio frequency interference and reduced power loss due to corona as well. Research continues both in exploring possible new conductor types (especially for cables) and, more particularly, in structures for supporting these wires.

Besides **all-aluminum-alloy conductors (AAAC)**, one important and frequently-used type of composite conductor is the **aluminum conductor steel-reinforced (ACSR)** type, which has a few center strands of steel surrounded by more strands of aluminum. The purpose of the steel strands is for structural strength—conductors must often cover spans of many feet—while the aluminum strands provide most of the current-carrying capacity. Conductors used in high-voltage transmission lines are typically ACSR; distribution lines with their shorter spans often use all-aluminum-alloy conductors.

UNITS, MEASUREMENTS, AND CONVERSION FACTORS

In the United States, it is still common practice to specify line dimensions in English units. Thus, length is usually given in miles (mi), conductor dimensions in inches (in), and conductor spacing in feet (ft). Accordingly, the equations for line parameters are often expressed in terms of these units. Some useful conversions involving metric and English units are as follows:

$$1 \text{ m} = 39.37 \text{ in} \tag{8.1}$$

$$1 \text{ mi} = 1.609 \text{ km} \tag{8.2}$$

$$1 \text{ km} = 0.6214 \text{ mi} \tag{8.3}$$

$$1 \text{ in} = 2.54 \text{ cm} \tag{8.4}$$

$$1 \text{ ft} = 12 \text{ in} \tag{8.5}$$

$$1 \text{ mi} = 5280 \text{ ft} \tag{8.6}$$

OBTAINING THE PARAMETER VALUES FOR THE PER-PHASE MODEL

Following are some useful ideas and equations for finding the parameters of the per-phase line model. *We will assume that the currents and voltages in the line are balanced and thus that only positive-sequence values are required.* Furthermore, we will assume the line to be **transposed**, which means that the phases are geometrically "rolled" at intervals of one-third of the line length. Therefore, each of the three phase conductors or composites occupies each conductor position for one-third of the line length. (Although typically assumed, transposition is not always needed or implemented in practice.)

Resistance

Conductor resistance is found from tables provided by manufacturers; on the PE exam, it probably will be given to you. (You can find tables for the characteristics, including the resistance of several conductor types, in most undergraduate power-systems texts, such as the two listed as references at the end of the chapter.) There may be very slight differences between the values of the line parameters given by different manufacturers or textbook authors. Especially in the case of resistance, this may be due to the calculation of the values at different temperatures. If a table is supplied with an exam problem, be sure to use the correct column with regard to temperature as stated in the problem.

If conductors are bundled, the per-phase resistance is found by dividing the per-phase resistance of one such conductor by the number of conductors in the bundle (usually 2, 3, or 4). The ac resistance of the conductor (which is of interest here) is greater than its dc resistance, because a phenomenon known as **skin effect** tends to force the current density to be greater near the outside of the conductor, in contrast to the uniform current density typical of dc currents. In fact, for large-diameter conductors, skin effect has a pronounced oscillatory behavior in terms of a *Bessel function* from the center to the outside of the conductor.

Values of resistance are typically given in Ω/mile in tables, but occasionally in Ω/meter.

Inductance

Because it is a more useful property, line inductance is usually expressed in reactance at fundamental frequency (60 Hz).

Shunt Capacitance

Being a shunt parameter, line capacitance is instead expressed as capacitive susceptance in siemens (S).

Equations will be given for inductive reactance and capacitive susceptance once some important prerequisites have been introduced, such as geometric mean radius, geometric mean distance, and bundling. Before doing this, though, the entries for a typical conductor in a table of conductor characteristics provided by a manufacturer will be examined to see what information is typically given, and identification of what is needed in particular from the table to calculate the line parameters will follow.

The Circular Mil and the Conductor Tables

Manufacturers provide tables of ACSR conductors' characteristics. We will look at some of the more important information given in such tables, including the three critical items needed for electrical parameter calculation. You can find such tables in the recommended references.

ACSR conductors were long ago given bird names, which are used conventionally in the United States. Thus, when a utility engineer speaks of a CARDINAL or FALCON conductor, for example, it is well known which specific conductor is meant.

Consider the frequently-used ACSR conductor denoted DRAKE. According to published tables, this conductor has a designation 795,000 because the cross-sectional area of the aluminum material in this conductor is 795,000 circular mils (cmil).

The **circular mil** is a unit of cross-sectional area that was very useful in pre-calculator times, because it removed the irrational factor π ($= 3.1416$) from the calculations. One circular mil is simply the area of a circle of diameter one mil ($= 0.001$ in), so one circular mil is equal to $\pi/4$ square mils, or

$$1 \text{ cmil} = \pi / 4 \text{ mil}^2. \tag{8.7}$$

This may be useful to know, just in case you have to perform this conversion.

Returning to the DRAKE conductor entry in the table, we would see other geometric information about the conductor, such as is having 26 aluminum strands, each of diameter 0.1749 inches, and 7 steel strands, each of diameter 0.1360 inches. Its overall outside diameter, Dia, is 1.108 inches, and it weighs 5770 pounds/mile. Its ac resistance, R, for "75% capacity" is 0.1288 Ω/mile and its geometric mean radius (GMR, explained below) at 60 Hz is 0.0375 feet. The current-carrying capacity of this conductor is stated in the table to be approximately 900 A.

The previous items are interesting and useful, but the three pieces of information we need most from the table for this DRAKE conductor for calculation of the parameters of Figure 8.1 are

- per-phase resistance r = 0.1288 Ω/mi,

- conductor diameter Dia = 1.108 in (so conductor radius r_c = Dia / 2 = 0.5540 in), and
- D_s = GMR = 0.0375 ft.

D_s or GMR, called the *geometric mean radius* of the conductor, is a property of the conductor that allows calculation of inductive reactance, whereas conductor radius r_c is used in the calculation of capacitive susceptance.

Example 8.1 illustrates the use of these characteristics.

Conductor Bundling

As mentioned earlier, conductors on EHV (extra high voltage) lines are often bundled by placing several identical composite conductors close to each other, usually no farther apart than two feet.

For a two-conductor bundle, we define a term

$$D_{SL} = (D_s\, d)^{0.5}, \tag{8.8a}$$

which replaces D_s, the conductor GMR, in inductance calculations where bundling is present. Here, d is the *bundle spacing*, the center-to-center distances between composite conductors in the phase bundle. Often, d = 1.5 feet in practice.

Similarly, for capacitive calculations, we will need the term

$$D_{SC} = (r_c\, d)^{0.5}. \tag{8.8b}$$

For a three-conductor bundle, in an equilateral triangle formation,

$$D_{SL} = (D_s\, d^2)^{0.333} \tag{8.9a}$$

and

$$D_{SC} = (r_c\, d^2)^{0.333} \tag{8.9b}$$

where, again, d is the center-to-center spacing of the composite conductors.

For the occasionally-encountered four-conductor bundle, in a square formation with the sides of the square equal to d,

$$D_{SL} = 1.091\,(D_s\, d^3)^{0.25} \tag{8.9c}$$

and

$$D_{SC} = 1.091\,(r_c\, d^3)^{0.25} \tag{8.9d}$$

where, again, d is the center-to-center spacing of the composite conductors.

Although the symmetrical bundle configurations given above are the ones encountered in practice, it is in fact a simple matter to find expressions for D_{SL} and D_{SC} in a bundle spacing of any arbitrary geometry, as given in the References.

In the absence of bundling, D_{SL} is replaced by conductor GMR from the table and D_{SC} is replaced by conductor radius r_c.

Geometric Mean Distance (GMD)

Useful in the calculation of both X_L and B_C is the idea of **geometric mean distance** (**GMD**, not to be confused with **GMR**). For a transposed three-phase line,

$$\text{GMD} = (D_{ab}\, D_{bc}\, D_{ca})^{0.333} \tag{8.10}$$

where D_{ab}, D_{bc}, and D_{ca} are the respective center-to-center phase spacings of the phase conductors. Note that while GMR is a function of only one (usually composite) phase conductor, GMD has to do with the spacings *between the phases*. Also, you should observe that in the case of bundled conductors, these centers may not lie within conductors at all but in the air space between conductors in a bundle.

Calculation of Inductive Reactance and Capacitive Susceptance

We now find the *per-mile* parameters (shown as lower-case letters) as

$$x_L = [0.1213 \ln (GMD / D_{SL})] \; \Omega/mi \qquad (8.11)$$

$$b_C = (3.3727 \times 10^{-5} / [\ln (GMD / D_{SC})]) \; S/mi \qquad (8.12)$$

and finally we can express the total parameters of Figure 8.1 as

$$X_L = [0.1213 \ln (GMD / D_{SL})] \, l \; \Omega \qquad (8.13)$$

$$B_C = (3.3727 \times 10^{-5} / [\ln (GMD / D_{SC})]) \, l \; S \qquad (8.14)$$

where l is the line length in miles.

Often we introduce the complex per-mile impedance and admittance

$$z = r + j \, x_L \; \Omega/mi \qquad (8.15)$$

$$y = jb_C \; S/mi. \qquad (8.16)$$

These complex numbers are used in Example 8.1.

Example 8.1

Six DRAKE conductors are used to make a three-phase line. There are two conductors per phase with a center-to-center spacing of d = 1.5 feet in each phase. The line is 80 miles in length. The conductors are in a horizontal configuration (as in the well-known H-frame structure) with center-to-center spacings between the three phases of 10, 10, and 20 feet. Find the parameters of Figure 8.1 for this line.

Solution

Because there are two conductors per phase, the resistance per phase is half the resistance of one conductor:

$$R = (0.5) \, (0.1288 \; \Omega/mi) \, (80 \; mi) = 5.152 \, \Omega.$$

Geometric mean distance, needed for inductance and capacitance calculations, is

$$GMD = (D_{ab} D_{bc} D_{ca})^{0.333} = [(10)(10)(20)]^{0.333} = 12.60 \; ft.$$

Because of the bundling, the modified GMR (called D_{SL}) and the modified radius D_{SC} must be calculated:

$$D_{SL} = (D_S \, d)^{0.5} = [(0.0375)(1.5)]^{0.5} = 0.2372 \; ft.$$
$$D_{SC} = (r_C \, d)^{0.5} = [(0.5540/12)(1.5)]^{0.5} = 0.2631 \; ft.$$

At this point, the line parameters of Figure 8.1 can be calculated as

$$X_L = [0.1213 \ln (12.60 / 0.2372)](80) = 38.55 \; \Omega$$
$$B_C = (3.3727 \times 10^{-5} / [\ln (12.60 / 0.2631)])(80) = 697.4 \times 10^{-6} \; S.$$

SHORT AND MEDIUM-LENGTH MODELS

The calculations of Example 8.1 gave the parameters of what is often called the **medium-length line model**, which is generally considered valid for line lengths up to 150 miles. *Note that, to find each parameter for the medium-length model, the per-mile value is simply multiplied by line length.*

A simplification can often be made for open-wire lines with lengths below about 50 miles. Called the **short-line model**, it simply neglects the shunt capacitive susceptance and employs series resistance and inductive reactance as in the medium-length model. Such simplifications were used more frequently before the advent of digital calculators, but may at times be useful in hand calculations if they apply.

For lines above about 150 miles in length, the **long-line model** is necessary. It is discussed in the section "Long Line Equations" later in this chapter.

ABCD PARAMETERS

Note that the per-phase circuit model (Figure 8.1) of a transmission line (or that of a two-winding transformer, for that matter) is actually a **two-port network**; it has two ends, one of which may be input and the other output. It is useful to represent such a two-port network by what are sometimes called **transmission parameters** or **ABCD parameters**, which are defined with regard to the two-port network model of Figure 8.2 as

$$\mathbf{V_S} = \mathbf{A}\,\mathbf{V_R} + \mathbf{B}\,\mathbf{I_R} \tag{8.17}$$

$$\mathbf{I_S} = \mathbf{C}\,\mathbf{V_R} + \mathbf{D}\,\mathbf{I_R} \tag{8.18}$$

where
- $\mathbf{V_S}$ = the per-phase sending-end voltage
- $\mathbf{V_R}$ = the per-phase receiving-end voltage
- $\mathbf{I_S}$ = the per-phase sending-end current
- $\mathbf{I_R}$ = the per-phase receiving-end current.

In matrix form,

$$\begin{bmatrix} V_S \\ I_S \end{bmatrix} = \begin{bmatrix} A & B \\ C & D \end{bmatrix} \begin{bmatrix} V_R \\ I_R \end{bmatrix} \tag{8.19}$$

The square matrix shown is known as the **ABCD matrix**. Note that the ABCD parameters are complex numbers; one (**B**) is an impedance, one (**C**) is an admittance, and two (**A** and **D**) are dimensionless.

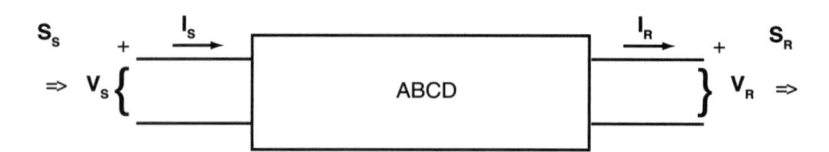

Figure 8.2 ABCD parameter model for a two-port network

Observe that, in the defining network of Figure 8.2, the sending-end current is directed *into* the network while the receiving-end current is directed *out*. This has decided advantages, some of which become evident when we show the load on the right side and the source on the left side of the model of Figure 8.2 as we usually do. One significant advantage is that we can **cascade** the models of two or more devices (such as two transformers separated by a transmission line) and easily find the equivalent. Because of the voltage polarity and current direction references, the sending-end quantities of one device become the receiving-end quantities of another. *The equivalent ABCD matrix of a group of cascaded elements is just the matrix product of the individual ABCD matrices in order from source to load.*

ABCD Models of PI and T Networks

ABCD modeling is not limited to transmission lines but can also be used in a wide variety of situations. For example, for the PI model of Figure 8.3a, we could show that

$$\begin{aligned} \mathbf{A} &= 1 + \mathbf{Y}_2 \mathbf{Z} \\ \mathbf{B} &= \mathbf{Z} \\ \mathbf{C} &= \mathbf{Y}_1 + \mathbf{Y}_2 + \mathbf{Y}_1 \mathbf{Y}_2 \mathbf{Z} \\ \mathbf{D} &= 1 + \mathbf{Y}_1 \mathbf{Z} \end{aligned} \tag{8.20}$$

and for the T-model of Figure 8.3b,

$$\begin{aligned} \mathbf{A} &= 1 + \mathbf{Y}\mathbf{Z}_1 \\ \mathbf{B} &= \mathbf{Z}_1 + \mathbf{Z}_2 + \mathbf{Y}\mathbf{Z}_1\mathbf{Z}_2 \\ \mathbf{C} &= \mathbf{Y} \\ \mathbf{D} &= 1 + \mathbf{Y}\mathbf{Z}_2. \end{aligned} \tag{8.21}$$

It is always true that the determinant of the ABCD matrix is unity; that is,

$$\mathbf{AD} - \mathbf{BC} = 1 + j0. \tag{8.22}$$

Because of this constraint, only three of the four ABCD parameters are independent; we can always rearrange Equation 8.22 to solve for one of them in terms of the other three. This has interesting consequences for the PI and T-models, each of which has only three unique parameters, as indicated in Figure 8.3. What this means is that a PI model can always be converted into a T-model, or vice versa, through the use of ABCD parameters. Noting that a *PI* is just a delta and a *T* is just a wye, in terms of the delta-wye conversions mentioned in Chapter 3, we see why this unique equivalence must be so.

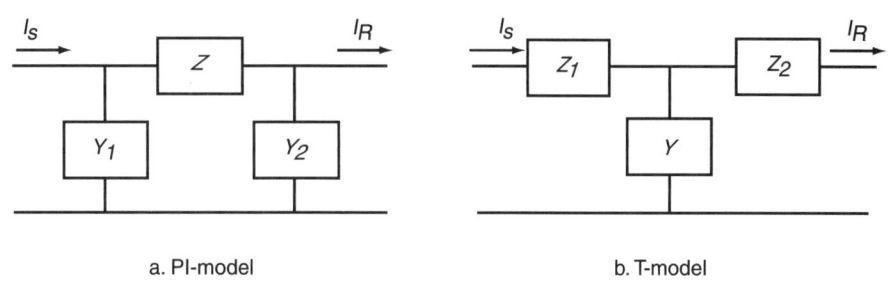

a. PI-model b. T-model

Figure 8.3 PI and T-models

Furthermore, for a **symmetric** network (that is, one that shows the same impedance whether we look into the sending or receiving side), it is always true that

$$\mathbf{A} = \mathbf{D}. \tag{8.23}$$

A transmission line is of course a symmetric network, and it is easy to find the ABCD parameters for its medium-length model by comparing Figures 8.1 and 8.3a and using Equations 8.20.

LONG-LINE EQUATIONS

For lines approaching 150 miles in length, the total parameters obtained by multiplying their respective per-mile values by line length begin to show some significant errors. The errors become more pronounced with increasing line length. That is because in actuality the resistance, inductance, and capacitance parameters are all *distributed* along a line and not lumped in the middle or at the ends as Figure 8.1 would suggest.

One way to visualize (or to actually solve for) the correct values is to envision the cascading of a large number of models like Figure 8.1 for short lengths of the line; for example, 200 such models, each modeling 1 mile of line, for a 200-mile line. This could be done with ABCD parameters but would involve a lot of calculation and would probably produce considerable round-off error.

Fortunately, we do not need to cascade a large number of models to try to represent this distributed behavior; compact equations for the equivalent can be obtained by calculus. The equations for these will simply be stated here, although derivations may be found in the recommended references.

To find the values of the parameters (called **exact** values) for the long-line model, we need to first introduce two complex-number quantities: **characteristic impedance**

$$\mathbf{Z_C} = (\mathbf{z}/\mathbf{y})^{0.5} \tag{8.24}$$

and **propagation constant**

$$\gamma = (\mathbf{z}\,\mathbf{y})^{0.5} \tag{8.25}$$

where **z** and **y** are the per-mile values of the series and shunt line parameters, respectively, obtained earlier.

Using these quantities, we would find the ABCD parameters in terms of *hyperbolic trigonometric functions* with complex arguments as

$$\mathbf{A} = \mathbf{D} = \cosh(\gamma\, l) \tag{8.26}$$

$$\mathbf{B} = \mathbf{Z_C} \sinh(\gamma\, l) \tag{8.27}$$

$$\mathbf{C} = (1/\mathbf{Z_C}) \sinh(\gamma\, l) \tag{8.28}$$

where l is the length of the line.

(If your calculator will not calculate these complex hyperbolic functions, please see Chapter 1 for equations for them in terms of simpler functions.)

WAVELENGTH

Wavelength is certainly more important in high-frequency communication work than in electric power engineering, but it has interesting implications for a low-frequency power line as well. The same concepts hold. To find the wavelength of a line, we need to express the previously-defined propagation constant in rectangular form as

$$\gamma = (\mathbf{zy})^{0.5} = \alpha + j\beta. \tag{8.29}$$

Then wavelength in meters is

$$\lambda = 2\pi/\beta = 2\pi/[\omega(LC)^{1/2}] \text{ m} \qquad (8.30)$$

where
- ω = the radian frequency
- L = the inductance of the line
- C = the capacitance of the line.

The wavelength is in MKS units, but converting it to miles at 60 Hz, we would find for a typical line that $\lambda \approx 3100$ miles.

THE LOSSLESS MODEL AND SURGE-IMPEDANCE LOADING

It is sometimes helpful to visualize the model of Figure 8.1 but without resistance. This leads to the lossless line model. We will not discuss this extensively except to point out some of its unique aspects and one useful performance measure to which it contributes.

If a line were ideal, its ABCD parameters would simplify greatly, becoming circular trigonometric functions with real arguments. However, an even more important piece of information can be gathered from this ideal line.

The characteristic impedance figured for a lossless line is called the **surge impedance**. Thus, for an actual line, the surge impedance is an approximation of the characteristic impedance. From Equation 8.24, it is evident that *the surge impedance is a positive real number*. When a line is terminated in a resistance equal to its surge impedance, we call this condition **surge-impedance loading (SIL)**. SIL is a useful measure, because of the fact that *at SIL, the voltage profile of the line is flat*. In fact, we could show that its voltage, current, and power profiles are all flat at SIL. That is, the magnitudes of those quantities are not different from one point on the line to the next for constant loading. This is in contrast to a line that is not loaded to SIL, whose profiles do change.

Because most lines are not likely to be terminated in SIL, it is evident that the voltage at the end of a typical line will likely be different at the sending and receiving ends. It was observed more than a century ago, for instance, that the voltage along a lightly-loaded line actually *increased* from source to load. This is known as the **Ferranti effect** and is due to the interaction of the series inductance and shunt capacitance, just as the inclusion of power-factor correction capacitors tend to cause the bus voltage in an actual power system to rise.

FULL-LOAD AND NO-LOAD PERFORMANCE CALCULATIONS

Regardless which model of the line is used, the ABCD parameters can be used to find parameters such as P and Q loss in the line, efficiency, voltage regulation, and voltage ratio. Because this is so frequently done, we will outline all the steps below, citing some equations with which you are already familiar, but perhaps with different notation.

The usual problem is to be given the load voltage and either *P* and *S*, or *PF* and *S*, at the load. From these, the receiving-end current I_R is calculated. Because total three-phase complex power S_{total} and line-to-line receiving-end voltage V_{RLL} are typically provided, the equations below will be given in those terms. Usually, the load is assumed to be full load for the desired condition.

Typically, we assume V_R to be the angle reference. The magnitude of receiving-end current can be calculated from

$$I_R = S_{total} / \left(\sqrt{3} V_{RLL}\right).$$

If Q_{total} and P_{total} are given, then

$$\phi = \tan^{-1}(Q_{total} / P_{total}).$$

Alternatively, if S_{total} and P_{total} are given, then

$$\phi = \cos^{-1}(P_{total} / S_{total}).$$

Then,

$$\mathbf{I_R} = I_R \angle -\phi.$$

Next, using Equation 8.19, the per-phase, V_S and I_S are found. At this point, the three-phase complex power at the sending end can be found as

$$S_{Stotal} = P_{Stotal} + j Q_{Stotal} = 3 \mathbf{V_S} \mathbf{I_S}^*. \tag{8.31}$$

From this the total complex line loss under load can be found as

$$\mathbf{S}_{loss\ total} = P_{loss\ total} + j Q_{loss\ total} = (P_{Stotal} + j Q_{Stotal}) - (P_{Rtotal} + j Q_{Rtotal}) \tag{8.32}$$

and the efficiency of the line may be expressed as

$$EFF = P_{Rtotal} / P_{Stotal}. \tag{8.33}$$

Similarly, **voltage ratio**, a parameter often used as one measure of line design effectiveness, is given in terms of voltage magnitudes by

$$VRATIO = V_S / V_R. \tag{8.34}$$

In order to find voltage regulation VREG, introduced in earlier chapters, we need to make **no-load calculations**, because VREG is given in terms of the full-load (FL) and no-load (NL) values of receiving-end voltage. To begin the no-load calculations, it is probably easiest to let the no-load value of V_S be the angle reference. We let its magnitude be equal to the value of V_S, calculated for the full-load case above:

$$\mathbf{V_{SNL}} = V_S \angle 0.$$

Then, from Equation 8.17, because the receiving-end current is zero,

$$\mathbf{V_{RNL}} = \mathbf{V_{SNL}} / \mathbf{A}. \tag{8.35}$$

And similarly, if desired, from Equation 8.18 we could find

$$\mathbf{I_{SNL}} = \mathbf{C} \mathbf{V_{RNL}}. \tag{8.36}$$

This tells us how much current per phase enters the energized end of the line with the other end open-circuited and gives the power-factor angle for the no-load case. (In general, this will be a very capacitive result for a high-voltage transmission line, and the sending-end current may be several hundred amps.)

If desired, the line P and Q loss under open-circuit conditions could be calculated in the same manner as for the loaded case above.

Finally (and this is the primary reason to perform the no-load calculations), the voltage regulation can be found from a combination of load and no-load case results as

$$\text{VREG} = (V_{RNL} - V_{RFL}) / V_{RFL}. \tag{8.37}$$

EFF and VREG are usually expressed in percent. Note that, like voltage ratio, they are dimensionless, real numbers.

RECOMMENDED REFERENCES

Glover, Sarma, and Overbye. *Power Systems Analysis and Design.* 4th ed. Thomson, 2008.

Grainger and Stevenson. *Power System Analysis*, McGraw-Hill Inc. 1994.

CHAPTER 9

Direct-Current Machines and Machine Basics

OUTLINE

BASIC CONCEPTS AND TERMINOLOGY 120

THE SEPARATELY EXCITED GENERATOR AND MAGNETIC SATURATION 121
Efficiency

SELF-EXCITED DC GENERATOR AND DC MOTOR CONFIGURATIONS 124

RECOMMENDED REFERENCES 124

DC machines are widely used in battery-powered applications, such as automotive applications (starter motors, windshield, and blower motors) and in small toys and other portable devices. They are often found in computers (e.g., for opening drive trays) and other places where very small, high-torque actuators are required. Some dc micromotors may be extremely small; other contemporary designs may employ novel features such as low-inertia printed-circuit rotor windings.

Because of the high torque and precise control capabilities of dc machines, a few large hoists and cranes use large dc motors, but most electric power-systems engineers are not likely to see large machines of this type. Also, it should be pointed out that, with the increasing sophistication of power electronics, many precise positioning systems are replacing more traditional dc machines with steppers—in fact, with ac types that are driven by switched signals.

You are not as likely to see problems relating directly to dc machines on the exam as you are to encounter problems relating to synchronous and induction (ac) machines, which may require a knowledge of other power-systems topics as well. Nonetheless, we include dc machines in this book for several important reasons.

First, dc machines are rather useful in illustrating some important machine concepts and equations. Also, the applications mentioned above suggest that dc machines are not likely to go away soon. Furthermore, there is one dc machine connection (the **universal machine**, found in small, high-torque, high-speed ac applications such as food processors and hand-held drill motors) that can be used with ac as well. Although we do not anticipate a test problem centering on such a device, it is a good idea to be prepared, because "power" studies have traditionally included electric machines.

For these reasons, we include this chapter as a brief introduction of the dc machine and, in fact, as a prelude to the discussion of ac machines in the following chapter. Because some concepts introduced here apply to ac machines, you should review this material at least once, with the idea of knowing where to look up information on these devices quickly. Furthermore, some performance measures such as voltage regulation and efficiency, which apply to other electrical devices as well, are illustrated in this chapter.

BASIC CONCEPTS AND TERMINOLOGY

Rotating machines are characterized by two basic parts: the **stator**, or stationary part of the machine, and the **rotor**, or rotating part. Windings, permanent magnets (where present), commutators, brushes, shafts, etc. are each associated with one of these. The terms *rotor* and *stator* apply to all rotational machine types.

Also, note that rotating machines are often characterized by the term generator or motor. This simply indicates mode of use. In **generator operation** a machine supplies electrical energy; that is, it converts mechanical energy into electrical energy. Under **motor operation** it converts electrical into mechanical energy.

The operation of all electromagnetic machines is based on some fundamental concepts relating to electrical and magnetic fields. Some of these concepts can be very effectively illustrated by the dc machine.

As we know, a conductor moving in a magnetic field will have a voltage induced in it. The voltage will be greatest when the conductor is normal (at right angles) to the magnetic flux, and least when it lies in the direction of the flux. At any instant of time, the magnitude of the voltage is proportional to the length of the conductor within the field, the velocity of the conductor with respect to the field, and the strength of the magnetic field.

This magnetic field may be due to a permanent magnet or may be produced and controlled by an **electromagnet**, which usually involves a coil of wire wrapped around a ferromagnetic material. The ferromagnetic structure of such an electromagnet is called the **field pole**; the wire coiled around the field pole is called the **field winding**.

The flux density in the ferromagnetic material of the field poles is related to the field current I_f by the **hysteresis curve** of the magnetic material and the air gap. The hysteresis curve is also called the **B-H curve** for its coordinates, which are magnetic flux density B and magnetic field intensity H. The flux density determines the magnitude of the generated voltage.

For example, a pair of conductors rotating in a magnetic field (shown at the instant the two fields are normal to each other in Figure 9.1a) will produce a voltage E_g caused by a field current I_f. We often collectively call these conductors the **armature**. *In a dc machine, the armature is customarily on the rotor and the field coil is on the stator for reasons of convenience; in the ac synchronous machine described in the next chapter, the reverse is true.* An approximate voltage output at the open-circuited armature of this simple, idealized device is shown in Figure 9.1c.

The moving armature is electrically connected with stationary circuits through a commutator arrangement. The **commutator** is a ring split into segments that is contacted by stationary brushes. Figure 9.1b shows a commutator of two segments; in practice, it usually contains many more, with the attached armature windings distributed around the rotor periphery. Because the brushes contact successive segments as the rotor turns, the commutator functions as a mechanical switch. In this capacity it is designed to function as a *mechanical rectifier*.

Also note that the drawing in Figure 9.1a represents a two-pole machine (the minimum number of poles possible because magnetic poles must occur in pairs). An actual dc generator or motor might have more pairs of poles (for example, many automotive starter motors have four poles). The connections from the armature conductors to the commutator segments and the number of pole pairs are important considerations in terms of the voltage and current ratings of the machine. Details of this lie outside the scope of this discussion but may be found in the machinery books referenced at the end of the chapter.

(a) Generator   **(b)** Commutator   **(c)** Voltage

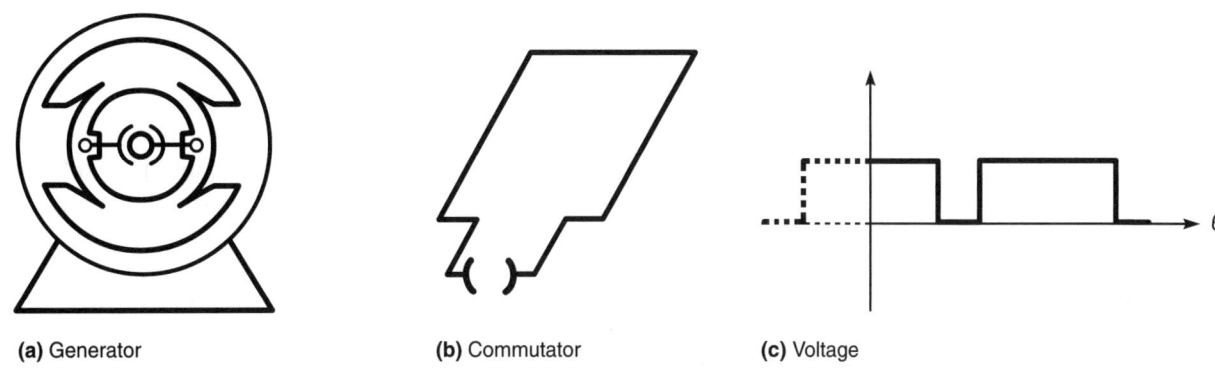

Figure 9.1 Simplified two-pole dc generator

Two important, fundamental equations to remember when analyzing dc machines are as follows:

$$E_g = k' \phi W_m \tag{9.1}$$

$$T_d = k' \phi I_a \tag{9.2}$$

where

E_g = the induced armature voltage
ϕ = the magnetic field flux
T_d = the developed torque (see Chapter 10 for a discussion of torque)
I_a = the armature current
k' = a constant dictated by machine design.
W_m = mechanical speed.

We will not investigate the use of these equations in detail, but they will help you review machine operation and might possibly help you to answer a test question.

THE SEPARATELY EXCITED GENERATOR AND MAGNETIC SATURATION

For simplicity, we consider the separately-excited dc machine first and assume that it is operating as a generator. By **separately excited**, we mean that the field coil is energized from a separate dc source and is not connected to the armature. The equivalent circuit for such a generator is shown in Figure 9.2a. The magnitude of the induced armature voltage generated is then a function of field current I_f and the rotational velocity at which the armature is driven, because all other parameters are fixed (see Equation 9.1, noting that the field flux is a function of field current). This induced voltage will be linearly proportional to rotational speed, but will be a nonlinear function of the field current because of the ferromagnetic material's nonlinear B-H curve (saturation curve).

With the field current supplied by a separate source, a typical plot of the generated voltage is given in Figure 9.2b. The solid curves of this figure constitute the open-circuit characteristic of the machine. Note that the voltage curve starts leveling off as the field current increases; this, of course, is due to the magnetic saturation of the ferromagnetic material, including the field poles, the armature, and the frame, or yoke, of the machine.

Generally the armature resistance, including that of the brushes (the brush resistance is actually nonlinear), is a relatively low resistance (perhaps a fraction of an ohm), whereas the resistance of the shunt field is relatively high (perhaps 25–100 ohms). If the armature current is small, the $I_a R_a$ drop is negligible. The implication is that the terminal voltage is very near the generated voltage E_g. This voltage is referred to as the **no load voltage** V_{NL}, because it would be observed at the generator's armature terminals if no load were present.

However, the purpose of a generator is to provide relatively large currents to a load, where the terminal voltage V_T now becomes the **full load voltage** V_{FL}. The difference between the load and no-load voltage leads to the concept of voltage regulation. The **percent voltage regulation** is defined as

$$\text{VREG} = (100\%)(V_{NL} - V_{FL}) / V_{FL}. \tag{9.3}$$

A similar term, **speed regulation**, is defined for *motor* operation as

$$\text{SREG} = (100\%)(n_{NL} - n_{FL}) / n_{FL} \tag{9.4}$$

where

n_{NL} = the no-load speed in rpm
n_{FL} = the full-load speed in rpm.

The two equations are presented together here because of their similarity. Both apply to other machine types where speed is not constant, and VREG is applicable to other situations as well.

(a) Equivalent circuit

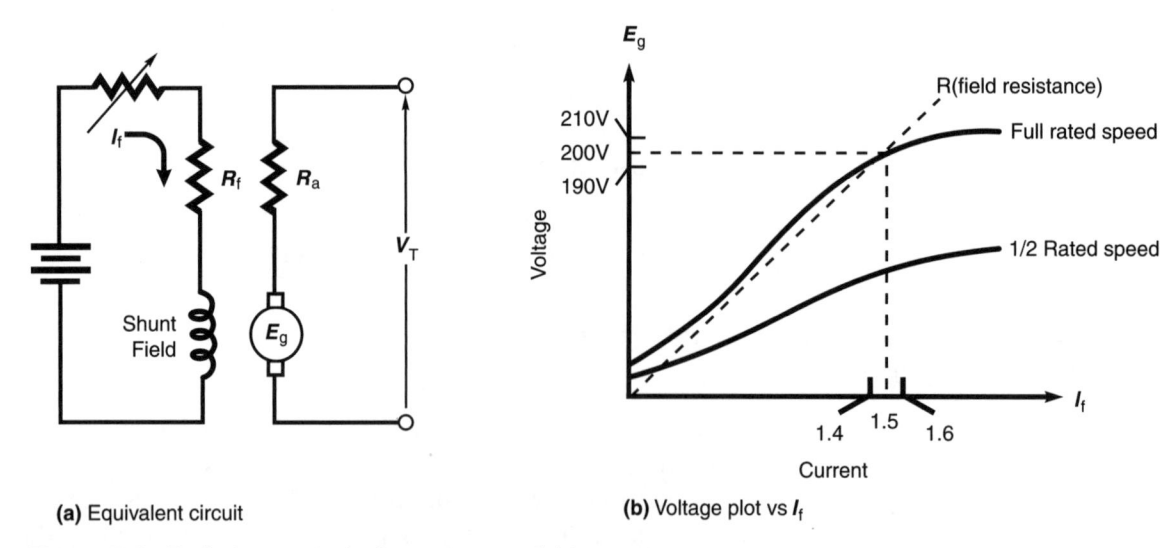

(b) Voltage plot vs I_f

Figure 9.2 Typical generated voltage curve vs. field current

Example 9.1

Assume a separately-excited dc generator is being driven by a prime mover at rated speed. The rated output voltage is to be 200 V when the rated armature current of 50 A flows to the load. If armature resistance is known to be 0.2 ohms, what value of field current I_f is necessary and what is the voltage regulation from no load to full load? For this machine, assume the open-circuit characteristic shown in Figure 9.2b.

Solution

The no-load voltage E_g minus the voltage drop in the armature resistance must be the full load terminal voltage, V_T; the no-load voltage is easily calculated to be

$$V_{NL} = 200 + (50 \times 0.2) = 210 \text{ V}.$$

Thus, from Equation 9.3, the voltage regulation is

$$\text{VREG} = 100 \, (210 - 200) \, / \, 200 = 5\%.$$

In other words, the measured voltage across the load resistance R_L is 200 V, but if the load resistance were removed, the terminal voltage would jump to 210 volts. From Figure 9.2b, the field current required is 1.6 A.

Efficiency

Recall that when there is a current flow in the presence of a magnetic field, a force is exerted on the conductor(s), and the torque on the armature conductors opposes the direction of the shaft rotation, as indicated by Equation 9.2. As the current from the generator increases, the counter torque increases, making the prime mover work harder. If the electrical power dissipated in the external load resistor is the power output, and the mechanical horsepower (from the prime mover) to the shaft is the power input, then one may easily determine the efficiency of the machine.

The calculation of efficiency has some interesting aspects. In Example 9.1, the electrical power delivered to the load was

$$P_L = V_T I_a = 200 \times 50 = 10 \text{ kW}$$

where I_a is the armature or output current.

Suppose it is known that the power taken from the prime mover is 16 hp at rated speed. From these values, the efficiency may be found. We first convert the mechanical power to watts:

$$P_m = 16(746) = 11936 \text{ W}$$

and then the efficiency may be found as

$$\text{EFF} = P_{out} / P_{in} = P_L / P_m = 10000 / 11936 = 0.838 = 83.8\%.$$

Once the losses within the machine (here, almost 2 kW) can be identified, one may begin to find a reasonable model for the dc generator. Two electrical losses are the power lost to provide for a magnetic field, which is $I_f^2 R_f$, and the power lost in the armature windings, which is $I_a^2 R_a$; these two losses are sometimes lumped together and called the **copper loss**.

Another loss may be found by driving the prime mover at rated speed with all electrical connections removed; the measured power taken from the prime mover by the electrical machine is called the **friction and windage loss**, F&W. There is still another loss that is difficult to measure. This loss is sometimes called the **iron loss** and is mostly due to both a hysteresis and eddy current effect caused by the armature rotating in the presence of a magnetic field. Efficiency may be found in terms of these losses:

$$\begin{aligned} \text{EFF} &= P_{out} / P_{in} \\ &= P_{out} / (P_{out} + P_{loss}) \\ &= (P_{in} - P_{loss}) / P_{in} \end{aligned} \quad (9.5)$$

where P_{loss} is the total of all the power losses in the device.

This equation applies equally well to other devices (whether ac or dc), of course.

SELF-EXCITED DC GENERATOR AND DC MOTOR CONFIGURATIONS

A dc generator can derive its excitation from the armature voltage; this leads to what are called the self-excited configurations. In the case of the **shunt self-excited generator**, the shunt field winding, which might have a variable resistance in series to control its current, is placed across the armature terminals. What gets this process started is the residual magnetism present in the core; as the generator is driven, this small magnetism leads to a small induced voltage in the armature which in turn causes a small field current, causing the armature voltage to rise more, and so on. The armature voltage stabilizes at some point dictated by the intersection of the saturation curve and the field resistance line as shown in Figure 9.2b.

In reality, many dc machines have both a shunt field winding (as discussed above) and, in addition, a series field winding on the same poles. The latter is a heavy winding of a few turns and is capable of carrying armature current. It is in fact connected in series with the armature when used. Several connection possibilities exist for such a machine. A **series self-excited generator** involves only the armature and the series winding in series; the shunt winding, if present, is simply left unconnected. Series generators have been used as portable arc-welder power supplies because the internal voltage is low under open-circuit conditions and begins to increase as current is drawn.

By employing both series and shunt field windings, the output of the generator (or the speed of the motor) can be better controlled. Using multiple, interconnected field windings in this manner is called **compounding**, and the resulting connection is called a **compound machine**. When the series field aids the shunt field, the configuration is called **cumulative compounding**; if the series field opposes the shunt field, it is called **differential compounding**. One reason for the versatility of dc machines in the aforementioned hoist and crane applications (as well as in traction applications of earlier times, such as for electrified trolleys) is the precise control available with the compound motor.

RECOMMENDED REFERENCES

For additional review reference material, almost any electric machinery text (even if a decade or more old) will be suitable. Two texts that are widely available and have appeared in a number of previous editions are:

Fitzgerald, Kingsley, and Umans. *Electrical Machinery.* 6th ed. McGraw-Hill, 2002.

Chapman, Stephen J. *Electric Machinery Fundamentals,* 4th ed. McGraw-Hill, 2003.

CHAPTER 10

Alternating Current Machines

OUTLINE

AC MACHINE TYPES AND GENERAL CHARACTERISTICS 126

SYNCHRONOUS MACHINES 127
The Revolving Field ■ Electrical vs. Mechanical Speed ■ Equivalent Circuit of a Synchronous Machine ■ Effect of Changing Field Current I_f ■ Torque

POLYPHASE INDUCTION MOTORS 138
Equivalent Circuit of an Induction Motor ■ NEMA Designs for Induction Motors ■ Speed-Torque Characteristics ■ Other Machine Parameters ■ Motors with Higher Starting Torque ■ Useful T Equations and Conversions ■ Mechanical Load Characteristics and Induction Motor Selection ■ Multi-Speed Motors ■ Effects of Voltage and Frequency Changes Upon Motors ■ Effects of Voltage Unbalance Upon Motors ■ Effects of Harmonics Upon Motors ■ Frequency Changers ■ Starting Methods for Large Induction Motors ■ Power-Factor Correction for Induction Motors ■ Induction Machine Examples

SINGLE-PHASE SYNCHRONOUS AND INDUCTION MOTORS 154

SOME OTHER MACHINE TYPES 155

RECOMMENDED REFERENCES 156

Not including some specialized, usually smaller types (such as steppers and others, which will not be discussed here), there are three major types of ac rotating machines:

- Universal or commutator machine
- Synchronous machine
- Induction machine

This chapter begins with a brief review of each of these major machine types. Following this summary are major sections on each of the two principal types of three-phase machines found in electric power systems: synchronous and induction machines. Fundamental operating characteristics, basic equations, and equivalent circuits are presented for each of these devices.

The chapter also includes a brief overview of single-phase ac machines. However, because three-phase machines are easier to understand than single-phase machines, and because questions on them are more likely on the exam, three-phase devices are discussed first.

AC MACHINE TYPES AND GENERAL CHARACTERISTICS

The **universal machine** is *very much like the series-connected dc machine* described in Chapter 9. It has an armature winding on the rotor and a field winding on the stator. Current is communicated between the two windings by means of a split-ring **commutator** and a corresponding set of **brushes**. Such machines have a *high power/weight ratio* and a *large speed drop with increasing load* and can be operated over a wide range. Some applications include specialized hoists and cranes and hand-held or table-top devices such as electric drills, saws, high-speed grinders, blenders, vacuum cleaners, and mixers. In such applications they are superior, not just because of their high power/weight ratio, but because they automatically adjust their speed downward when more torque is demanded. These machines are capable of high speeds—in fact far higher speeds than the other machine types when operated at power frequencies (of 50 or 60 Hz). Unfortunately, they have relatively complicated windings and produce arcing at the brushes even under normal operating conditions, as well as experiencing brush and commutator wear. In contrast, synchronous machines and induction motors are more robust and simple and are much more likely found in larger sizes in ac power systems. Universal machines will not be covered further in this chapter.

A **synchronous machine** is one in which *the shaft speed is proportional to the frequency of line currents* under normal operation. Because utilities maintain line frequency relatively constant, the synchronous motor can be considered a constant-speed device. Most large generators are synchronous machines. Synchronous motors are used in applications where constant speed is critical; single-phase synchronous motors, for instance, find wide application in clocks and timers. Large synchronous motors, sometimes called **synchronous condensers**, can provide precise power-factor correction while delivering mechanical energy as well.

The operation of an **induction motor**, an asynchronous (non-synchronous) device, depends upon the induction of currents in the rotor by magnetic action from the stator, as the name implies. *Speed is actually a function of load but in certain ranges is nearly constant as load varies*. Although usually employed as motors, induction machines can also be used as generators and find wide application in asynchronous (variable-speed) situations such as wind farms.

Some of these design features can be combined; for example, **reluctance motors** start as induction motors but run at synchronous speed. Larger synchronous motors may be outfitted with **amortisseur (damper) windings** on the rotor, which serve a similar purpose, allowing the machine to start as an induction motor while also damping out transients caused by loading variations. Motor designers have been quite resourceful in developing a wide variety of hybrid designs.

Induction and synchronous motors may have any convenient (but even) number of poles (2, 4, 6, etc.) and have been made in one-, two-, or three-phase design. Three-phase synchronous and induction motors (and even two-phase machines) are easier to understand than single-phase ones, so we discuss three-phase designs first. However, many of the concepts presented in connection with three-phase machines apply to single-phase designs as well.

Large electric-utility generators are usually synchronous machines. Their output frequency is determined by the number of poles and how fast the machines are driven. However, as mentioned above, induction machines can be used to generate ac power also. Induction generators derive their excitation from the ac line and generate power at line frequency, without the requirements of being driven at synchronous speed and requiring a separate dc excitation.

The following sections expand upon the characteristics, construction, and use of synchronous and induction machines.

SYNCHRONOUS MACHINES

A synchronous machine is defined as one in which the speed of shaft rotation is directly proportional to the frequency of armature currents under steady-state conditions. Thus, in electric utility applications, in which frequency is usually very tightly controlled, synchronous machines are constant-speed machines.

Synchronous and induction machines share a number of characteristics, so we will consider both from time to time in this section. However, there are major differences as well, such as in rotor design and speed-torque behavior. To confuse the issue, synchronous motors are not self-starting per se, and to make them self-starting, rotor windings called **damper windings** or their equivalent are installed, which cause them to start like induction motors.

The concept of the synchronous machine (and the induction machine) dates to Nikola Tesla in the 1880s. By the late 1890s, both of these machine types were widely used in ac power systems.

Like other machines, the synchronous machine is defined as (and behaves as) a **generator** when converting mechanical to electrical energy, and as a **motor** when converting electrical energy to mechanical. A synchronous generator is sometimes called an **alternator**. (In fact, on a smaller scale, the automotive alternator is really a three-phase synchronous machine, with three-phase currents rectified to dc by diodes embedded in the case. Driven by the engine, it provides dc power through this group of diodes and a regulator to charge the battery and power auxiliary devices such as headlights and ignition system.) Synchronous machines can be made to operate equally well as motors or generators.

The rotors of synchronous machines may be permanent-magnet, wound, or a cage variety in which some of the teeth are cut away to leave pole saliencies (the latter is sometimes referred to as reluctance design). There are also **hysteresis motors**, usually small single-phase units in which the rotor is made of a smooth, magnetically-hardened steel alloy.

Larger synchronous machines (such as utility generators in the MVA range) will either have a cylindrical (round) rotor or a salient-pole rotor. A **salient-pole rotor** looks like the one shown in Figure 10.1b. It is basically a ferromagnetic core with well-defined (salient) poles on which a coil carrying direct current to control the magnetic field is usually placed. A **cylindrical rotor**, in contrast, is basically a cylindrical, solid steel forging with slots milled parallel to the axis; the winding is placed in these slots and thus functions like the winding on the salient-pole rotor. Machines with many poles (as for low-speed applications such as hydroelectric generation) typically employ salient-pole rotor design; large generators with two or four poles are typically cylindrical-rotor.

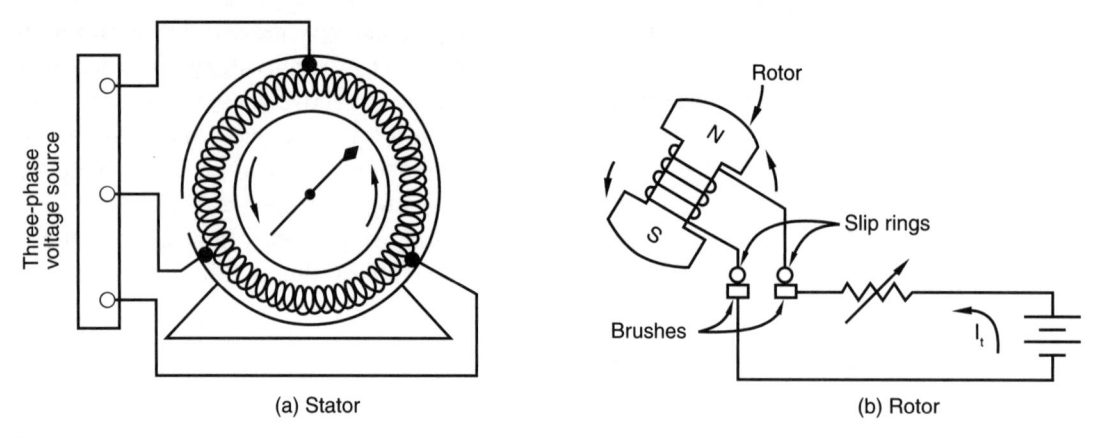

Figure 10.1 Symbolic representation of synchronous-machine stator (showing revolving magnetic field as double arrow) and salient-pole rotor

Except for some smaller (kW range) induction generators used in wind farms, most large generators found in electric utility systems are synchronous machines. In fact, synchronous generators have been built as large as approximately 1500 MVA. Virtually all of these larger machines are three-phase machines, although small (fractional-horsepower) synchronous machines are used widely in clocks, timers, and rotation sensors.

Although the polyphase synchronous machine is probably the simplest machine in concept (that of a dc magnetic field rotating in the presence of a set of stator conductors in the usual configuration), it is not the most frequently-used large machine in terms of number of units constructed. That distinction goes to the *induction motor*, which is discussed later in this chapter.

The Revolving Field

Synchronous machine and induction motor operation is based upon the development of a **revolving (or rotating) magnetic field** within the machine. It was Tesla's brilliant perception of this in the 1880s that led to the development of both machine types.

The revolving field is probably best explained with respect to motor operation. Figure 10.1a shows, somewhat symbolically, a three-phase, balanced-wound stator to which three-phase, balanced voltages of constant frequency are applied. Under these balanced conditions, the stator winding set produces a magnetic field of constant strength which rotates at a fixed speed. This field's speed of rotation is defined as **synchronous speed** n_s. For a given line frequency, the synchronous speed of a motor becomes less as the number of poles increases:

$$n_s = 120 f / p \qquad (10.1)$$

where
- f = the frequency of the ac armature currents, in Hz
- p = the number of (rotor or field) poles
- n_s = the synchronous mechanical speed in revolutions per minute (rpm).

This resulting revolving field can be envisioned somewhat analogously as a permanent magnet of bar shape spinning about its center so that the north (N) and south (S) magnetic poles pass by the stator poles regularly. As mentioned previously, this revolving field rotates at a mechanical speed n_s with respect to the stator. If a permanent-magnet rotor (or an electromagnet rotor, supplied by direct current, as shown in Figure 10.1b) were placed near this stator-induced revolving field, and this rotor were brought close to synchronous speed by some outside prime mover, the rotor would tend to lock into step with the stator field. In fact, this is the principle of the synchronous machine's operation—the magnetic interaction of a stator field and a rotor field which "track" each other while both traveling at synchronous speed and thus produce rotational torque.

Let's examine Figure 10.1 closely again to emphasize this. The resulting magnetic field of the stator has a maximum in the direction of the arrow and physically rotates, as mentioned above, at synchronous speed n_s as per Equation 10.1. The two-pole, salient-pole rotor coil shown in Figure 10.1b carries dc field current which it receives through brushes and slip rings. If the rotor is somehow brought up to near synchronous speed, it is easy to see that the two magnetic fields will attract each other and cause the rotor to travel at synchronous speed also.

Another very helpful illustration of the stator-induced revolving field can be made using software[*] that animates this process in slow motion, by performing calculations repeatedly at small time steps to calculate the stator field mmfs along each stator coil axis and the total air-gap mmf. A snapshot of the revolving field generated in this manner is shown in the upper left of Figure 10.2, where the motion is frozen at approximately the one o'clock position. Also shown in the figure are the sinusoidal stator currents in the lower part of the figure and the stator-coil mmf values which they produce along each stator coil axis (these appear in the upper right with arrows). With animation, we would see the mmfs pulsate along the stator coil axes, varying in magnitude and direction (much like the movement of the cylinders in an internal-combustion engine) and we would also see the net rotating field-vector, the tip's trace of which is a circle for balanced applied voltages or currents. The sinusoidal currents themselves move like an oscilloscope trace during animation, and the vertical center line on that trace indicates the point at which the calculation is made.

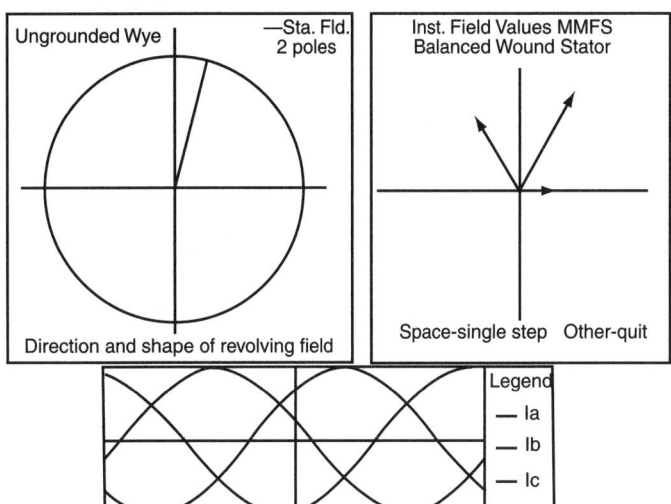

Figure 10.2 Simulation's snapshot of revolving field generated by balanced applied voltages

---

[*] Figures 10.2, 10.9, 10.10, and 10.11 were generated by the interactive, animated machine-demonstrator and analysis software MACHDEM. This program animates machine behavior graphically for induction and synchronous machines subject to balanced and unbalanced fundamental-frequency excitation. It also provides calculation of input current, torque, power, symmetrical components, unbalance factor, etc. It is available from the author of this book at the Klipsch School of Electrical and Computer Engineering, New Mexico State University, Box 3-O, Las Cruces, NM 88003, phone 575-646-3834, *e-mail hsmollec@nmsu.edu*.

Even though you cannot see the animation in the screen image here in the book, envisioning the process can help clarify how the synchronous machine works. We might say that in motor operation, as in the figure, the stator field drags the rotor around at synchronous speed. The rotor is not shown in Figure 10.2 for reasons of clarity, but if it were requested during the animation, you would see it turning in the upper left at the same speed as the stator field but lagging that field by a constant angle, called the **torque angle**. We will consider the torque angle in more detail later in this chapter.

For the same machine in generator operation, exactly the opposite happens. The spinning rotor electromagnet, driven by some prime mover, causes a three-phase set of sinusoidal voltages to be developed in the stator, which results in three-phase voltages at the stator terminals. This process is easy to envision, because it follows Faraday's law for induced voltages. In a sense, the rotor field drags the stator revolving field along in this case, again yielding a constant torque angle.

Some important ideas introduced so far are summarized in the following:

- The speed of a synchronous or induction machine connected to a power line of constant frequency is related to the number of poles. The speed of a synchronous machine is constant at steady state, while that of the induction machine is nearly constant in the typical range of operation, but less than synchronous speed for motor applications.

- In all ac induction or synchronous machines, *the number of rotor poles must equal the number of stator poles per phase* in order to deliver power uniformly. A machine having p stator pole projections per phase is designated a p-pole machine. Of course, that machine must have p poles on its rotor, so it is usually easier to count the rotor poles.

- *The number of poles is always even.*

- Unlike an induction machine, *a synchronous machine delivers uniform torque only at synchronous speed.*

Electrical vs. Mechanical Speed

Equation 10.1 gives the speed at which a synchronous-machine rotor would turn in terms of number of poles and electrical frequency. However, synchronous machines of a number of different speeds (e.g., 3600, 1800, or 1200 rpm) may all be attached electrically to the same system. We sometimes introduce the concept of **electrical speed** n_{elec} to unify these. Electrical speed is related to mechanical speed as

$$n_{elec} = (p/2) \, n_s. \tag{10.2}$$

Combining Equations 10.1 and 10.2 gives an interesting result:

$$\begin{aligned} n_{elec} &= (p/2) \, n_s \\ &= (p/2)(120 \, f / p) \\ &= 60 \, f. \end{aligned}$$

From this we see that electrical speed is always the same for a given system frequency, regardless of the number of poles, and that electrical and mechanical speeds are equal only for the two-pole machine. At 60Hz, that electrical speed is obviously 3600 rpm.

Let's explore this idea a little further. As we saw earlier, magnetic poles must occur in pairs, so p is always an even number in a conventional, symmetrically-built machine. Thus, from Equation 10.1, the highest speed possible (corresponding to the minimum number of poles, p = 2) for 60 Hz operation is 3600 rpm; the next lowest speed would be half that, 1800 rpm for a four-pole machine, followed by 1200 rpm for a six-pole machine. Therefore, for a given frequency of stator currents and restricting our attention to machines of only a few poles, there are just a few values of synchronous speed available. In fact, two- and four- pole synchronous machines predominate in generator applications involving high-speed prime movers such as steam turbines. The 1800 rpm speed (corresponding to a four-pole machine) is a particularly convenient speed for many applications. Hydroelectric generators, whose turbines convert energy best at lower speeds, may have in the range of 40 poles.

Constructing a machine with a greater number of poles means that this magnetic field must make more revolutions for a single mechanical revolution, because an electrical cycle is completed for each movement of a rotor pole from a stator a-phase pole to the next a-phase pole of like polarity. For example, a four-pole machine contains twice as many stator pole projections as a two-pole machine would for the same number of phases, and two electrical (or magnetic) cycles will be completed in moving the rotor through a complete mechanical revolution for this four-pole machine (see Equation 10.2).

Example 10.1

A certain three-phase synchronous machine is designed for 25 Hz use and is stated to have six poles. Find synchronous speed, number of pole projections on the rotor, and number of pole projections on the stator.

Solution

From Equation 10.1, synchronous speed is

$$n_s = 120 (25) / 6 = 500 \text{ rpm}.$$

Because it is a six-pole machine, the machine must have six rotor pole projections (this would be true regardless of the number of phases).

Because it is a three-phase machine, there must be this many stator poles per phase, or a total of (3)(6) = 18 stator pole projections.

Example 10.2

For a typical car alternator having 14 rotor poles, how many electrical cycles of stator output voltage are produced by one mechanical revolution of the rotor? If the rotor is driven at 4000 rpm from the belt and pulley system coupling it with the car's engine, what frequency of stator currents will it produce?

Solution

From Equation 10.2, there are 14 / 2 = 7 electrical cycles produced by one mechanical revolution.

From Equation 10.1 the resulting stator frequency would be

$$f = n_s p / 120 = 4000 (14) / 120 = 467 \text{ Hz}.$$

This frequency is almost an octave above middle *C* on the piano and is in about the middle of a typical woman's singing range, so we would hear this sound very clearly (and annoyingly) if coupled into the car's audio system. This points out the need for the large power-supply filters inherent in good automotive audio gear.

Equivalent Circuit of a Synchronous Machine

Just as with the transformer, we would like to find a simple circuit model that adequately represents the synchronous machine over the range of operation of interest while preserving information on all of the effects with which we are concerned. Under steady-state conditions, a synchronous generator is usually modeled by an ideal sinusoidal voltage source in series with an impedance, which for large machines is very highly inductive. This per-phase model has been found to represent the synchronous machine well under normal operating conditions. It is shown in Figure 10.3a. This is of course a Thevenin model. Note the impedance Z_s, which we call **synchronous impedance**, and the **internal or generated voltage** E_f (the subscript f indicates its origin by induction from the field current in the rotor). Although it results from the field current, E_f is a sinusoidal phasor quantity in the model because it is a sinusoidal voltage induced in the stator winding. Its magnitude is a function both of (synchronous) speed and the dc field current of the rotor.

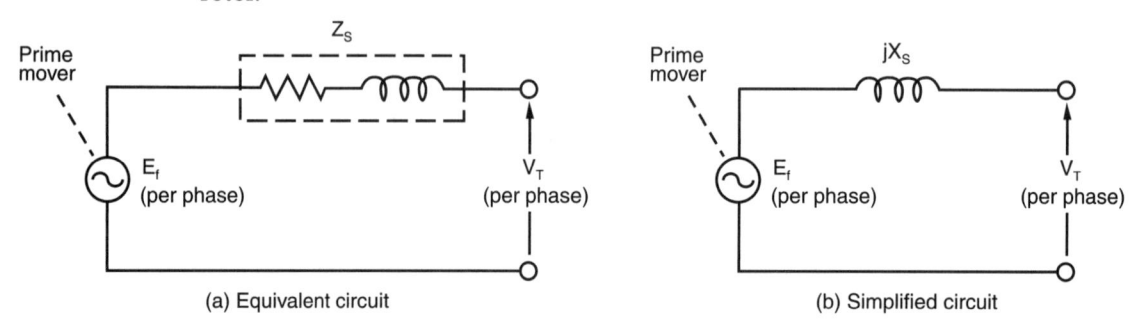

Figure 10.3 Equivalent circuit for a synchronous generator

The synchronous impedance is due, in part, to the magnetizing and leakage flux caused by both the field and armature currents. In the case of large machines, the resistive portion is considerably smaller than the reactance and, without much loss of accuracy, we can usually assume that

$$|Z_s| \approx X_s,$$

which reactance is called **synchronous reactance**. This simplification leads to the circuit of Figure 10.3b.

As we saw, the synchronous machine may be operated as either a generator or as a motor. Under generator operation, current is assumed directed out of the positive terminal of the circuit model, and P and Q are assumed directed out of the machine into the circuit to which it is connected. For motor operation, the opposites of these directions are assumed. Q may in fact move in the opposite direction to P, as we will see in several of the following examples.

Example 10.3

Consider a synchronous generator attached to a typical RL load, as shown in Figure 10.4a. Often one considers the terminal voltage V_T constant and if necessary finds the internal voltage E_f required to support V_T, usually assuming a wye connection, for a given load. Assume that the rated line voltage in the three-phase circuit is 15 kV, the synchronous reactance of the synchronous generator is 10Ω/phase, and a load impedance of 20 + j20Ω per phase is present. Both generator and load are connected in a wye configuration. Compute the line current and the power output of the machine, as well as the internal voltage and the power-factor angle and torque angle.

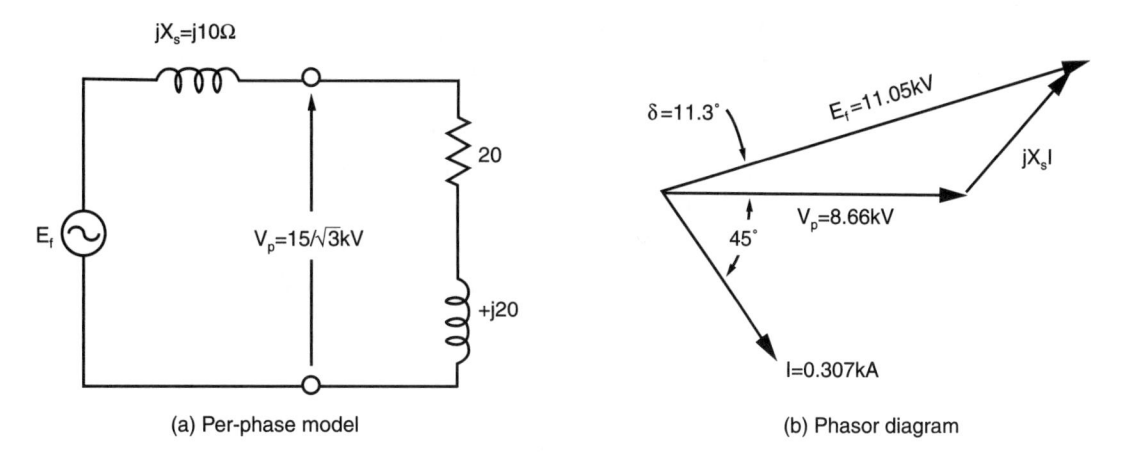

Figure 10.4 Equivalent circuit and phasor diagram for example problem: synchronous generator supplying RL load

Solution

The circuit model and phasor diagram are shown in Figure 10.4. First, we need to introduce two angles of interest in synchronous-machine problems, as follows:

The **power-factor angle** is the angle by which the terminal voltage V_T (sometimes called phase voltage V_P) leads the current I. You will recognize this angle, which we designated by the letter φ (phi) in earlier chapters. It has the same significance here that it had before; the cosine of this angle is simply the power factor.

The **torque angle** or **power angle** is a new concept. It is the angle by which the internal voltage E_f leads the terminal voltage V_T or V_P in a generator (or lags it in a motor). We usually designate this angle by the letter δ (delta).

It is evident in this RL circuit of Figure 10.4a, even before calculations are made, that the internal voltage E_f will lead the terminal voltage, and the current into the load will lag the terminal voltage. Because it is customary to specify voltages in line-to-line form, but to perform calculations on a line-to-neutral basis, the 15kV will be understood as a line-to-line value and must be converted into a line-to-neutral voltage for the per-phase model by dividing by the square root of 3 (see Chapter 5).

Assuming the terminal voltage to be the zero-degree reference, the phasor line current is

$$\mathbf{I} = \mathbf{V}_p / (20 + j20) = (15/\sqrt{3} \angle 0°) / (20\sqrt{2} \angle 45°) = 0.307 \angle -45° \text{ kA},$$

from which the per-phase value of E_f can be found as

$$\mathbf{E}_f = j X_s \mathbf{I} + \mathbf{V}_p = (j10)(0.307 \angle -45°) + 8.67 \angle 0° = 11.04 \angle 11.3° \text{ kV}$$

and the output power in the per-phase model is

$$P_{out} = V_p I \cos \varphi = (8.67 \times 10^3)(0.307 \times 10^3) \cos 45° = 1.88 \times 10^6 \text{ W} = 1.88 \text{ MW}.$$

The total output power is three times the per-phase power:

$$P_{out\,total} = 3 P_{out} = 3 \times 1.88 = 5.63 \text{ MW}.$$

Furthermore, the two angles of interest are obviously

$$\delta = 11.3°$$

and

$$\varphi = -45°.$$

The phasor diagram corresponding to this condition is shown in Figure 10.4b. Study this phasor diagram and note how it would change if the load had been RC (i.e., leading power factor). Also note the similarity between the calculations above and those for the transformer. To emphasize an important point, recall that calculations for balanced three-phase circuits are almost always made on a per-phase basis, assuming wye connection.

Effect of Changing Field Current I_f

We mentioned earlier that the magnitude of the internal voltage of the synchronous machine is a function of the dc field current I_f. Suppose our machine is operating as a *motor* and we change I_f. What will be the effect upon the machine? As long as the rotor does not fall out of step (which would be possible if I_f were reduced below some small value, thus resulting in insufficient rotor magnetic field), the rotor turns at synchronous speed and the mechanical load attached to the motor does not know the difference. However, the current and both the angles φ and δ would change as E_f would change in both magnitude *and* angle in the phasor diagram.

This leads to a very important point: *by changing the dc field current we can change the power factor that the machine presents to the ac network.* For high values of field current, the machine (whether operating as motor or generator) is called **overexcited** and produces Q. For low values of field current, the machine operates in the **underexcited** mode and absorbs Q. For some particular value of I_f, which changes with machine load, the machine looks like a unity power factor circuit to the network.

The resulting change in power-factor characteristics is often represented by the well-known *V curve*. A ***V curve*** is a plot of armature current magnitude versus dc field current; the "V" in the designation refers to the shape of the curve. Typically a family of *V* curves, for different values of load, is generated from loading tests. An example of such a family of curves is shown in Figure 10.5. You may recall taking data to produce such curves in a machine laboratory course.

You can see that the unity power factor locus (which corresponds to the dashed line connecting the minimum points of the *V* curves) is not a vertical or even straight line but has a rather significant upward shift to the right. Similarly, curves of 0.8 leading and 0.8 lagging power factor are also shown as dashed lines. Together, the solid curves and dashed lines of Figure 10.5 provide much information regarding the operation of this machine.

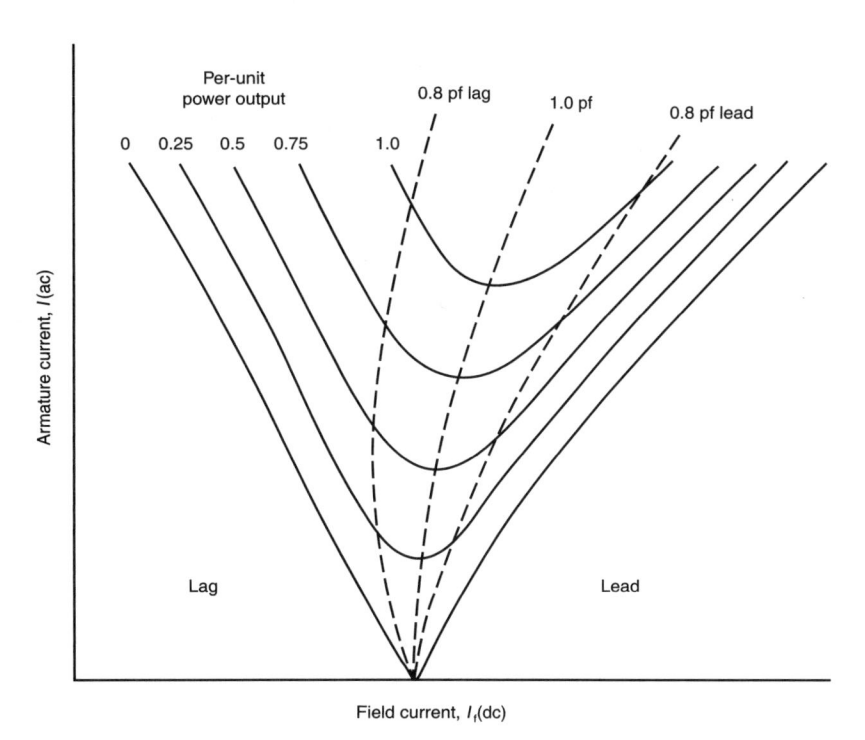

Figure 10.5 Typical family of **V** curves with three lines of constant power factor shown

As an example of using the adjustable power factor feature of synchronous motors via the field current, consider Example 10.4.

Example 10.4

Assume that an existing plant, operating at full capacity, requires 100 kVA at 208 V and has a lagging power factor of 0.8. The local power company charges a higher rate to its commercial customers if their power factor falls below a certain lagging value. The plant manager has been asked to increase production by buying machinery that can be driven by a 50 kW synchronous motor with a power factor adjustable from 0.71 leading to 0.71 lagging at full load. Neglecting losses, what is the kVA rating of the motor and what would the overall power factor be if the manager bought the new machine and operated it at 0.71 leading power factor?

Solution

This problem is similar to the power-factor correction problem addressed in Chapter 4, but much easier. The range of operating power factor stated for this machine is somewhat wide since synchronous motors are often operated at power-factor values no lower than 0.8. Because of this, the effect of power-factor correction will become very evident here. Let's present a graphical solution here in terms of what is sometimes called the *power triangle*. The given and calculated quantities are shown in Figure 10.6.

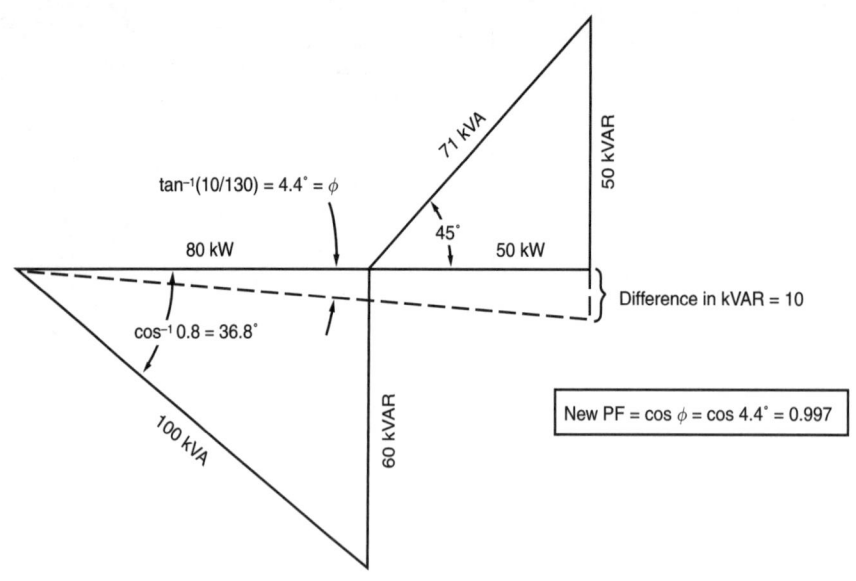

Figure 10.6 Power triangles (showing P and Q values and power factor values) for Example 10.4

The kVA of the new machine, when operating at its limit of either 0.71 leading or 0.71 lagging power factor, can be found by dividing the real power by the power factor of this composite load:

$$\text{kW} / \text{PF} = 50 / 0.71 = 70.4 \text{ kVA}.$$

The total kW required from the utility is of course the sum of what is currently required plus that required by the newly-added synchronous motor: 80 + 50 = 130 kW. The added synchronous machine is to be operated at its maximum leading (0.71) power-factor limit, so that is indicated on the diagram as well. It can be seen that the new motor requires 71 kVA (this figure may be useful in specifying the motor's protection, because it relates directly to the current). The addition of this power-factor correction reduces the reactive demand of the plant to a total value of 10 kVAr, as shown.

The PQ relations are shown in Figure 10.6. The total power triangle solution gives the new power factor of 0.997 lagging. In practice, this amount of correction may be excessive, but it might be useful later to offset the Q required by future inductive loads. Recall that the VAr consumption or supply of a synchronous machine can be continually adjusted by varying the dc field current.

Example 10.5

Consider the synchronous alternator whose per-phase equivalent circuit is shown in Figure 10.4. The circuit is altered by replacing the 20 + j20 Ω load by an RC load of impedance 20 − j20 Ω. Taking the per-phase terminal voltage as reference, determine the magnitude and phase of the internal voltage E_f, with respect to the reference.

Solution

The per-phase current is

$$\mathbf{I}_p = \mathbf{V}_p / \mathbf{Z}_L = (15{,}000 / \sqrt{3}\,)\angle 0° / (20 - j20) = 0.3062 \angle +45° \text{ kA}$$

and so

$$\mathbf{E}_f = j\mathbf{X}_s \mathbf{I} + \mathbf{V}_p = (j10)(0.306 \angle 45°) + (15/\sqrt{3}\,)\angle 0° = 6.85 \angle 18.4° \text{ kV}.$$

Note that a lower field current is required in this case than for the same machine supplying the RL load. Here the machine is **underexcited**; it must absorb Q from the load. This is evident by P and Q balance, because an RC load must supply Q. One danger in the underexcited mode, when the field current is reduced significantly, is the possibility of the machine losing synchronism, or falling out of step, as discussed earlier.

Example 10.6

Two wye-connected, three-phase synchronous machines are connected in parallel at their armature terminals. The per-phase equivalent circuit, with additional information, is shown in Figure 10.7. One machine is acting as a source (generator) and the other as a load (motor). The terminal voltage V_T or V_P (which of course is the same for the two machines) is 400 V (phase-to-neutral) and the synchronous reactances are $X_{Gen} = 0.5 X_{Mot} = 0.4\ \Omega$. Furthermore, the dc field currents of the machines are adjusted such that the system is operating at unity power factor. Determine the internal motor voltage E_{Mot} if it is known that the current I is 100 A. The internal resistances are all negligible.

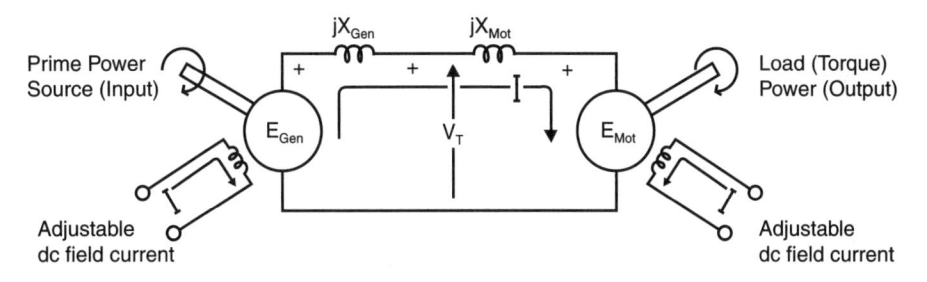

Figure 10.7 Equivalent circuit for a synchronous generator connected to a synchronous motor

Solution

This is a typical problem in steady-state synchronous operation and emphasizes that a synchronous machine may act as either a motor or a generator. By looking at the equivalent circuit, you can see that it reduces to a simple ac circuit problem-containing two ideal sources and impedances. A simple application of Kirchhoff's voltage law yields

$$\mathbf{E}_{Gen} = jX_{Gen}\mathbf{I} + jX_{Mot}\mathbf{I} + \mathbf{E}_{Mot}.$$

Similarly,

$$\mathbf{V}_T = jX_{Mot}\mathbf{I} + \mathbf{E}_{Mot} \quad \text{or} \quad \mathbf{E}_{Mot} = \mathbf{V}_T - jX_{Mot}\mathbf{I}.$$

Because unity power factor at the machine terminals is specified, we know that V_T and I are in phase. Letting the terminal voltage (and, also in this case, the armature current) be the angular reference,

$$\mathbf{E}_{Mot} = 400\angle 0° - j(0.4/0.5)(100\angle 0°) = 400 - j80 = 408\angle{-11.3°}\ \text{V}.$$

Note that many problems relating to power and machinery, such as Examples 10.4–10.6, can be solved as straightforward steady-state ac circuit problems. In the case of three-phase devices, be careful to distinguish between line-to-neutral and line-to-line voltages and between three-phase and single-phase P and Q (see Chapter 4).

Torque

If you are required to make calculations involving torque on the PE exam, remember that *power is speed times torque*. Conversions are necessary if non-MKS units are used; they are shown in Equations 10.10–10.12 in the section "Useful Torque Equations and Conversions" later in this chapter.

POLYPHASE INDUCTION MOTORS

Like other machines, an induction motor (we usually refer to it as a motor because that is the most frequent way it is used) operates from the interaction of two magnetic fields: a stator magnetic field and the magnetic field of the rotor. As the name implies, *the rotor field is induced magnetically from the stator*, in contrast to the synchronous-machine rotor to which a direct current is applied. Induction motors are of two types, distinguished by the form of the rotor: cage (or squirrel cage) rotor and wound rotor. We will concentrate primarily on the simpler and more frequently encountered cage-rotor machine, although most of the theory holds for either type.

An induction machine can be distinguished physically from a synchronous machine by its rotor; in theory, the stators are virtually the same. The rotor construction determines whether the rotor locks into step with the stator's revolving field (synchronous-machine behavior) or rotates at a slightly lower speed when motoring (induction-machine behavior). A **cage rotor** essentially consists of a number of conducting bars, short-circuited at the ends, on a laminated core. A **wound rotor** contains a winding similar to that of the stator; the windings are brought to the stator by means of slip rings and either shorted directly or through resistance. There are no external rotor connections to a power source for a cage rotor. In either device, voltage is applied only to the stator winding, and currents induced by transformer-like action flow in the short-circuited rotor. These ideas will all be made clear in the following.

To summarize: the two common types of induction machine, distinguished from each other by rotor type, are:

- cage rotor machine, and
- wound-rotor machine.

The vast majority of single- and three-phase motors (especially above about one-eighth hp but smaller than 1 MVA or so) used in ac systems are induction motors, because they are simple and reliable and meet most load demands well. Most of these machines are cage-rotor machines because of the simplicity of this device. Examples (both single-phase and polyphase) include most fan and well-pump motors, appliance and shop-tool motors (having speeds below 3600 rpm at 60 Hz), and barbecue-spit motors.

Recall that the speed of the revolving field set up by the three-phase set of stator currents flowing in a three-phase machine rotor was given as

$$n_s = 120 \, f / P. \quad (10.1)$$

The speed of the revolving field is the same as what we found for the synchronous machine, which is to be expected, because the *stator* construction of the two machines is virtually identical.

We emphasize again that for a *synchronous* motor, operating under steady-state conditions, *shaft or mechanical speed n_m is equal to n_s*. For an *induction* motor, the shaft will always turn at a *lower* speed than synchronous when the machine is motoring.

We now define **slip speed**, the difference between the speed of the stator revolving field and the mechanical speed n_m of the induction-machine rotor, by

$$n_{slip} = n_s - n_m. \tag{10.3}$$

A more important and useful parameter for the induction machine is **slip**, or really **per-unit slip** s:

$$s = (n_s - n_m) / n_s \tag{10.4}$$

from which we can express mechanical speed as

$$n_m = (1 - s) n_s. \tag{10.5}$$

Note two extreme conditions of motor operation:

locked rotor or standstill: $(n_m = 0)$ for which $s = 1$
synchronous speed: $(n_m = n_s)$ for which $s = 0$

Of course, we recognize that the induction *motor* cannot operate at synchronous speed, because at that speed there is no relative motion between the rotor conductors and the stator revolving field; therefore, there is no torque. Thus, for *motoring* operation, it is always true that $n_m < n_s$. If, however, we were to drive the shaft of the induction machine with a prime mover, we could make $n_m > n_s$. We will show that under such conditions the machine will actually behave as a *generator*.

The slip s is obviously dimensionless but is usually expressed in percent. It is typically on the order of 0.02–0.05 (that is, 2–5 percent) under load, or less than 0.01 for very light or no-load conditions, in a typical motor.

Based upon the conditions described above, the **frequency of induced rotor currents** can be found as

$$f_r = s f. \tag{10.6}$$

We could show that *the induced currents in the rotor will in turn produce a rotor revolving field. This revolving field interacts with the revolving field produced by the stator to develop machine torque.*

In order to begin exploring the nature of this torque, we will now develop an equation for the speed of the rotor-produced revolving field with respect to a stationary reference frame.

Let n_{rf} be the speed of the induced rotor field *with respect to the rotor*. We see that

$$n_{rf} = s n_s \tag{10.7}$$

because the speed of a revolving field is proportional to frequency.

The speed of the induced rotor field with respect to a stationary reference frame, n_{rs}, can be expressed as

$$n_{rs} = n_{rf} + n_m$$

and thus, using Equation (10.5),

$$n_{rs} = sn_s + (1 - s)n_s = n_s. \tag{10.8}$$

This important conclusion shows that *regardless of rotor (mechanical) speed, the stator and rotor revolving fields always travel at the same speed with respect to a stationary reference.* Thus, at any given value of load and speed, there is a constant torque angle between the rotor and stator fields for balanced stator voltages. It is this property that ensures the three-phase induction machine constant torque for a given mechanical load when balanced currents are applied to the stator and which also ensures that the machine will have a net starting torque. (This result also clarifies why short-circuited damper windings are placed on the rotor of the *synchronous* machine to make it self-starting. The windings work just like an induction machine rotor winding.)

Like a synchronous machine, an induction machine may have any even number of stator poles per phase, with two being the minimum. Synchronous speed, and thus mechanical speed, is dependent upon the number of poles and the applied frequency, as Equation 10.1 showed. Also, from what was discussed in connection with the synchronous machine, recall that synchronous speed n_s is really synchronous *mechanical* speed, because it is the physical speed at which the magnetic field travels with respect to a stationary reference frame. Again, in the case of the induction machine, this is the speed of the revolving field, *not* the speed of the rotor.

Also, it is important to point out that a wound-rotor machine has a rotor winding that is a mirror image of the stator winding and thus must conform to the number of stator poles and phases. In contrast, the simpler cage rotor develops its own required number of magnetic poles during operation.

An induction motor cannot deliver torque at synchronous speed. It is easy to see why. Recall that, at starting (zero speed), s = 1, but at synchronous speed (which the induction motor cannot attain), *s* would be zero. At synchronous speed, therefore, the rotor has no relative motion with respect to the rotating stator field; that is, there is no *slip* of the rotor with respect to the stator field, and thus no currents would be induced in the rotor to produce a counter-torque. Compare this with the synchronous machine, which has a rotor that does rotate at synchronous speed; in the synchronous motor the revolving stator field locks in magnetically to the rotor and drags it along at constant, synchronous speed.

It may thus be helpful to think of induction and synchronous motors as complementary to each other; the synchronous motor can deliver uniform mechanical power at only one speed (synchronous speed) for a given electrical frequency, while the induction motor can deliver power at any practical speed *except* synchronous speed (although it performs best within certain speed ranges). In fact, an induction motor whose shaft is driven faster than synchronous speed becomes a generator, and slip is negative. Such generators are sometimes used in wind-power systems and other variable-speed applications, because as *asynchronous* (non-synchronous) machines their output electrical frequency is that of the power system to which they are connected. In such variable-speed applications, a synchronous machine would require a frequency converter.

Equivalent Circuit of an Induction Motor

As we did for other devices, we would like to find a sufficiently comprehensive but simple circuit model for the induction machine, because that will allow us to analyze the machine and find power, torque, line current, efficiency, and a number of other parameters at different operating conditions. Without any derivation, we show the per-phase equivalent circuit for a typical induction motor in Figure 10.8. In a more exact representation, the shunt branch consisting of b_n and g_n would be placed between r_1 and x_2', but (as with the transformer) we can often move it to the source side as shown here, in order to simplify the calculations.

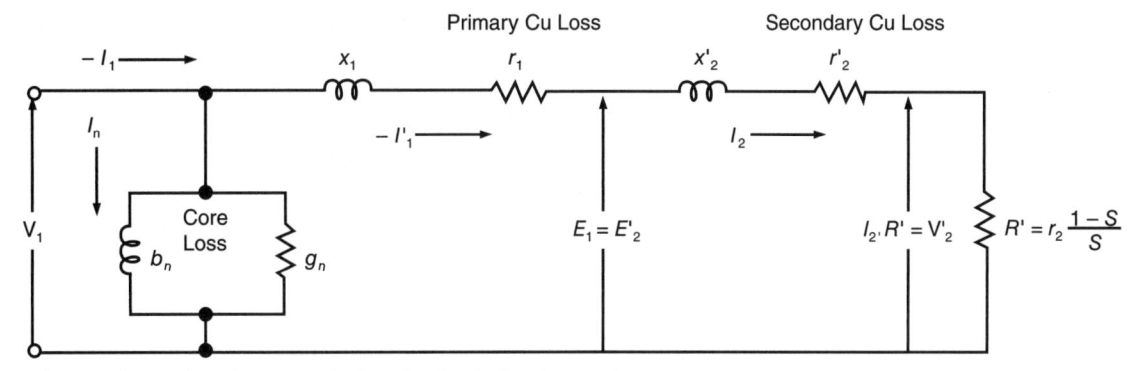

Figure 10.8 Per-phase equivalent for the induction motor

This circuit model emphasizes that, unlike the synchronous machine, the induction machine always looks RL to the source supplying it; it can only consume (not supply) Q and, under motoring conditions, it can only consume (not supply) P.

Most of the model parameters in Figure 10.8 are evident. Assume all are referred already to the stator. (This referral from rotor to stator is like that of a transformer but is complicated by the difference in rotor and stator frequencies. We will not explore this mathematically but will simply state the results. The point at which rotor and stator meet in the circuit is where $E_1 = E_2'$. The prime indicates that the particular parameter or variable has been referred to the stator.) These model parameters, all in per-phase, include the following:

- r_1 stator resistance
- x_1 stator leakage inductance
- r_2' referred rotor circuit resistance
- x_2' referred rotor circuit reactance
- g_n core loss conductance
- b_n core magnetizing susceptance

The similarity to the transformer models in Chapter 7 is evident. Each of these parameters has basically the same significance that it had in the transformer model. Here, the stator is the "primary" and the rotor is the "secondary". Further, we should note that V_1 is the per-phase terminal voltage and I_1 is the input current. (Sometimes the terms V_a and I_a are used for these terminal quantities, but the subscript "1" here was used on purpose to indicate the positive-sequence model as per Chapter 12.)

One more parameter needs to be mentioned: the resistance

$$R' = r_2'(1-s)/s \qquad (10.9)$$

at the right side of the model. This parameter, which has no counterpart in the model of the transformer, models the **developed power** supplied to the rotor. From this developed power, mechanical losses (windage and friction) are subtracted to find the output power of the motor.

Observe that if we drive the motor faster than synchronous speed (e.g., by using a wind turbine), slip becomes negative and thus R' becomes negative. In theory, a negative resistance would supply P, and this is exactly what happens; the induction motor functions as a *generator*. It must still receive, from the three-phase network to which it is connected, the equivalent of what for the synchronous machine was known as *excitation*. The difference here is that this excitation, which provides the magnetic field needed for machine operation, is a three-phase ac signal rather than a dc current as for the synchronous machine.

In keeping with typical procedure, the drawing of the circuit model of Figure 10.8 begins with the part of the device (the stator windings) connected to the source on the left and ends with the load or output on the right.

Although we will not cover this, all the parameters of Figure 10.8 (except slip s, of course) can be found from tests that parallel those used in finding the transformer parameters in Chapter 7.

NEMA Designs for Induction Motors

In the United States, induction-motors are typically manufactured to conform to several NEMA (National Electrical Manufacturers Association) *designs*, the most familiar being designs A, B, C, and D. Probably most motors in manufacture are Designs A and B. These designs differ in terms of performance measures such as slip, speed-torque characteristics, and efficiency. Table 10.1 summarizes these NEMA motor designs in terms of starting current and starting torque.

Table 10.1 NEMA Motor Designs

NEMA Design	Starting Torque	Starting Current
A	normal	normal to high
B	normal to high	low
C	high	low
D	very high	low

We will primarily consider the Design B motor, which is widely used in HVAC (heating, ventilating, and air conditioning) applications for such loads as fans, blowers, and pumps, as well as for other general purposes. Design A motors are quite rare; Designs C and D are occasionally found. Note that these "design letters" have a different meaning from the "code letters" discussed in Article 430.7 of the National Electric Code, which range from A to V and indicate kVA per horsepower for locked-rotor conditions.

Speed-Torque Characteristics

The speed torque characteristics of the induction motor can be found from the equivalent circuit model. Usually we assume the application of a fixed per-phase voltage and evaluate the developed torque versus slip for a number of slip values between 0 and 1 and graph the result.

A typical speed-torque curve for a Design B motor is thus simulated in Figure 10.9. Although we often refer to this as a speed-torque curve, it might better be called a *torque-slip curve*. In general, induction motor operating characteristics are plotted versus slip, as will be seen.

Because the concept of **torque** is so important in dealing with induction motors, we pause to define it here. Torque is defined as force acting through a moment arm. A good illustration is a socket wrench having a handle of one foot radial length to which one pound of force is applied at the end. *In most motor work in the United States, mechanical torque T is expressed in lb-ft.* For the socket wrench example, the torque would be T = 1 lb-ft. Equations 10.10–10.12 define and provide conversions for torque.

The shape of the curve of Figure 10.9 is very interesting and we now examine the important points and regions.

Starting torque is the torque value corresponding to zero speed, where of course s = 1. Starting torque is approximately (neglecting the mechanical losses) the torque available, with rated voltage applied to the motor, to accelerate a shaft load. It is good that starting torque is greater than full-load torque in the figure, because the machine will usually need to accelerate a load from rest, and the torque requirement may be greater at starting than under running conditions.

Full-load torque is the value of torque at rated load. This point would be found on the right side of the curve at a very low (several percent) value of slip. For this Design B motor, it is less than starting torque. Note that, because the curve has a high slope in this region, only a small variation of slip (and thus of speed) occurs in response to a relatively large excursion of torque demanded from the motor. This is an advantage, because it helps a motor to "ride through" substantial momentary torque excursions. The steepness of the curve in this region leads some to consider that the induction motor is a constant-speed machine (especially in comparison with the universal motor discussed at the beginning of the chapter), but of course it is evident that this is not truly the case; at least, the speed is not sufficiently "constant" to qualify the machine for use in applications such as timers.

Maximum (pullout) torque, often called **breakdown torque**, is the maximum value of the curve in Figure 10.9. Note that this torque value, for the Design B machine, is significantly higher than either starting torque or full load torque. This, also, is an advantage, allowing the machine to supply considerably higher torque than the rated value in order to meet momentary demands. Of course, we recognize that trying to run the motor constantly at this value would quickly result in high currents and overheating, as will be evident from the input-current curve for the same machine, shown in Figure 10.10.

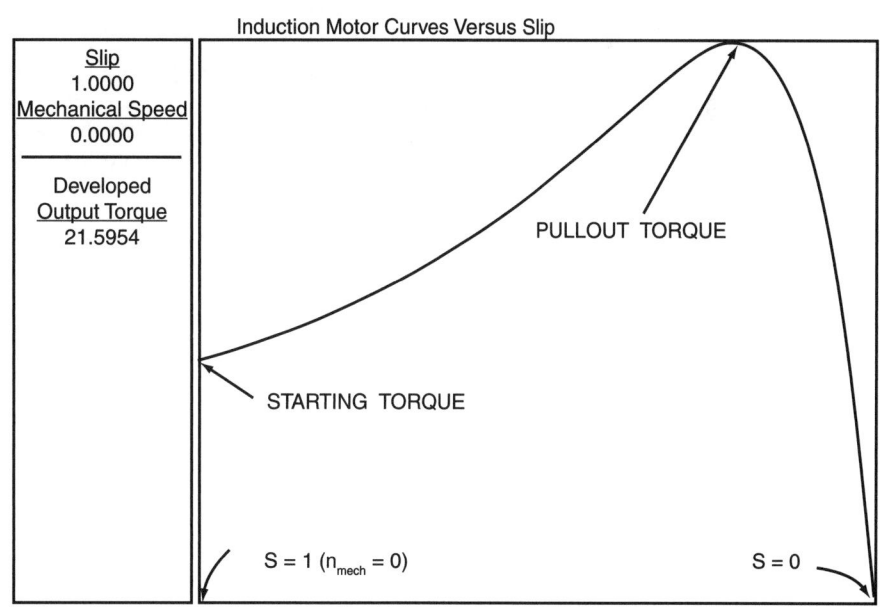

Figure 10.9 Speed-torque curve for typical induction motor

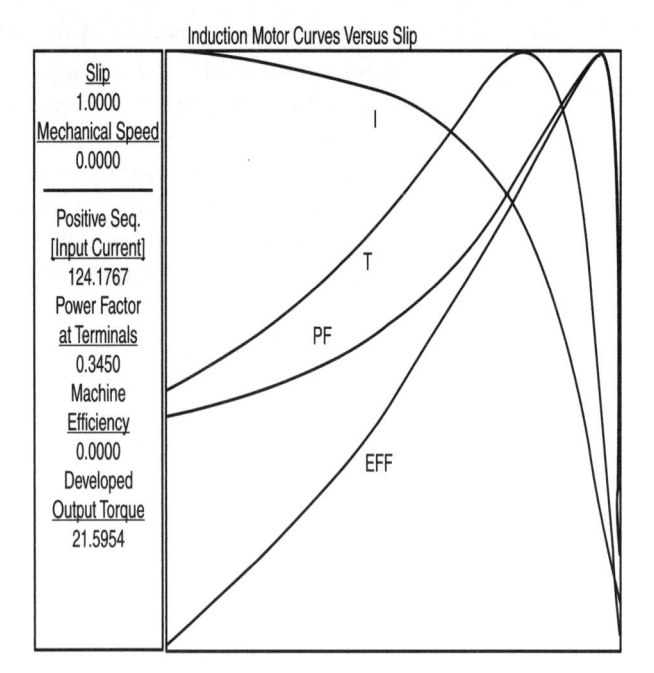

Figure 10.10 Motor parameters in terms of speed or slip

Other Machine Parameters

Numerous other operating parameters can be calculated in terms of slip for a given machine. Besides torque, simulated curves of input current, power factor, and efficiency are shown for a typical Design B induction machine in Figure 10.10. From these we can see the following:

- Starting current is much higher than full-load current and decreases rapidly as the machine nears full-load (rated) speed. A rule of thumb is that starting current can be approximated as six times full-load current.

- Power factor is low upon starting and increases to a much higher value near rated output. This is very fortunate, because a motor is often operated near rated load. However, the power factor is still on the order of 85 percent (lag) at rated output and thus may need correcting.

- Efficiency is highest near full load, which of course is also a decided advantage. Of course, these advantages are obtained through careful design of the machine.

The first two points above provide the motivation for **reduced-voltage starting**. Because of its low (and very highly inductive, as evidenced by the low power factor) impedance at starting, the machine imposes a heavy burden on the typical system, which may result in voltage flicker, sags, dips, or other problems, including tripping of protective devices. Motor starters can be designed to reduce these problems. Some discussion of motor starting will be presented later in the chapter.

Motors with Higher Starting Torque

As we have seen, a typical induction motor, when running at full load, operates at a slip of just several percent; that is, at a speed close to and a little under synchronous speed. In fact, Design A, B, and C motors typically have less than 5 percent slip under full load. Design D motors might have slip values up to about 13 percent.

Motors of NEMA Designs C and D have higher starting torques than those of Designs A and B, which can be an advantage in some specialized applications. However, they have higher full-load slip, which means a greater speed drop as load is increased, as well as lower efficiency. Curves for a representative high-slip machine are shown in Figure 10.11.

One could show analytically that *the speed-torque characteristics of cage-rotor induction machines are primarily a function of rotor resistance and thus a function of rotor construction (such as cross section of the conducting bars and their depth from the rotor surface). Increasing the effective rotor resistance from that of the Design B motor yields higher-slip designs.* In fact, this is how we obtained the simulated curves for the machine of Figure 10.10. For a wound-rotor machine, the effective resistance in the rotor circuit can be increased directly by the user to achieve the same effect.

Note that in the higher-slip machine, whose response is plotted in Figure 10.11, the maximum torque point on the curve is near the locked-rotor (starting) point. Such a motor might be used to drive a rock crusher, where very high initial values of torque are needed. In fact, rotor resistance can usually be increased to the point where maximum torque actually occurs at starting. Again, this increase in starting torque comes at a price: lowered efficiency and poorer speed regulation.

Induction Motor Curves Versus Slip

Slip
1.0000
Mechanical Speed
0.0000

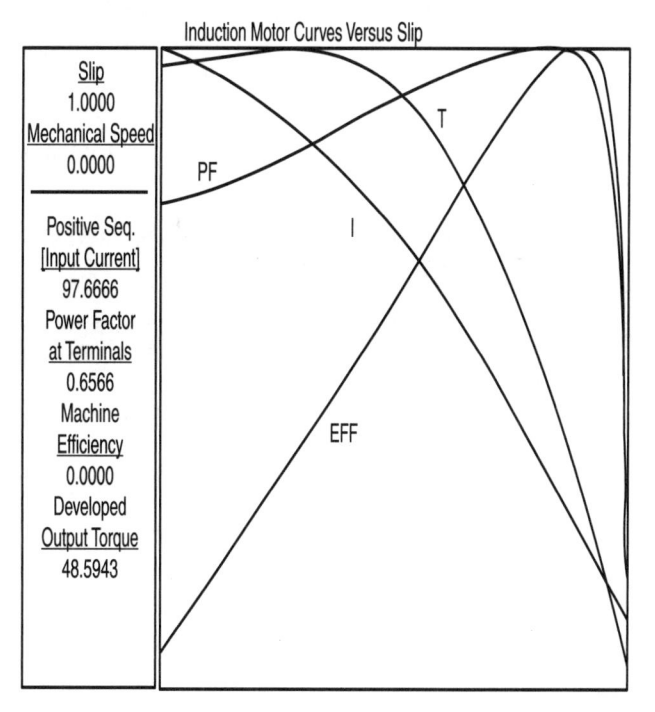

Positive Seq.
[Input Current]
97.6666
Power Factor
at Terminals
0.6566
Machine
Efficiency
0.0000
Developed
Output Torque
48.5943

Figure 10.11 Speed-torque curve for high-slip induction motor

Useful Torque Equations and Conversions

Recall that *power is equal to torque times speed*. The basic equation is

$$P = T' \omega \qquad (10.10)$$

where
P = the mechanical power in watts
T' = the torque in newton-meters
ω = the mechanical speed (angular velocity) in radians per second.

More useful equations for those applying motors in the United States are

$$P = n T / 7.04 \qquad (10.11)$$

and

$$HP = n T / 5252 \qquad (10.12)$$

where
P = the mechanical power in watts
HP = the mechanical power in horsepower
n = the mechanical speed in revolutions per minute (rpm)
T = the mechanical torque in lb-ft.

These equations are a consequence of the fact that

$$1 \text{ revolution} = 6.283 \text{ radians}$$

and

$$1 \text{ horsepower} = 746 \text{ watts}.$$

Mechanical Load Characteristics and Induction Motor Selection

Mechanical loads are sometimes classified according to speed-torque characteristics. They are then grouped into four general categories:

- **Constant torque**—torque remains constant within operating speed range of the load. Also called constant torque/variable horsepower. Conveyors and hoists often fall into this category.

- **Constant power**—torque is inversely proportional to speed. Also called constant horsepower/variable torque. Circular saws and lathe drives are good examples of this kind of load.

- **Torque proportional to speed**—power is proportional to square of the speed. Roller devices are examples, such as the calendar, which presses cloth or paper.

- **Torque proportional to the square of speed**—power proportional to third power of speed. Fans, blowers, and other air-handling devices, and some pumps, are often assumed to fall into this category.

In practice, a given load often has several of these characteristics, but one will usually predominate. A number of loads actually have a characteristic that lies between the last two, and for purposes of motor selection, are treated similarly.

Knowledge of the mechanical characteristics of the load is important to proper selection of the motor and its electrical auxiliaries. For instance, the load characteristic dictates how the electric current drawn by the motor heightens with increasing load. The characteristics above become especially important when selecting multi-speed motors.

Motors are selected to meet the following criteria:

- Starting and running requirements of the load
- Constraints of ambient conditions
- Constraints of the power supply (available voltage, frequency, system stiffness, etc.)
- Duty cycle

The usual objective is to minimize the physical size and weight, horsepower rating, and cost, including cost of auxiliaries such as motor starters and protection. Reliability and ease of maintenance may also be issues of concern. Efficiency, of course, is a matter of concern, because higher efficiency reduces operating cost, which can be high over the life of a large motor.

A motor must be able to accelerate its load from zero to rated speed, in a reasonable time, without overheating or tripping protective devices. Sometimes, in order to provide a safety factor or for anticipated increase in load, an application is **overmotored** by selecting a substantially larger motor than needed. This is often tempting to do, because going to the next motor size may only incur a small additional cost. However, *when oversized motors are selected, the starting current and cost (of motor and perhaps starter and protective devices) are greater, and perhaps more importantly, the power factor during running is lower.*

Multi-Speed Motors

Multi-speed motors are designed as one of the following, in order to meet the load requirements just discussed:

- *Constant-torque motors*: horsepower rating at each speed is directly proportional to the speed
- *Constant-horsepower motors*: have the same horsepower rating at each speed
- *Variable-torque motors*: used on loads such as centrifugal pumps and fans where the horsepower requirement decreases more rapidly than the square of the reduction in speed

Effects of Voltage and Frequency Changes Upon Motors

Three-phase motors are usually designed to operate successfully (although not optimally) under the following extreme ranges of conditions:

- Where the supply voltage magnitude is within ±10 percent of nominal
- Where the frequency of the supply voltage is within ± 5 percent of nominal
- Where the sum of these variations does not exceed ± 10 percent (provided that the constraint on frequency is not exceeded)

In general, motors and other devices do not operate as well when their voltage and frequency are not close to nominal (nameplate) values. Furthermore, such factors as voltage unbalance and harmonics, the latter being prevalent today because of electronic loads, mitigate against optimal operation and motor life. Usual symptoms include excessive heating, reduced efficiency, poor starting, noise, reduced life, operation of protective devices, and perhaps even burnout.

Effects of Voltage Unbalance Upon Motors

An unbalance in the phase voltages applied to a three-phase motor usually manifests itself in excessive heating within the motor. Voltage unbalance may be easily estimated by taking the three line-to-line voltage readings, figuring their average, and using the equation below:

percent voltage unbalance =
(maximum deviation of voltage readings from average) / average

(See Chapter 12 for a more complete discussion and examples of voltage unbalance.)

A small amount of voltage unbalance at the terminals of the motor may be accompanied by a large unbalance in the full-speed line currents (perhaps as much as approximately ten times the voltage unbalance) and a slight decrease in torque and full-load speed. Temperature rise will increase substantially. The *Standard Handbook for Electrical Engineers* cites, for instance, a 25 percent increase in temperature rise for a 3.5 percent voltage unbalance. It is probably best to try to keep the voltage unbalance at the motor terminals to 2 percent or less.

Effects of Harmonics Upon Motors

As discussed in Chapter 14, the notion of **harmonics** provides a way of representing periodic, steady-state distortion in the power system. Capacitive-discharge power supplies in computers, televisions, and other modern electronic devices and (more important to us) *solid-state motor drives*, which are commonly found in industry, can create significant harmonic problems. In general, harmonic components in the voltage supply to the motor cause adverse effects similar to those caused by voltage unbalance. Undesirable effects include excessive noise, heating, and premature failure. A motor can absorb a certain level of harmonic components, but its operation may be impaired.

Harmonics can sometimes be *detected* by comparing the readings of different types of voltmeters or by measurement of neutral currents, but the level of harmonic content is difficult to ascertain using simple instrumentation. Furthermore, the neutrals of three-phase motors are nearly always ungrounded or the windings are connected in delta, making neutral current zero. *A harmonic analyzer, oscilloscope, or other sophisticated piece of equipment may be required to properly assess the problem.* Fortunately, these devices are becoming smaller, more user-friendly, and to some extent less expensive and thus more popular. A good power-quality analyzer with oscilloscope mode and wave-capturing capability can now be obtained in a size that is smaller and lighter than a large textbook. These instruments provide an accurate profile of the harmonics present and can even be left in place for periods of time to record abnormalities.

As mentioned earlier, power-electronic devices are often associated with the presence of harmonics in the system. Some comments on effects and mitigation of harmonics are presented in Chapter 14.

Frequency Changers

Power electronic circuits known as **cycloconverters** can be used today to alter electrical frequency if necessary. However, it has been known for a century or more that a three-phase *wound-rotor* induction motor can be reconnected as a **frequency changer**, in which the frequency of the output currents is different from that of the input currents. In the usual case, voltages of line frequency are applied to the stator of the machine, and the shaft is driven by an external prime mover whose speed can be adjusted and held constant. Output is taken from the rotor circuit via its slip rings. Equations 10.1, 10.5, and 10.6 allow us to calculate the mechanical speed at which the machine must be driven in order to supply a rotor frequency f_r in terms of the frequency f of stator currents.

For an example of the use of the frequency-changing property of the wound-rotor induction motor, see Example 10.10 later in this chapter.

Starting Methods for Large Induction Motors

Starting current for a typical induction motor might be on the order of six times or more of the rated full-load current, and the current might remain close to this level for several seconds while the motor is being started (see, for instance, Figure 10.10). Higher values of mechanical load aggravate the problem by requiring longer starting times. Starting a motor by connecting it directly across the line is known as **across-the-line starting** or simply **line starting**. This method of bringing a motor on-line may be acceptable for relatively small machines but can present problems when used with large ones.

To reduce the impact on the system, a large motor is often buffered from the line during starting. Most such methods involve reducing the voltage to the machine during starting and are thus called **reduced-voltage starting** methods. They include the following:

- **Resistor or reactor starting**: A resistor or inductive reactor is inserted in series with each line to the motor. This impedance is switched out after a timed interval or when the motor has reached a certain percentage of rated speed. At least two contactors (motor relays) are required, three if there are two steps of impedance insertion. Reactors are preferred to resistors because they provide impedance (to reduce line currents) without extensive I^2R heating as would be characteristic of resistors.

- **Autotransformer starting**: An autotransformer is inserted between the source and motor during starting. This has the twofold advantage that motor currents can remain higher (for torque production) than line currents because of the transformer action, and line currents can be equivalently reduced. Sometimes autotransformers are placed in only two of the phases, because they are used only during starting and unbalance is not an issue.

- **Part-winding starting**: This requires a motor having several *sets* of stator windings, each rated for line voltage; one set is energized for starting and the second set is connected later. Because the first set remains in during running, the switching strategy is relatively simple. This is not truly "reduced voltage" starting but has a similar effect upon the power system.

- **Star-delta (or wye-delta) starting**: The motor is started in wye and the winding terminals are then rearranged so that it runs in delta. This method is very popular in Europe and for loads such as compressors in the United States. It requires the motor to have both ends of each phase-winding to be accessible and rated for delta connection and in addition demands a relatively complicated switching strategy.

- **Electronic starting (soft starting)**: Here, electronic devices are used to start the motor. Extensive choice of voltage ramping and acceleration characteristics may be available by means of feedback. Like other power-electronic circuits, such methods have the disadvantage that they may cause harmonic and other problems in other upstream loads (see Chapter 14).

Power-Factor Correction for Induction Motors

Recall that most ac power-system loads, whether residential, industrial, or commercial, are characterized by lagging power factor. These include induction motors, solenoids and other inductive devices, fluorescent and high-energy-discharge lighting systems employing magnetic ballasts, etc. Two of the few exceptions, which have nearly unity power factor, are incandescent lamps and radiant heating. Very few loads have leading power factor.

Although lighting may account for a significant share, probably the largest concentration of load in an industrial plant is induction motors. Most such motors have the following characteristics:

- Reactive power taken by an induction motor is relatively constant over its normal range of load, increasing slightly with increasing load.

- Real power increases nearly linearly with load (mechanical) power.

- Apparent power (VA or KVA) thus increases with load.

- *Power-factor is extremely low at light loads.*

Figure 10.12 illustrates these ideas for a typical induction motor.

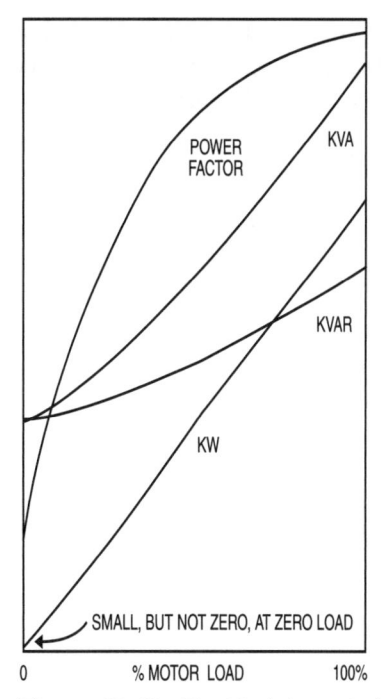

Figure 10.12 Electrical characteristics of typical induction motor in terms of load.

Because induction motors operate at power-factor levels that are typically lower than desired, it is often essential to correct the power factor of large units. From Figure 10.12 we see that *power-factor correction for induction motors is straightforward and can provide desirable results over a wide range of motor operation.* You may wish to review the principles of power-factor correction as discussed with examples in Chapter 4 before we discuss this further.

A further motivation for power-factor correction occurs when a drive is for some reason **overmotored**, in which case the motor operates at a fraction of its rating. It can be shown that operating an induction motor at lower values of load compared with rated load results in a considerably lower power factor (perhaps in the range of 0.5 to 0.8 lagging). With a large number of such motors on line, the power factor at a plant bus could be quite low. This is in addition to the problems of high current and very low power factor during motor starting, which tend to depress bus voltages and might cause problems for other nearby loads.

Utilities often penalize an industry for low power factor, because (as noted in Chapter 4) more amps are required per kilowatt delivered at low power factor, and this higher current results in transmission and distribution losses, voltage drop, and flicker problems.

Because reactive power consumption of an induction motor is nearly constant, this device is an ideal application for power-factor correction, as will be shown here. The power factor of a motor can be easily corrected over a wide range of loading by placing a capacitor at its terminals. Suppose, for example, the capacitor is chosen to correct the power factor of the motor in Figure 10.12 to 95% lagging at no-load conditions. The curves of Figure 10.13 result. (A stiff source was assumed, such that the capacitor does not significantly affect the voltage at the motor, and thus it does not affect the motor load. Actually, the installation of this capacitor reduces the current in the line from the power system and in fact would typically *raise* the voltage at the motor.) The combination of the motor and capacitor is sometimes treated, and switched, as a unit. We should note, as in a previous example, that induction motors are not corrected to unity or leading power factor values in practice. This is to avoid overexcitation and resonance problems.

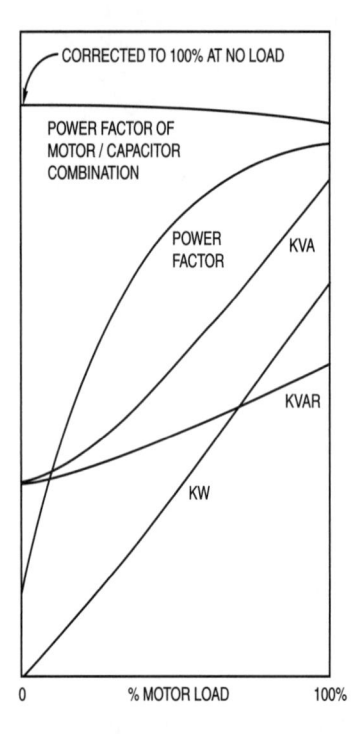

Figure 10.13 Motor of Figure 10.12 corrected to 95% lagging power factor at no load

Although motors may be corrected to other lagging power-factor values, the example here makes clear the ease and effectiveness of power factor correction for the induction motor.

Keep in mind that the curves shown here and elsewhere in this chapter are for typical example motors and might not accurately represent a given application completely.

Induction Machine Examples

Example 10.7

A certain 60 Hz induction motor has four poles and operates at a slip of 0.03 (= 3 percent) at rated load. Find synchronous mechanical speed and actual mechanical speed.

Solution

For this machine, the speed of the revolving magnetic field (= n_s and thus synchronous mechanical speed) would be

$$n_s = 120 \times 60/4 = 1800 \text{ rpm}$$

and the actual shaft speed would be

$$n_m = (1 - 0.03)(1800) = 1746 \text{ rpm.}$$

Example 10.8

A 60 Hz, three-phase induction motor operating at full load has a slip of 0.04. With the load removed, the slip reduces to 0.002, where the speed is measured to be slightly under 1200 rpm. Estimate the full-load speed in rpm.

Solution

Here, because synchronous speeds for a 60 Hz machine are 3600 rpm (for two poles), 1800 rpm (for four poles), 1200 rpm (for six poles), etc., it is obvious this is a six-pole machine. Furthermore, note from Figure 10.9 that the torque in terms of speed is approximately linear (with a negative slope) near full speed.

For a slip of 0.002, this machine would have a speed of

$$n_m = (1 - 0.002)\,1200 = 1197.6 \text{ rpm (this checks the no-load condition)}.$$

At full load (for a slip of 0.04), the speed would be

$$n_m = (1 - 0.04)\,1200 = 1152 \text{ rpm}$$

which, as you would expect, is lower than what it would be at a smaller value of load.

Example 10.9

The efficiency of a typical induction motor may be expressed as

$$\text{EFF} = P_{out} / P_{in} = P_{out} / (P_{out} + \text{all losses}).$$

The losses usually considered are the electrical winding losses (I^2R), the magnetic core losses (P_{core}), and the friction and windage loss ($P_{F\&W}$).

A particular 60 Hz, cage-rotor induction motor is known to produce a shaft output power of 5 hp at a speed of 1750 rpm. Determine (a) the slip when producing 5 hp, and (b) the power input if the electrical loss is 200 W, the core loss is 100 W, and the friction and windage loss is 150 W.

Solution

(a) Because the machine is rated for 60 Hz service, the synchronous speed n_s must be 3600 rpm, 1800 rpm, 1200 rpm, or some other speed corresponding to an even value of p in Equation 10.1. The actual speed (under load) is slightly less than n_s. It should be obvious for this motor that n_s is 1800 rpm (i.e., it is a four-pole machine). The slip is given as follows (see Equation 4.3):

$$s = (n_s - n_m) / n_s = (1800 - 1750) / 1800$$
$$= 0.028 \text{ or } 2.8 \text{ percent (a typical value, incidentally)}.$$

(b) The output power must be converted from hp to W. The power input is given by

$$P_{in} = P_{out} + \text{all losses} = 5 \times 746 + 200 + 100 + 150 = 4180 \text{ W}.$$

Incidentally, the efficiency would be 3730 / 4180 = 0.892 = 89.2 percent, which, again, is a typical value for an induction machine.

Example 10.10

A six-pole, three-phase wound-rotor induction motor is to be used as a frequency changer. Currents of 60Hz are applied to the stator, and we wish to obtain 50Hz currents from the rotor. At what speed should a prime mover drive the machine?

Solution

From Equation 10.6, the required slip must be

$$s = f_r / f = 50 / 60 = 0.8333 \ (= 83.33 \text{ percent}).$$

Synchronous speed of the stator field must be

$$n_s = 120 \, f/p = 120 \, (60) / 6 = 1200 \text{ rpm}.$$

Thus, from Equation 10.5,

$$n_m = (1 - s) \, n_s = (1 - 0.8333)(1200) = 200 \text{ rpm}.$$

Note that as the desired rotor frequency becomes closer to the stator frequency, the speed at which we must drive the shaft decreases. For instance, if we desired a frequency of 59 Hz at the rotor, the shaft speed would need to be 20 rpm.

SINGLE-PHASE SYNCHRONOUS AND INDUCTION MOTORS

Synchronous and induction motors can be made in single-phase and two-phase configurations as well. Two-phase applications, widely used in positioning systems, are not often found in large ac power contexts (although it is noteworthy that the early induction machines built by the device's inventor, Nikola Tesla, were two-phase, as were the original Niagara Falls synchronous generators). The main differences between polyphase (higher phase order than one) induction machines and single-phase ones are the methods of getting these machines to start and to develop a rotating flux phasor.

Single-phase machines are called for, typically, when it is difficult or inconvenient to provide three-phase service. This includes homes, small commercial enterprises, and in situations where single-phase service is generally more convenient. Although single-phase machines are not found in large sizes in most industrial applications, because of their widespread use they should be considered at least briefly. Most ac machines other than universal machines of less than one horsepower rating (**fractional-horsepower machines**) are single-phase induction machines (typically where load torque within a narrow speed range is desired) or synchronous machines (where truly constant speed is required, as in electric clocks and timers).

In general, two- and three-phase machines behave very much alike from the standpoint of starting. Recall that two- and three-phase induction motors are inherently self-starting and directional, while two- and three-phase synchronous machines are not self-starting. (They are usually fitted with a short-circuited *damper* or *amortisseur* winding on the rotor to emulate an induction machine under starting conditions.) *Single-phase synchronous and induction machines are neither inherently self-starting nor directional*; something artificial must be done to shape the magnetic field in order to make the motor accelerate a load from rest and decide the direction in which it will turn.

Popular single-phase motor types include shaded-pole and split-phase designs, named for the way in which they modify the air-gap field. A motor in a six inch tabletop fan, for instance, probably has **shaded-pole** construction (heavy shorted turns diagonally opposite each other on the stator core which reshape the flux under parts of the pole faces in order to give net torque). A one-half HP motor purchased for a home table saw, by contrast, is probably a **split-phase** machine with starter windings displaced in position (and electrically shifted in phase as well) from the main windings. The starter windings are removed from the circuit by a centrifugal switch once the motor has accelerated to near rated speed (if the centrifugal switch fails in the closed position, the starter winding will likely burn out if an integral protector or external protective device does not operate).

Some single-phase induction motors (particularly split-phase) employ an external capacitor to provide additional phase-shifting of the starter-winding current during starting; they are called **capacitor-start** motors. When a capacitance is left in the circuit, they are called **capacitor-run** motors.

Regardless of which technique is used for starting, most of these single-phase induction motors essentially have one thing in common—they effectively have two revolving flux phasors rotating in opposite directions until running. Then as they accelerate, one or the other flux phasor predominates in the rotational direction of starting. From then on, the machine acts much like the corresponding three-phase machine. As in the case of polyphase machines, whether the single-phase machine is synchronous or induction in behavior depends upon the rotor structure.

The same equations relating to synchronous speed apply regardless of the number of phases. For single-phase induction machines, the equations involving slip (as developed earlier in the chapter) still apply also. Questions involving slip for single-phase machines might be expected on the examination.

SOME OTHER MACHINE TYPES

For the sake of completeness, we end this chapter with brief mention or recapitulation of a few other machine types besides the synchronous and induction machine. Some of these are mentioned or described in Chapter 9. Such types of machines, usually in the smaller sizes, are found in a wide variety of applications. Often they do not fit neatly into specific categories.

DC machines (see Chapter 9) may be of the permanent-magnet type, in which the stator field is provided by a permanent magnet and the rotor winding is supplied power through a mechanical switching device consisting of a split-segment **commutator** and slip rings. Such machines are often found in electric toys, inexpensive tape recorders, videocassette rewinders, etc. Larger dc machines usually have windings on both rotor and stator. They may have two or more field windings (on the stator) as well as the armature (rotor) winding. These windings may be interconnected in series or shunt, or in **compound** (where both series and shunt field windings are used). By controlling the relative values of the currents and voltages in these field windings, a wide variety of speed-torque characteristics may be obtained.

Many applications require the use of a high-speed motor with a high value of *speed regulation* (i.e., a significant drop of speed with increasing load; speed regulation is analogous to voltage regulation). Some examples of such applications are hand-held drill motors, electric mixers, blenders, and other small high-speed appliances. The motor used here is essentially a series-connected dc motor, commonly called a **universal motor** because it can be used with dc or ac. The universal motor is, of course, a single-phase machine. Sometimes this device is simply called a **commutator motor**, because, unlike other ac machines, it contains a commutator.

In recent years, **stepper motors**, which are usually driven by synchronized, electronically-generated pulses, have found wide application in positioning devices such as computer printers, and robotics. You can often distinguish a de-energized stepper motor by the **cogging**; as the rotor is turned it seems to seek specific rest positions because of the interaction of an internal rotor magnet and the salient stator poles.

Probably the most fertile field for unique design has been the area of small (fractional-horsepower) single-phase induction and synchronous machines and novel types of small commutator machines. From about the 1920s onward, many different designs, often widely different in appearance, have been marketed, culminating in (for example) today's very inexpensive, simple, and reliable ac clock and timer motors. Many of these applications are, of course, being supplanted by digital electronic devices.

As with transformers, it is generally true that electric machines of a given horsepower rating have become successively smaller, lighter, more efficient, and more reliable over the years, as better shaping of the magnetic field and better insulating, conducting, and core materials are used. Some modern designs for high-speed, low-inertia motors actually dispense with the ferromagnetic rotor core.

A device related to motors is the linear actuator called the **solenoid**, which consists essentially of a ferromagnetic plunger free to move within a coil. Solenoids are typically used where straight-line (**translational**) mechanical motion is required (e.g., to move the clutch mechanism in a clothes washing machine). **Rotary solenoids** operate similarly but produce rotary motion without continuous rotation. In order to make solenoids efficient for operation with sinusoidal ac currents, **shading coils** (heavy short-circuited turns) are often placed on the stator as is done with the shaded-pole induction or synchronous motor.

It is probably unlikely that these machine types will be included on the PE exam. If they are, it is often sufficient to develop a simple circuit model and solve for the unknowns or to use standard equations relating torque, power, and so on.

RECOMMENDED REFERENCES

For additional review reference material, almost any electric machinery text (even if a decade or more old) will be suitable. Two texts that are widely available and have appeared in a number of previous editions are:

Fitzgerald, Kingsley, and Umans. *Electrical Machinery.* 6th ed. McGraw-Hill, 2002.

Chapman, Stephen J. *Electric Machinery Fundamentals.* 4th ed. McGraw-Hill, 2003.

Machinery and power-systems texts often contain useful reviews of steady-state ac problems as well, usually in an early chapter or an appendix.

For specific, detailed information on single-phase machines, see the classical books by C. G. Veinott, including:

Veinott, C.G. *Fractional- and Subfractional-Horsepower Electric Motors.* McGraw Hill, 1970.

Veinott, C.G. *Theory and Design of Small Induction Motors.* McGraw Hill, 1959.

CHAPTER 11

Power-Flow Study

OUTLINE

FUNDAMENTALS OF POWER FLOW STUDY 158

CIRCUIT THEORY 160

CALCULATION OF LINE FLOWS 163

RECOMMENDED REFERENCE 164

The term **power flow** or **load flow** denotes a system study, carried out in the sinusoidal steady state, to calculate certain unknown quantities in terms of network impedances (or admittances) and certain specified operating parameters. A power-flow study is normally conducted as a planning study, in which it is desired to determine the effects of anticipated loadings and generation schedules, device outages, and other changes in system structure or operation. Power-flow studies can be made for small or large systems and sometimes include the low-voltage portions of the system as well. Because electric utilities are interconnected, some system models for power flow contain 3000 or more buses.

Power-flow study is a fundamental component of much utility analysis and power system performance and is listed explicitly as an exam topic in the NCEES specifications. Thus it is worthwhile to be familiar with this material. Additionally, the equation development in this chapter provides another, more comprehensive, review of SSAC circuit solution.

In this chapter, we will examine the basics of power-flow study. Power-flow algorithms used by electric utilities are much more sophisticated than this and allow inclusion of tap-changing and phase-shifting transformers, series and shunt line compensation, dc lines with associated electronics, partition of the system into zones, etc. They typically use fast algorithms that exploit system sparsity and sometimes include optimization techniques.

FUNDAMENTALS OF POWER FLOW STUDY

Typically, the *known* or specified quantities in a power flow study include:

- impedances of all lines and transformers,
- *P* and *Q* values for all loads, specified in MW, MVAR at nominal voltage, and
- magnitudes of generator bus voltages (in per unit).

Unknowns to be calculated, and the units in which they are most frequently expressed, include:

- angles of generator bus voltages (in degrees),
- magnitudes (in per-unit) and angles of load bus voltages,
- real and reactive power flows in the lines (MW and MVAR), and
- (sometimes) total real and reactive power losses in the transmission system (MW and MVAR).

Most power-flow study *outputs* display at least the following quantities:

- All bus voltages in per-unit, with angles,
- Line *P* and *Q* flow out of each bus,
- Generator *P* and *Q* values,
- Bus loads (*P* and *Q*) and also shunt-device *P* and *Q* values if such devices are present, and
- Possibly total generation and total load *P* and *Q* values for the system.

Figure 11.1 shows a single-line diagram of a simple four-bus network containing two generators and two loads. We will use this network to illustrate the basic principles of power flow, including the use of the bus admittance matrix Y_{bus} (which was introduced in Chapter 4). As is typically done in a single-line diagram, we indicate loads by arrows and machines by circles, and each bus is numbered. The network is a three-phase system, and balance is assumed as is typical in power-flow studies involving transmission networks. The usual conventions of voltages being understood as line-to-line values and MVA values being understood as total MVA are followed.

Note that the *P* and *Q* values for the loads are shown directed *out* of the passive network, while those for the generators are assumed directed *into* the passive network. This reflects the usual understanding of those quantities. For nodal analysis, currents are assumed directed *into* the buses, so we need to introduce a negative sign for the stated load *P* and *Q* when setting up the nodal problem. This again points out the need for careful attention to reference directions and polarities.

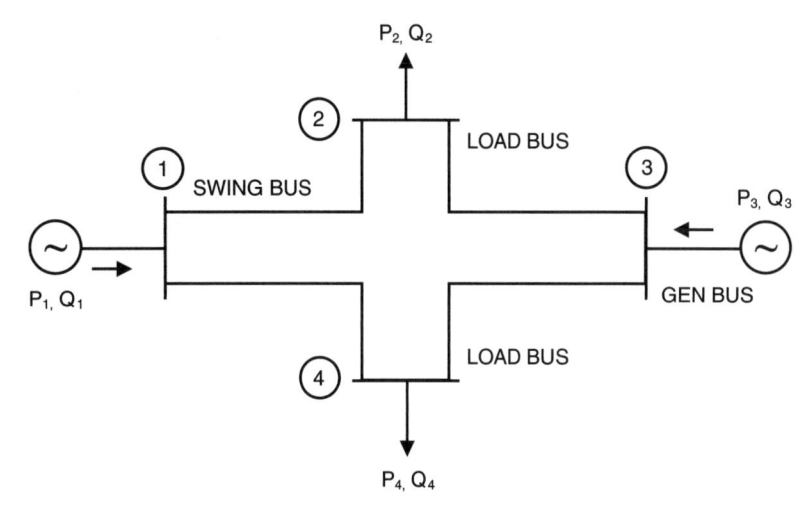

Figure 11.1 Four-bus sample system for power-flow study

In power-flow studies, we divide the buses in the network into three major types. Four quantities (P, Q, voltage magnitude V, and bus-voltage angle δ) are of interest at each bus. Table 11.1 shows the names of these bus types and indicates what is known and what must be determined at each type of bus.

Table 11.1 Values for an example four-bus network

Bus Type	Known	Unknown
Load bus	P, Q	V, δ
Generator Bus	V, P	Q, δ
Swing (or slack) bus	V, δ	P, Q

The concept of a **swing bus**, introduced for convenience in the calculations, is necessary because we don't know system losses prior to the study and thus cannot specify the real power into the system at every generator bus. We can, however, specify the real power at every bus except one. That one bus is called the *swing bus* and is usually selected to be an important bus at which generation is present. Because P is specified for each of the other buses having generation, and P is specified at each load bus, in a numerical sense the swing bus "takes up the slack" in supplying the difference of total generation and total load power; that is, system real-power loss. This is computationally convenient, although in actuality we recognize that the generator at the swing bus is just another generator supplying P and Q into the system, where it flows according to electrical laws.

Calculation of the generators' Q outputs is also an important part of the study because this partially determines generator capability; recall that increasing Q increases generator armature current (just like increasing P does) and, therefore, increases generator heating and the internal voltage drops.

Probably the most important outputs from a power-flow study are the bus voltages and the line P and Q flows. In order to calculate the line P and Q flows, a power-flow algorithm needs the magnitude of each bus voltage V and its angle δ. From Table 11.1 we see that those two values are known at the swing bus, so nothing has to be calculated at that bus. We also see that at each generator bus only the angle of the voltage has to be found, and at each load bus both voltage magnitude and angle are to be found.

CIRCUIT THEORY

To illustrate the power-flow calculation procedure, we will use the simple system of Figure 11.1 which contains only lines, generators, and loads. The interconnected transmission network containing the transmission lines (as well as transformers and other passive devices in actual power networks) constitutes a *passive* system, to which generators and loads are to be tied. Both loads and generation are treated as bus injections *into* the system, and thus both behave mathematically as *active* devices. This means that neither generator circuit models (such as used in short-circuit study) nor load impedances need to be specifically known. Of course, in most cases power actually flows out of the system into a load, so again it is important to emphasize that the sign convention must reflect this.

At each bus n, the current *entering* the system can be specified in terms of the phasor bus voltages and some (complex-number) elements Y_{nm} of the bus admittance matrix as

$$\mathbf{I}_n = \sum_{m=1}^{NBUS} (\mathbf{Y}_{nm} \mathbf{V}_m) \tag{11.1}$$

where NBUS is the total number of buses.

Also, the complex power injected into the system at any bus n is

$$S_n = P_n + jQ_n = \mathbf{V}_n \mathbf{I}_n^* \tag{11.2}$$

as shown in Figure 11.2.

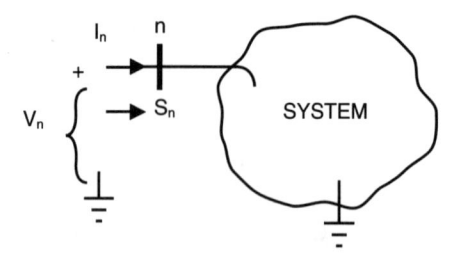

Figure 11.2 Relation of quantities at a bus

Solving Equation 11.2 for $\mathbf{I}_n^*$, the conjugate of bus injection current, gives

$$\mathbf{I}_n^* = (P_n + jQ_n)/\mathbf{V}_n = S_n/\mathbf{V}_n$$

or

$$\mathbf{I}_n = S_n^*/\mathbf{V}_n^*. \tag{11.3}$$

Setting Equation 11.1 equal to Equation 11.3, we have

$$\mathbf{I}_n = S_n^*/\mathbf{V}_n^* = \sum_{m=1}^{NBUS} (\mathbf{Y}_{nm} \mathbf{V}_m). \tag{11.4}$$

Removing the diagonal term from the summation yields

$$S_n^*/\mathbf{V}_n^* = \sum_{\substack{m=1 \\ m \neq n}}^{NBUS} (\mathbf{Y}_{nm} \mathbf{V}_m) + \mathbf{Y}_{nn} \mathbf{V}_n, \tag{11.5}$$

which can be solved for V_n as

$$V_n = (1 / Y_{nn})[S_n^* / V_n^* - \sum_{\substack{m=1 \\ m \neq n}}^{NBUS} (Y_{nm} V_m)]. \qquad (11.6)$$

Note that in Equation 11.6, V_n appears on both sides of the equation. This equation represents an **iterative form**, and is the basis of what is called the **Gauss-Seidel power-flow algorithm**. The usual solution procedure for this kind of equation involves substituting a value of V_n^* into the right-hand side, solving for V_n, again substituting the new value, and so on until the most recently-calculated value equals (approximately) the next previous value. Because there is more than one bus voltage to be calculated in our problem, we calculate them successively, updating the appropriate values at each calculation, as will be illustrated below.

To start the process, values of magnitude and angle must be assumed at all the buses where one or the other, or both, are not known. From a numerical-analysis standpoint (especially for a large system), a good estimate of the bus voltages may be critical and the success or accuracy of the numerical process may hinge on this. Fortunately, using the per-unit system, we can make a good intelligent guess, because in a typical stable system most bus voltages will be fairly close to 1 pu. Thus, at each load bus, we guess a voltage of $1\angle 0°$ pu. Furthermore, at each generator bus we will include the specified magnitude to which the generator will hold the voltage and assume $\delta = 0°$ for each (collectively this is known as a **flat start**). Now let us see, specifically, how to handle each type of bus and why we defined the different types of buses in the first place.

To illustrate the algorithmic process of performing a power-flow study, we will now examine the buses of Figure 11.1 in order, noting what we do and do not need to calculate at each bus. Each of the steps of the process within an iteration corresponds to one bus; because there are three types of buses, there are three basic types of steps.

Step 1—Bus 1 (the swing bus in this example):

As mentioned earlier, one generator bus in the system is defined as the swing bus. At this bus we specify voltage magnitude and angle, so it becomes the "angle reference" for the circuit problem. Because the generated real power is specified at each of the other generator buses, we can't specify it at the swing bus when we don't yet know how much P will be lost in the lines.

Because we know the voltage at the swing bus, no calculation is needed there.

Step 2—Bus 2 (a load bus in this example):

Because Bus 2 is second in the list, we proceed to it next. A load bus is one at which there might be some load, but at which there is no generation. We assume that this load is known in terms of real and reactive power. As an interesting aside, *it is the specification of P and Q at the load buses (rather than the load impedance) that causes the basic power-flow iterative Equation 11.6 to be nonlinear*. The network model is of course assumed linear, as indicated by Figure 11.1, but the complete equation for bus voltage is in nonlinear form. Such equations are often easier to solve by iterative methods, especially in a multivariate situation, and this is exactly what we do.

With the assumed bus voltages and the values of Y_{nm} from Y_{bus}, and noting that the load S_n at bus n is known, it is possible to substitute directly into Equation 11.6 and determine a "new" V_n in the case of a load bus. For example, for the four-bus example system of Figure 11.1, we would use the given and assumed data to compute a new V_2. This value of V_2 (magnitude and angle) replaces the assumed values in all calculations to follow. With this new value of V_2 we now move to Bus 3 and try to do the same thing.

Step 3—Bus 3 (a generator bus in this example):

Note that at a generator bus we know P_n but we do not know Q_n, and thus we cannot apply Equation 11.6 directly yet. Before using Equation 11.6, Q_n must first be determined, which we can do as follows:

$$Q_n = \text{Im}\{S_n\}$$
$$= \text{Im}\left\{V_n \left(\sum Y_{nm} V_m\right)^*\right\} \qquad (11.7)$$

Using the newly-calculated value(s) of voltage (so far, in the first iteration, just V_2) and the assumed values of voltage as well as values from Y_{bus}, we can now determine Q_n. Now $S_n = P_n + jQ_n$ is known at this generator bus (P_n was specified and Q_n was just calculated).

At this point we substitute the result for S_n into Equation 11.6 and solve for a new V_3. This will give new values of both magnitude and angle. However, remember that the magnitude of the voltage at each generator bus was assumed known at the beginning of the study (it can be controlled locally at an actual plant). Thus, we merely keep the specified magnitude value and replace the assumed angle with the newly calculated angle; that is, let

$$V_3 = \left|V_{3\text{ given}}\right| \angle \delta_{3\text{ new}}. \qquad (11.8)$$

This value of V_3 replaces the old value for the calculations to follow.

Step 4—Bus 4 (another load bus in this example)

We now move to Bus 4 and repeat what we did at the other load bus (Bus 2) to solve for a new V_4. This calculation uses the new values of V_2 and V_3.

An *iteration* has now been completed. However, we probably do not yet have accurate values for the calculated voltage magnitudes and angles. To find more accurate values of these, we begin another iteration by returning to Bus 2 (remember that no calculations were needed at Bus 1, the swing bus) and proceeding through all the relevant steps again to complete this second iteration.

We continue doing this until we reach **convergence**; that is, when each of the newly-calculated values at each bus is nearly equal to that calculated for the previous iteration. To determine convergence numerically, the real and imaginary parts of each bus voltage must satisfy the following criteria:

$$\left|\left|\text{Re}\{V_{new}\}\right| - \left|\text{Re}\{V_{old}\}\right|\right| = \zeta \qquad (11.9)$$

$$\left|\left|\text{Im}\{V_{new}\}\right| - \left|\text{Im}\{V_{old}\}\right|\right| = \zeta \qquad (11.10)$$

where ζ is the **convergence tolerance**, and "new" and "old" are interpreted to be the most recent and next most recent calculated values, respectively. Because we often want bus voltages accurate to at least two decimal places in the output, from experience we might choose a tolerance $\zeta = 0.00001$.

When *all* bus voltages satisfy both Equation 11.9 and Equation 11.10, we have finished calculating the bus voltages and can now calculate line flows and losses using simple, standard circuit methods in connection with the circuit models of the appropriate elements (see below).

The numerical procedure outlined here is a variant of what numerical analysts call the *Gauss iterative method*. However, because we typically use each newly-calculated value as needed for subsequent calculations *within the same iteration without waiting to use these values in the next iteration*, the procedure is technically called the **Gauss-Seidel** iterative procedure.

Because it is simple and easy to understand and program, the Gauss-Seidel method was applied to early digital-computer algorithms for power flow, and it is often used to illustrate the numerical process for finding system voltages. However, it was soon found that this method can require significant computational time and may even fail (diverge) for larger systems. Other methods have been used with often better result, including the **Newton-Raphson** method, which is based upon a Taylor series. Decoupled and fast-decoupled modifications of this latter method have led to much faster algorithms which can still preserve computational accuracy and which can adapt to unusual network situations.

As mentioned earlier, actual power-flow algorithms provide much more sophisticated capabilities and enable the modeling of off-nominal transformers, interties with other utilities, and dc interfaces.

CALCULATION OF LINE FLOWS

The preceding section shows that the calculation of bus voltages, in general by an iterative method, is central to the power-flow algorithm. After bus voltages have been calculated, the final calculations for line P and Q flows are very straightforward.

Consider the per-phase π model of a transmission line (presented in Chapter 8) as shown in Figure 11.3. We will call this line "line mn" because it spans buses m and n as its endpoints. Note that if the endpoint phasor voltages V_m and V_n are both known, and assuming that we know the impedance and admittance parameters of the line, we can immediately find the current I_{nm} flowing into this line from bus n as

$$\mathbf{I}_{nm} = (\mathbf{V}_n - \mathbf{V}_m) / \mathbf{Z}_{nm} + (\mathbf{y}_c / 2)\mathbf{V}_n \qquad (11.11)$$

from which the real and reactive power entering the line at bus n are found as

$$P_{nm} + j Q_{nm} = \mathbf{V}_n \mathbf{I}_{nm}^*. \qquad (11.12)$$

The real and reactive power values $P_{mn} + j Q_{mn}$ entering the line at bus m can be found from similar equations. (As at the other end, one or both of these may be negative, except that the real power entering a line cannot be negative at both ends at the same time. This is because a line cannot generate P.) Then, the P and Q **loss** in the line can be found as the difference between the entering quantity and the leaving quantity:

$$(P + jQ)_{\text{loss in line nm}} = (P_{nm} + P_{mn}) + j(Q_{nm} + Q_{mn}) \qquad (11.13)$$

Additionally, the **efficiency** of the line can then be found, if desired, as

$$\text{Eff} = P_{out} / P_{in} = (-P_{mn}) / P_{nm}. \tag{11.14}$$

Note that a line will always consume P, but it may consume or produce Q depending upon the relative values of the Q associated with the inductive and capacitive parts of the line model. For instance, a lightly-loaded, high-voltage line is likely to supply net Q to the system, because the Q that it produces is proportional to the square of voltage while the Q that it absorbs is proportional to the square of current. In contrast, a highly-loaded line is likely to be a net consumer of Q, implying that the Q produced by its capacitive behavior is less than that consumed by its inductance.

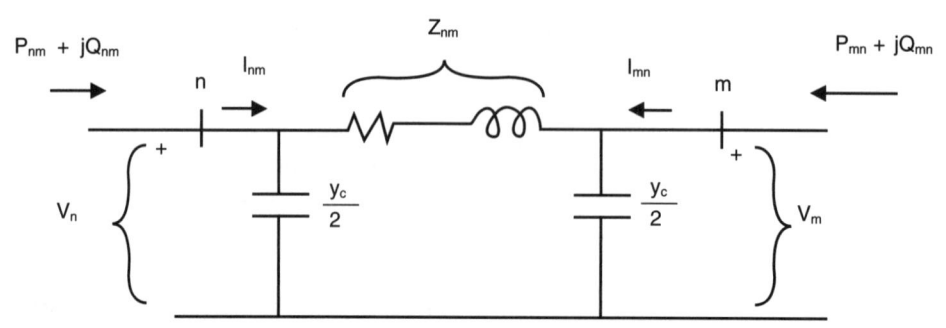

Figure 11.3 Transmission line model

RECOMMENDED REFERENCE

Glover, Sarma, and Overbye. *Power Systems Analysis and Design*. 4th ed. Thomson, 2008.

CHAPTER 12

Symmetrical Components and Unbalance Factors

OUTLINE

THE METHOD OF SYMMETRICAL COMPONENTS 165

PHASE UNBALANCE FACTORS 170
Approximate Voltage Unbalance Factor

RECOMMENDED REFERENCES 171

The method of symmetrical components has been in use for nearly a century as a means of simplifying unbalanced three-phase problems. It allows many problems with unbalanced loads or unsymmetrical short circuits to be solved easily. Furthermore, it provides a justification for using per-phase models to solve *balanced* three-phase problems.

Calculation of symmetrical components requires matrix multiplication (or the equivalent) involving complex-number quantities. Some fundamental ideas concerning the meaning and use of the symmetrical-component transformation are presented in this chapter. A brief discussion of the application of the method to unbalanced faults is given in Chapter 13.

At the end of this chapter, we introduce the idea of **phase unbalance factor**, which, in one form of calculation, involves the use of symmetrical components.

THE METHOD OF SYMMETRICAL COMPONENTS

The method of symmetrical components involves resolving a set of unbalanced three-phase vectors, such as phasor voltages, currents, or magnetic fluxes, into *three sets of three components each*, as follows:

- **Positive sequence** set: a balanced set of abc sequence vectors
- **Negative sequence** set: a balanced set of acb sequence vectors
- **Zero sequence** set: a set of three identical residual vectors

Note that two of these sets of fictitious, transformed quantities are *balanced* sets. This gives rise to the name of the method.

From these three sets (containing a total of nine vectors), we choose the "a" quantity from each set. These three resulting components are then called, respectively, the **positive-, negative-, and zero-sequence components** of the original unbalanced three-phase set. Collectively, they are called **symmetrical components** or **sequence components**. For voltages, for example, they would be designated V_1, V_2, and V_0 respectively.

Most authors facilitate the notation by defining the **rotational operator a**, which is a vector of unity magnitude and angle 120°. In rectangular form,

$$a = -\frac{1}{2} + j\frac{\sqrt{3}}{2} \tag{12.1}$$

and its square would thus be

$$a^2 = -\frac{1}{2} - j\frac{\sqrt{3}}{2}, \tag{12.2}$$

which of course represents a vector of unity magnitude at 240°. (Recall that squaring a vector involves squaring the magnitude and multiplying the angle by two.)

The **symmetrical-component transformation matrix A** is defined as

$$A = \begin{bmatrix} 1 & 1 & 1 \\ 1 & a^2 & a \\ 1 & a & a^2 \end{bmatrix}, \tag{12.3}$$

which in rectangular form is

$$A = \begin{bmatrix} 1 & 1 & 1 \\ 1 & -0.5 - j0.866 & -0.5 + j0.866 \\ 1 & -0.5 + j0.866 & -0.5 - j0.866 \end{bmatrix} \tag{12.4}$$

The **inverse transformation matrix** is easily remembered; it looks like the transformation matrix A but with a one-third premultiplier and with the a and a^2 terms changing place:

$$A^{-1} = \frac{1}{3}\begin{bmatrix} 1 & 1 & 1 \\ 1 & a & a^2 \\ 1 & a^2 & a \end{bmatrix}. \tag{12.5}$$

To check this inversion and refresh your matrix multiplication skills, you might want to multiply A into its inverse or vice versa; in either case the result is the identity (or unity) matrix, as shown below.

$$AA^{-1} = A^{-1}A = \begin{bmatrix} 1 & 0 & 0 \\ 0 & 1 & 0 \\ 0 & 0 & 1 \end{bmatrix} \tag{12.6}$$

The phase (abc) and symmetrical—or *sequence*—component phasors are related by

$$I_{abc} = A\, I_{012} \qquad (12.7a)$$

and

$$I_{012} = A^{-1}\, I_{abc} \qquad (12.7b)$$

where

I_{abc} = the vector of abc phase currents

I_{012} = the corresponding set of sequence currents.

Similar relations hold for voltages or magnetic fluxes.

The following examples will illustrate some of the important properties of symmetrical components.

Example 12.1

Find the symmetrical components for the set of slightly unbalanced phase currents given in polar form as

$$I_{abc} = \begin{bmatrix} 10 \\ 12e^{j230°} \\ 11e^{j118°} \end{bmatrix} \text{A}$$

or, equivalently, in rectangular form as

$$I_{abc} = \begin{bmatrix} 10 \\ -7.7135 - j9.1925 \\ -5.1642 + j9.7124 \end{bmatrix} \text{A}.$$

Solution

The symmetrical (or sequence) components are found as

$$I_{012} = A^{-1} I_{abc} = \begin{bmatrix} -0.9592 + j0.1733 \\ 10.937 - j0.8226 \\ 0.0222 + j0.6493 \end{bmatrix} \text{A}$$

or, in polar form,

$$I_{012} = \begin{bmatrix} 0.9747 e^{j169.76°} \\ 10.9679 e^{-j4.30°} \\ 0.6496 e^{j88.04°} \end{bmatrix} \text{A}.$$

Notice that the positive sequence predominates in this example, because the original set was of abc sequence and nearly balanced. This observation is more evident from examining the polar form of the result.

Example 12.2

Now, let's switch I_b and I_c of Example 12.1 to get basically an acb sequence. Find the symmetrical components of this new case.

Solution

Here the vector of phase currents becomes

$$I_{abc} = \begin{bmatrix} 10 \\ 11e^{j118°} \\ 12e^{j230°} \end{bmatrix} \text{ A}$$

and the symmetrical components are calculated to be

$$I_{012} = \begin{bmatrix} 0.9747e^{j169.76°} \\ 0.6496e^{+j88.04°} \\ 10.9679e^{-j4.30°} \end{bmatrix} \text{ A.}$$

Note the strong negative sequence and weak positive sequence, just as we might have guessed. Also observe that the zero-sequence current has not changed. This case is akin to reversing two of the phases to obtain a reversed set of rotational sequences, as we do when we want to reverse the direction of a three-phase motor. Because the zero-sequence is not in reality a rotational sequence like the others, it is not affected.

Example 12.3

Find the symmetrical components of the set of *balanced* abc-sequence currents given as

$$I_{abc} = \begin{bmatrix} 10e^{j5°} \\ 10e^{j-115°} \\ 10e^{j125°} \end{bmatrix} \text{ A.}$$

The set of currents in this example happens to be displaced 5° from the reference, but that of course does not change the fact that it's balanced, as the calculation of its corresponding symmetrical components will show.

Solution

The symmetrical components are found as

$$I_{abc} = \begin{bmatrix} 0 \\ 10e^{j5°} \\ 0 \end{bmatrix} \text{ A.}$$

As you might have guessed, *only the positive sequence is active here, and its value is equal to the "phase a" value*. This is true for any balanced abc-sequence set, because there is no way for a negative-sequence or zero-sequence component to exist in such a case.

Example 12.4

We could have given examples, just as easily, for the calculation of I_{abc} in terms of I_{012}, in this case using Equation 12.7a. The use of this equation appears in Example 12.4.

Now, let's use symmetrical components to find the currents on one side of a delta-wye transformer, given the currents on the other side. To help emphasize the relevant concepts and compare the results, we will assume a 1:1 transformation ratio of line-to-line voltages (or, equivalently, assume that all the voltages are in per-unit).

Suppose that the sequence currents into HV (delta) side of the transformer are known to be

$$I_{012HV} = \begin{bmatrix} 0 \\ 300e^{j50°} \\ 60e^{j10°} \end{bmatrix} \text{ A},$$

which are in reality created by the phase currents

$$I_{abcHV} = \begin{bmatrix} 348.11e^{j43.64°} \\ 244.48e^{-j74.82°} \\ 315.99e^{-j179.22°} \end{bmatrix} \text{ A}.$$

Find the phase currents out of the low-voltage side.

Solution

We want to find the phase currents leaving the wye side. This can be done easily with symmetrical components, because it is true that for a delta-wye or wye-delta transformer built to American standards, the positive-sequence currents out of the LV side lag those into the HV side by 30°, while the negative-sequence currents do the reverse and thus lead by 30°.

First, the **sequence** currents leaving the LV (wye) side are found by modifying the corresponding positive- and negative-sequence values known from the other side to reflect this standard convention:

$$I_{012LV} = \begin{bmatrix} 0 \\ 300e^{j(50-30)°} \\ 60e^{j(10+30)°} \end{bmatrix} \text{ A}$$

Phase currents leaving the LV (wye) side are then found as

$$I_{abcLV} = AI_{012} = \begin{bmatrix} 356.97e^{j23.3°} \\ 295.55e^{-j111.5°} \\ 256.95e^{j148.6°} \end{bmatrix} \text{ A}.$$

The use of symmetrical components makes such problems such as Example 12.4 very easy despite the unbalances. The only difficulty is a purely mathematical one: multiplying a complex-number matrix by a complex-number vector, but this can be done on most scientific calculators or even by hand if necessary.

Symmetrical components are particularly useful in modeling unbalanced faults (or other unbalanced situations), such as phase-to-ground, phase-to-phase, two-phase-to-ground, and open-phase short circuits and loads. A brief discussion of this appears in the next chapter.

It is wise to verify the preceding examples on your own calculator to make sure you can manipulate these matrix quantities properly, either as matrices or as individual phasor combinations.

PHASE UNBALANCE FACTORS

A useful numerical measure of the amount of unbalance in a set of three-phase vectors is the phase unbalance factor, or simply the **unbalance factor,** defined by the Institute of Electrical and Electronics Engineers (IEEE) in terms of sequence components. The unbalance factor is defined as the magnitude of the negative-sequence quantity divided by the magnitude of the positive-sequence quantity. Thus, for current,

$$UF_{IEEE} = \frac{|I_2|}{|I_1|} \qquad (12.8)$$

For Example 12.1, for instance,

$$UF_{IEEE} = \frac{|0.6496|}{|10.9679|} = 0.0592 = 5.92\%.$$

Note that the unbalance factor is easy to find once we know the symmetrical components.

This unbalance factor can be calculated for voltages, currents, fluxes, etc. It relates to some power-quality issues (see Chapter 14) but is logically discussed here because of its dependence upon symmetrical components. To help understand the nature of this unbalance factor, it may be helpful to think of an induction motor as the load. Recall that positive-sequence currents tend to drive the rotor of a motor in the desired or reference direction, while negative-sequence components tend to push it in the opposite direction. Even in cases of relatively high unbalance, positive-sequence effects typically predominate, but a small negative-sequence component may cause extensive heating.

Approximate Voltage Unbalance Factor

However, another unbalance factor is often used to express unbalance in a set of line-to-line voltages. It was developed to be used with line-to-line voltage magnitudes only, as measured in the field with a hand-held voltmeter, for instance. In most practical cases, it provides a very good approximation to the unbalance-factor calculation above.

This **approximate unbalance factor** is defined as

$$UF_{approx} = \frac{\text{(maximum deviation of voltage from average)}}{\text{(average of voltage readings)}}. \qquad (12.9)$$

In most conventional cases, the value obtained by this method is close to that obtained using the more involved calculation for UF_{IEEE}, which requires symmetrical components.

Example 12.5

The three line-to-line voltage magnitudes are measured at one point in a three-phase system to be 435, 459, and 427 V. Find the approximate unbalance factor corresponding to those voltages.

Solution

The average of the measured voltages is

$$(435 + 459 + 427) / 3 = 440.33,$$

and thus the maximum deviation from the average is in connection with the second voltage given. The unbalance factor becomes

$$UF_{approx} = (459 - 440.33) / 440.33 = 0.0424 = 4.24\%.$$

Voltage unbalance is an important issue with motors, as increased unbalance leads to increased heating (with possible burnout) and reduced levels of performance. The 4.24% unbalance factor calculated here suggests an unbalance of sufficient magnitude to be addressed.

RECOMMENDED REFERENCES

Most general university texts on electric power systems contain a chapter or section on symmetrical components. Two recommended texts are:

Grainger and Stevenson. *Power System Analysis*. McGraw-Hill Inc., 1994.

Glover, Sarma, and Overbye. *Power Systems Analysis and Design*. 4th ed. Thomson, 2008.

The classic text in symmetrical components, which is still useful and makes interesting reading, is:

Wagner and Evans. *Symmetrical Components*. McGraw-Hill Book Company, 1933.

A more recent (and perhaps more readable) small book that gives numerous applications is:

Myatt. *Symmetrical Components*. Pergamon Press, 1968.

CHAPTER 13

Generator Transient Behavior, Short-Circuit Study, and Power System Protection

OUTLINE

FAULTS AND FAULT PROTECTION 174
The Importance of Generator Modeling in Fault Studies ■ Transient Behavior of a Synchronous Generator Subjected to a Fault ■ Use of Subtransient Model ■ Rationale for Short-Circuit Simulation Studies ■ Simplified Procedure for Three-Phase Short-Circuit Study ■ Multiple Sources of Fault Current ■ Unbalanced Faults

POWER SYSTEM PROTECTION 185
IEEE Standard Device Numbers

RECOMMENDED REFERENCES 187

No power system (or other system) can perform at all times in a totally desired, ideal, steady-state manner. The expected probability of a system's functioning as intended is expressed partly by the concept of **reliability** (see Chapter 15). Many things can impact the reliability of a system. One of these is the *fault*.

In this chapter, we will discuss typical faults and illustrate how they can be analyzed. Generally, a **fault** is any unwanted, usually adverse occurrence that prevents the system from operating in a desired steady-state manner. Usually when we think of faults we have a **shunt fault** or **short-circuit** in mind, and because this is usually the most frequent and most severe type of fault, we will concentrate on it here. (Open-phase faults, or even simultaneous faults, can in fact be simulated as well, but these are not expected on the exam.) **Overcurrents** can also pose a threat and are of concern to system protection but are usually not of the magnitude of short-circuit currents—at least not in high-voltage (transmission) systems. (The term *overcurrent* is often understood to mean a current somewhat higher than allowable load current but probably not as great as the current that would flow as a result of a nearby fault.) This chapter will thus primarily address shunt faults, with some consideration of overcurrents in connection with system protection.

The topic of short-circuit study relates particularly to synchronous generators, as well as to the power system in general, and is of course involved with the protection of these components. Thus, the brief discussion of system protection and relaying which we present fits logically at the end of this chapter.

FAULTS AND FAULT PROTECTION

Faults may be *shunt* or *open*. A shunt fault involves the contact of two or more conductors, which might include ground or neutral. *Typical shunt faults include line-to-ground, line-to-line, two-line-to-ground* (or *double-line-to-ground*), *and three-phase*. **Open faults** are typically considered to include the loss of either one (usually) or two conductors. One frequent cause of an open-line condition is the operation of a fuse.

Faults may be *balanced* or *unbalanced* (or, as sometimes described, *symmetrical* and *unsymmetrical*). Of the shunt faults, only the three-phase fault is balanced. A balanced fault can be analyzed by using a per-phase (positive-sequence) equivalent circuit. It is the easiest fault to simulate numerically (and often the most severe), and thus we concentrate on it here. A three-phase fault may or may not involve ground; under ideal conditions no ground current would flow even if a ground connection were present. All other faults are unbalanced (or unsymmetrical) and are usually solved by application of the method of *symmetrical components* (see Chapter 12).

Faults in a power system may be *intermittent* (temporary) or *permanent* (often called "bolted", because they act like a piece of busbar bolted across the conductors and typically are of negligible impedance). Faults on the outdoor portion of the system are usually weather-related in origin. An example of an intermittent fault is the single-phase fault produced when a tree blows against one phase-conductor in a rainstorm, and either moves or burns away shortly after, leaving the system able to have power restored without a visit to the fault location. An example of a bolted fault is a damaged power-line structure in which one or more phases are brought together in contact with each other or ground, and for which the problem requires some intervention to repair.

Power systems have sensing devices including current and potential transformers (see Chapter 15) which send signals to relaying points. These signals are evaluated and compared to determine if a fault is present. Historically this was done by mechanical relays; these tasks are usually performed today by sophisticated digital computer-like devices. Regardless of what kind of relaying is used, even many simple-looking protection schemes are in reality sophisticated. For example, a *differential protection scheme* for a transmission line involves sensing the current into and out of the line; deviations can suggest a fault involving the conductor in question. A similar strategy can be employed for transformer protection.

Two major classes of protective devices exist to *clear* faults in the power system: **fuses** and **circuit breakers**. Breakers can be opened and closed many times; they are designed to interrupt expected fault currents (which may be much greater than typical load currents) and thus can be opened under load currents as well. Fuses are single-operation devices in which the fusible material melts during a fault or sustained overload. Equipment, such as transformers, is sometimes protected by fuses as a "last resort", because fuses rarely malfunction. However, a fuse requires manual replacement, while a breaker can simply be reclosed. Especially in distribution (lower-voltage) systems, an attempt is sometimes made to trip a fault by opening a breaker; if the fault remains on reclosing, sometimes the breaker is kept closed to burn a downstream fuse with the intent of confining a permanent fault to a smaller part of the system and thus controlling the geographical extent of the outage.

As the previous discussion implies, once a fault is detected, a complex *protection* strategy comes into play automatically. For instance, when a fault is detected on a transmission line in the differential scheme mentioned above, signals are sent to the breakers at the ends of the line to trip. In the high-voltage transmission network, all of this may happen within three cycles of the waveform or less; longer response times are usually allowed in the lower-voltage system. After a predetermined short time, an automatic attempt is often made to **reclose** the breakers to restore the system. If the fault was intermittent, the system would probably have been restored by this operation; if the fault is permanent, such as caused by the downing of a conductor, then the protection scheme locks out the breaker in question, and a crew must be dispatched to correct the problem at the fault location.

The strategy above is designed to remove the faulted device quickly. If this does not happen, *backup relaying* should automatically come into play. It is easy to see that the design of a total protective strategy for even a small part of a power system is complex, with redundancies intentionally built in. More will be said about such topics at the end of the chapter.

The Importance of Generator Modeling in Fault Studies

Fault analysis in an electrical power network is very much related to the synchronous generator(s) in the system. Typically it is the generators that contribute most to short-circuit currents, and because of their expense and criticality it is often the generators that require the highest level of protection in the system. Because of the way a large machine responds to a fault, a particular model of the generator is needed, as will be shown below. It is worthwhile to revisit the synchronous generator, which was discussed in Chapter 10, because we will need to extend the steady-state modeling of that device to account for faults and other deviations from normal behavior.

Recall that, under normal, steady-state operating conditions, a generator is typically modeled on a per-phase basis by an ideal source in series with a reactance called the *synchronous reactance* X_s, as in Figure 10.3. In Chapter 10 we assumed that this reactance was constant and that it was much greater than stator resistance for large machines, so for many initial studies not involving losses we ignore stator resistance. We often refer to machine reactances in terms of the **direct axis**, which is the primary axis of the rotor magnetic field between the N and S poles in Figure 10.1b. Thus, we often find X_s referred to as X_d, the **direct-axis steady state reactance**. This definition fits in well with the definitions of other machine reactances that will appear in the following.

Also, it is well to emphasize the difference in how a generator is modeled for power-flow studies (as discussed in Chapter 11) and the way in which we will need to model it for short-circuit studies. For power-flow studies, which typically represent the system under some kind of acceptable (if not desirable) steady-state condition, a generator is usually modeled simply as a P and Q injection into the system at the bus of connection. In other words, for many steady-state power studies, it is sufficient to regard a generator simply as a source and to look no deeper into it beyond its terminals. But under sudden large load changes or short circuit conditions, other effects become important, necessitating the use of a particular internal generator model. This will be examined below.

Recall from Chapter 10 that a three-phase synchronous machine consists of stator containing an armature winding (which is connected to the ac system) and a rotor carrying a field winding to which direct current is applied. Synchronous machines are designated by rotor type. The cylindrical rotor (also called *round rotor*) machine is the simpler of the two and is usually used if the prime mover is a high-speed device (1800 or 3600 rpm at 60 Hz), such as a steam turbine. The other is the salient pole type, usually used if the prime mover is slower and the number of poles is high (say 72 poles at 100 rpm). A cylindrical rotor is simply a solid steel cylinder into which slots are milled parallel to its axis to contain the field (dc) winding. In a salient-pole machine, the poles are well defined and a winding usually appears on each pole, as in Figure 10.1b.

In general, the cylindrical-rotor machine is simpler to model, and the simplified approach we show here is more suited to that machine type. Detailed modeling of salient-pole machines is not likely to appear on the exam and is beyond the scope of this chapter. However, the short-circuit analysis fundamentals presented here will provide a good approximation to machine and system behavior, to a fault for most situations.

Transient Behavior of a Synchronous Generator Subjected to a Fault

Under short-circuit or other transient conditions, a sudden large current becomes present in the stator windings. Magnetic fluxes begin to change in the machine and the normal steady-state behavior is altered. The changes in the mmf values in the core could cause a large current to be induced in the rotor, especially so for a salient-pole machine. This in turn has an effect upon the mmf of the machine core. Because the core is so massive, however, the change in the magnetic field is slow in penetrating the core. The result of all this is an effective change in the inductance of the machine as seen by the external power system.

If we consider only the "symmetrical component" of sinusoidal generator output current immediately following a fault (i.e., if we do not consider the asymmetry in the wave caused by the initial switching), we would see the typical time plot of short-circuit current shown in Figure 13.1a. Note that the current reaches relatively high (but progressively decreasing) values within the first few cycles, and then its rate of decrease is smaller, finally (if the fault were left connected that long) reaching a steady state. The *envelope* of this behavior is shown in Figure 13.1b.

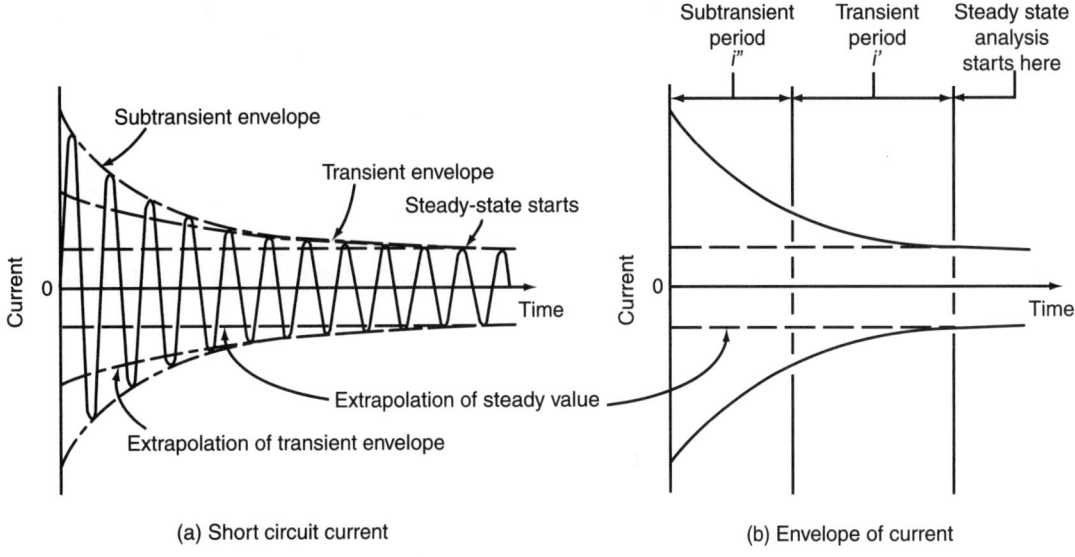

Figure 13.1 Typical short-circuit current for a synchronous machine

(a) Short circuit current

(b) Envelope of current

The damped sinusoidal response shown in Figure 13.1a can be better understood by drawing two envelopes for the symmetric wave, plus the steady-state envelope, all of which are shown in Figure 13.1a. These ranges are described as the **subtransient**, **transient**, and **steady-state** ranges, respectively. Because in fault studies we are usually interested in worst-case currents, and in fact want to remove the fault within a short time after occurrence, we are most interested in the *subtransient* range. (By contrast, *transient stability analysis*, not covered in this book, would be most concerned with the *transient* range.) Knowing some of these ideas might help you to answer a question or two on the exam.

To summarize, the high armature current within a few cycles after a short circuit occurs may be designated by three time periods:

- The *subtransient* period, lasting only several cycles with an instantaneous current designated i"

- The *transient* period which follows this, with an instantaneous current designated i'

- The *steady-state* period which would theoretically follow, if the fault were left in place for that long a time period (it won't, because protective devices will act before this time)

The current wave is of course continually changing throughout the whole transient, as Figure 13.1a shows, and its rms value is continually changing in each of its first two defined periods. Of course the "steady state" period here is untenable, because it would represent a still-high value of fault current which would likely cause extensive equipment damage. However, the breaking down of the current wave into these three periods proves to be very useful in simplifying the analysis of the faulted machine.

During a short circuit, the net transient may be analyzed by defining a **subtransient reactance as X_d"** for the first part of the curve in Figure 13.1, a **transient reactance X_d'** for the mid part of the curve, and a steady-state reactance X_d for the remainder. These correspond, of course, to the three ranges mentioned above. Engineers usually make a simplifying assumption, somewhat justified in practice for practical purposes, that these reactances are constant. Because we are most interested in the first of these periods when analyzing short-circuit response, we concentrate on subtransient reactance.

Use of Subtransient Model

It might seem from the machine response of Figure 13.1 that a detailed, time-domain transient-type analysis must be done any time we wish to calculate the anticipated short-circuit currents resulting from a particular fault. In fact, a simplification is usually made which allows this problem to be well approximated by steady-state analysis.

To do this, each generator (and large motor) is represented by a per-phase model which includes an ideal internal constant-voltage sinusoidal source in series with the direct-axis subtransient reactance. This is usually simply called the **subtransient reactance** and designated X_d'' as above, or simply X''. This reactance models the generator quite effectively during the first few cycles after the fault and allows the problem to be solved by steady-state network theory. The ideal voltage source in this generator model is sometimes called the *voltage behind subtransient reactance*.

It should be noted that synchronous generators are characterized by other parameters, too, including time constants and resistances. However, the subtransient reactance is the most important parameter for short-circuit study, and because the X/R ratio of the machine's armature circuit may be 30 or more, stator resistance can usually be neglected for short-circuit studies. This leads to the simple generator circuit model shown in Figure 13.2b (the figure actually illustrates two generators tied to a common plant bus). An example using this model will be presented, following a discussion for the short-circuit study's rationale.

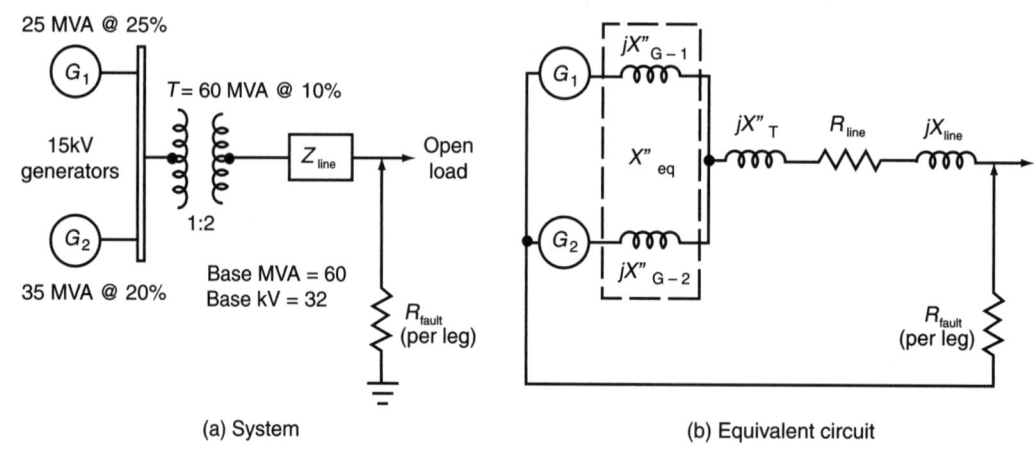

Figure 13.2 Subtransient problem arrangement

Rationale for Short-Circuit Simulation Studies

Fault analysis (short-circuit study) is performed upon an electric power system for the primary purpose of determining expected short-circuit currents in lines, transformers, generators, etc. which would occur as a result of certain expected fault types. These results are used by the power systems engineer to specify device ratings, select fuses and other protective devices, set relays, etc. Importantly, the results of a short-circuit study allow us to:

- correctly select circuit breakers and other devices so that they will successfully *withstand* expected fault currents in the event of a short circuit,

- select protective devices such as fuses and circuit breakers that will be able to *interrupt* these fault currents safely,
- set breakers and relays to *respond* properly when such faults occur, and
- *coordinate* the protective devices in order to isolate the fault to the smallest possible area and minimize the impact upon the total system.

From this brief introduction, you can see that synchronous-machine response to a short circuit is an extensive topic and that protective strategies can become very sophisticated. However, a straightforward approach can be taken to obtain values of short-circuit currents. Now that we have presented the motivation for short-circuit study and some of the related ideas and terminology, we will concentrate on some simple short-circuit modeling and analysis issues. Remember that we do not know where or when a fault will strike; we can only consider some likely faults and plan for their amelioration. In practice, a utility might perform fault simulations for all of the prominent buses in the system and check to make sure that the worst cases are covered.

Theoretically, any type of fault can be simulated at any point on a power network. In practice, the faults of greatest interest are usually **three-phase (usually the most severe)** and **single-phase-to-ground (usually the most frequent**; one utility reported that about 85% of its faults were line-to-ground). In fault studies, we usually simulate faults at identified buses. This is mathematically easier, and, furthermore, we can simulate a bus at nearly any point that interests us. Another justification for this approach is that bus faults are often more severe than faults somewhere on a line, and thus conservative results are obtained from the study. Some more sophisticated programs, however, allow the simulation of a fault at any distance along a line, and this enables us to make our simulation match actual faults and in so doing learn more about where they occur.

We will first consider three-phase faults; these are balanced and thus can be solved using a per-phase equivalent (i.e., the positive-sequence network). Also, we will assume a small, easily-sketched network, which might in fact represent the Thevenin equivalent of a much larger network and illustrate the principles through hand calculation. For discussions of fault analysis in large systems, and in fact in analysis software, some technique involving the *bus impedance matrix* of Chapters 3 and 4 is usually used (see for instance the reference listed at the end of the chapter for more detail).

Simplified Procedure for Three-Phase Short-Circuit Study

A potential exam problem may involve constructing a simple system model from a single-line diagram or word problem and solving for fault currents in particular devices for a fault at a specified bus. Accordingly, we have outlined the following steps to construct and analyze the model for a small system. Such a system may include several buses of particular interest and the Thevenin model of a much larger and more complicated network.

Step 1. Begin with an understanding of the network in words or (better) a single-line diagram.

Sketch the resulting per-phase network. Determine all device impedances, which you will usually want to express in per-unit on system base. Remember to use the change-of-base equation where necessary. The simplest models of the relevant devices are as follows:

- *Transmission lines* and *transformers* are usually modeled as single series impedances. Shunt capacitances of transmission lines and the shunt branch of the transformer model are usually neglected, because these typically have negligible effect on short-circuit current calculations. Series resistance is often neglected in line and transformer models, because the X/R ratio of a transmission line or transformer is often much greater than unity (a possible exception is a cable, where resistance may affect the solution significantly).

- *Synchronous machines* (particularly generators and large motors) are modeled by an ideal voltage source in series with an impedance, usually the direct-axis subtransient reactance X_d''. As shown earlier in this chapter, this reactance represents the behavior of the machine during the first few cycles of a short circuit, which is the period of highest fault currents and is, therefore, of greatest concern. The subtransient reactance is the most important machine parameter to use when calculating short-circuit effects.

- *Loads*, particularly *non-rotating* loads such as lighting, are usually neglected, because they do not contribute as sources to short-circuit currents. *Large motors* can contribute to short-circuit currents, and they can be modeled like synchronous generators. Sometimes an equivalent load will be given as a synchronous-machine model, in which case it should be included.

These approximations give results which are sufficiently accurate for the purpose of most short-circuit studies and remove much of the mathematical complexity. A good rule for test takers, though, is to properly use whatever relevant data is given in a problem statement. If resistances are given, the answer was probably figured in terms of resistances by the problem's author. On the other hand, the developer of the problem might be testing your ability to identify and use only the data relevant to the problem. Always examine the problem statement very carefully and use good engineering common sense!

The system is usually assumed to be at nominal (1.0 pu) voltage prior to the fault, although sometimes a voltage of around 1.05 pu or so is indicated because that value may better describe the voltage at the generator bus. With pre-fault (load) currents neglected, each bus can be assumed to be at the same voltage before the fault. Fault currents in the devices can then be found from circuit analysis.

Special mention should be made of the *delta-wye transformer*. Even if such a device is assumed ideal, there is no way to find a primary current in phase with any given secondary current because of the 30° phase shift which the device introduces under balanced conditions. However, this matter is probably not of concern for the PE exam; it may be sufficient to know that under ANSI standards typical of American transformer manufacture, positive (abc)-sequence currents or voltages are shifted +30° in going from the low-voltage side to the high-voltage side, while for the negative (acb)-sequence (needed for unbalanced fault analysis only) the reverse is true.

Step 2. Find the Thevenin equivalent or other simplified network behind the faulted bus.

Note that the Thevenin voltage will typically be 1.0 pu or whatever the stated nominal was, unless prefault currents are included.

Step 3. Using standard circuit techniques, calculate the current in the fault and then work backwards and calculate the fault current in any desired location using basic circuit analysis.

Although having the Thevenin equivalent as viewed from the fault point allows us to calculate the current *in the fault* immediately, it may be necessary to leave the Thevenin equivalent to calculate whatever device current is desired. (Just use whatever appropriate circuit analysis technique will allow you to get the current in the device under consideration.)

Note that bus voltages during the fault may also be calculated by this procedure. The voltage at a bus may range anywhere from zero (if that was the faulted bus) to approximately nominal.

Step 4. If it is desired to take pre-fault current into account, but if it was omitted from the calculations above, combine it appropriately with the calculated fault current to find the total fault current.

As this is a linearized problem, the principle of *superposition* applies and you can simply add pre-fault currents to short-circuit currents at the corresponding points. Remember to add these quantities as phasors, not simply magnitudes, and be careful that the reference directions are treated properly. In some problems, pre-fault current can significantly affect certain results; use it if it is given and is numerically significant.

Example 13.1

Note: As is typical in fault studies, this brief example makes use of per-unit values (see Chapter 6) in the analysis. *Because problems on the PE exam often require facility with per-unit methods and per-unit conversions at the least, working through this example with these issues in mind is good practice.*

Assume that two three-phase generators of the same voltage rating, but different in MVA rating and subtransient reactance, are tied together to feed a load through a step-up transformer as shown in Figure 13.2a. The transmission line impedance from the transformer secondary to the load point is known to be $2 + j10\ \Omega$. We have shown the information on the figure with a notation that is used by some engineers to give you a different perspective that you may need for the exam.

A three-phase fault to ground with a very low resistance of $0.5\ \Omega$ in each leg of a balanced wye occurs as shown. We desire to know the subtransient current in the fault itself and in each generator. For this example, the load is assumed open-circuited, so you can neglect pre-fault load.

Solution

The per-phase model of Figure 13.2b is first drawn. We will use this equivalent circuit (simplified a bit later) to calculate the short-circuit current once the per-unit impedances are calculated. You may wish to review Chapter 6 at this time if you have not made per-unit and base calculations recently.

It is necessary to assume two base quantities in one part of the circuit, from which all other base quantities can be calculated as needed. We will choose

$$\text{MVA}_{base} = 60\ \text{MVA}$$

and

$$\text{kV}_{base} = 32\ \text{kV}$$

on the high voltage (sometimes called the *high tension*) side of the transformer.

Next we convert the individual subtransient reactance of each generator to the base reactance of the system. For this we apply the change-of-base Equation 6.9 to each generator as follows:

$$X_{G-1} = 0.25(60\text{MVA} / 25\text{MVA}) = 0.6 \text{ pu}$$
$$X_{G-2} = 0.20(60\text{MVA} / 35\text{MVA}) = 0.343 \text{ pu.}$$

Note that no information is given for the transformer turns ratio. Thus we make the natural assumption that it is 15/32. For this reason, it was not necessary to correct the pu impedances of the generators for voltage, only for MVA.

The equivalent subtransient reactance of the two generators in parallel is then

$$X_{Geq} = X_{G-1} \| X_{G-2} = [0.6 \cdot 0.343 / (0.6 + 0.343)] = 0.2182 \text{ pu.}$$

Here the symbol $\|$ means the parallel combination of impedances, for which we have used the standard product-over-sum equation.

To help understand the problem, we draw the equivalent circuit of Figure 13.3 and numerically label this equivalent generator reactance.

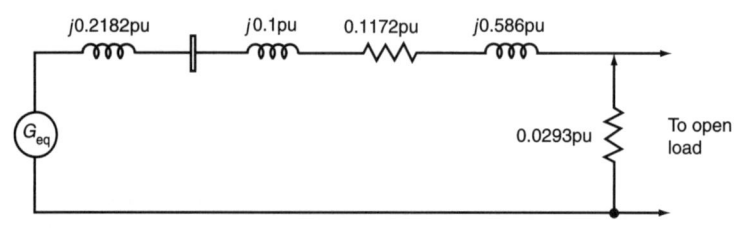

Figure 13.3 Reduced equivalent circuit of Figure 13.2b

Now examine the transformer. Because it is rated at the base values, no application of the change-of-base equation is necessary, and

$$X_{transformer} = 0.1 \text{ pu on system base.}$$

Consider next the transmission line. To convert the actual line impedance and the ground fault resistance to per-unit values, one must first find the base impedance *in that part of the circuit*:

$$Z_{base} = V_{base} / I_{base} = kV_{base}^2 / \text{MVA}_{base}$$
$$= 32^2 / 60 = 17.0667 \text{ }\Omega$$

from which

$$Z_{line} = Z_{actual} / Z_{base} = (2 + j10) / (17.0667) = 0.1172 + j0.586 \text{ pu.}$$

Similarly,

$$Z_{fault} = (0.5 + j0) / 17.0667 = 0.0293 \text{ pu.}$$

The subtransient current in the fault is computed by referring to Figure 13.3. We first find the total per-unit impedance seen by the source:

$$Z_{total} = (0.1172 + 0.0293) + j(0.2182 + 0.1 + 0.586)$$
$$= 0.1465 + j0.9042 \text{ pu} = 0.916\angle 80.8° \text{ pu.}$$

The current in the loop of Figure 13.3 is then

$$I = 1.0\angle 0 / 0.916\angle 80.8° = 1.091\angle -80.8° \text{ pu.}$$

The base current in the part of the network near the fault is

$$I_{baseHV} = 60{,}000/[(\sqrt{3})\,(32)] = 1082.5 \text{ A},$$

and thus the actual value of the current in the fault is

$$I_F = (1.091\angle{-80.8°})(1082.5) = 1181.3\angle{-80.8°} \text{ A}.$$

In the generator portion of the circuit,

$$I_{baseLV} = 60{,}000/[(\sqrt{3})\,(15)] = 2309.4 \text{ A}.$$

The per-unit currents out of the parallel generators may then be found; they are in inverse proportion to their per-unit impedances:

$$I_{G-1} = [0.343\,/\,0.6 + 0.343)]\,(1.091\angle{-80.8°})(2309.4) = 916.6\angle{-80.8°} \text{ A}$$
$$I_{G-2} = [0.6\,/\,0.6 + 0.343)]\,(1.091\angle{-80.8°}) = (2309.4) = 1603.4\angle{-80.8°} \text{ A}.$$

Just as a check, we calculate the fault current into the transformer low-voltage (generator) side:

$$I_{TLV} = (1.0912\angle{-80.8°})(2309.4) = 2520\angle{-80.8°} \text{ A}$$

The total of the fault currents in the two generators should equal this value:

$$916.6\angle{-80.8°} + 1603.4\angle{-80.8°} = 2520\angle{-80.8°} \text{ A},$$

so this checks.

In this example, calculations were made for a fault that was far from the generators. Because there are no sources near the assumed fault point, the fault current will be less than would be present if the fault were closer to the generator(s).

Example 13.2

Suppose that the fault in Example 13.2 had instead been at the generator bus. Assuming no fault-current contribution from the right side of the network in Figure 13.2, the currents in the generators will be

$$I_{G-1} = 1.0\,/\,jX_{G-1} = 1\,/\,j0.6 = -j1.667 \text{ pu}$$
$$I_{G-2} = 1.0\,/\,jX_{G-2} = 1\,/\,j0.343 = -j2.916 \text{ pu}.$$

And the total fault current into the bus would be

$$I_F = -j1.667 - j2.916 = -j4.583 \text{ pu}.$$

Or, in actual amperes,

$$I_F = (-j4.583)(2309.4) = -j10{,}585 \text{ A}$$

which, as expected, is considerably greater than the current out of the generator plant for a fault at the load end of the line.

Multiple Sources of Fault Current

Suppose there had been a large generator (or even a large motor, which could be considered to be a source of short-circuit current) at the load end of the line. How would Example 13.2 have been different?

As mentioned in the instructions for short-circuit study above, it is often best to obtain the Thevenin network as seen from the fault point. This will account for the multiple sources. The current in the fault can then be found immediately by dividing the pre-fault voltage by the Thevenin impedance. One can then work back to determine the fault contributions for the parts of the circuit on the sides of the fault.

In the case of a large system too complicated to draw and solve by hand, the bus impedance matrix Z_{bus} can be used. Recall from Chapter 4 that the diagonal terms of this matrix are just the Thevenin impedances at the buses. Thus, you can see that if Z_{bus} is available, all we need is one diagonal term from it in order to calculate the fault current at the faulted bus. Fault currents in other lines can then be calculated from other terms of Z_{bus} in terms of very simple equations, regardless of the size of the network. The Glover text listed in the Recommended References section at the end of this chapter shows you how to do this in detail.

Unbalanced Faults

As mentioned earlier, the three-phase fault is the easiest to simulate, because it is balanced and thus requires only the per-phase network, which in Chapter 12 was called the *positive-sequence network*.

Other types of shunt faults besides the three-phase fault that might occur, in roughly the frequency expected, are

Line-to-ground (L-G)
Line-to-line (L-L)
2 Line-to-ground (2L-G)

Solution of any of these unbalanced fault types is more complicated than that of the balanced fault. However, the procedures are straightforward and require just a few more mathematical steps if the **method of symmetrical components** of Chapter 12 is used.

The method of symmetrical components dates to about 1918 and has been used ever since, because it drastically simplifies the procedure for handling unbalanced three-phase circuits.

In symmetrical-component terms, the three-phase fault solution such as in Examples 13.1 and 13.2 involves only the positive-sequence network. *Short-circuit study for an unbalanced fault essentially involves at least two, and sometimes all three, sequence networks, interconnected in a unique way to represent the type of fault in question.* Before describing the models for each of these fault types, we should consider some specific, distinguishing properties of the negative- and zero-sequence networks.

The **negative-sequence network** looks just like the positive-sequence network topologically except that it contains no sources, if we assume a balanced system prior to the fault. Therefore, the negative-sequence network for Examples 13.1 and 13.2 would look exactly like Figure 13.3 but with the equivalent voltage source replaced by a short circuit. Once the positive-sequence model is built, the negative-sequence network is thus very easy to obtain. Except for the notable exception of rotating machines, the positive- and negative-sequence impedances of most devices (including lines, transformers, and non-rotating loads) are identical. This is of course because such devices behave exactly the same if presented with a balanced acb, as opposed to abc, sequence.

Although no negative-sequence sources exist in the "traditional" three-phase system operating in abc sequence, negative-sequence *currents* can exist in network elements (including machines) as a result of unbalanced faults. In fact, a damaging condition of unbalance in rotating machines known as *negative-sequence heating* is called that, because the prominent effect there can be found in terms of the negative-sequence current as the unbalanced currents are resolved into their sequence components.

The **zero-sequence network**, which is *required only for faults involving ground*, is often somewhat different in appearance, topologically similar to the positive-sequence network as far as the location of components is concerned but containing breaks or ties to reference depending upon the connections of the respective device neutrals.

A very important point to remember in unbalanced analysis is that *the sequence networks are always interconnected at the point of unbalance (here, the fault point)*. Each fault type requires a particular interconnection of these networks, as follows:

- To model a L-G fault, all three sequence networks are interconnected at the fault point in series-aiding, like cells in a flashlight.

- To model a L-L fault, the positive- and negative-sequence networks are connected in parallel at the fault point, like two batteries of identical voltage would be connected to achieve higher capacity battery.

- To model a 2L-G fault, all three sequence networks are connected in parallel, with their reference buses tied together.

Once the sequence-network model has been constructed for the particular type of fault as described above, sequence currents in the respective networks are calculated for whatever device we want to find the fault current in. The set of sequence currents for the desired device is then converted back to phase currents using the symmetrical-component transformation.

The current in the fault itself can be expressed in terms of one or more sequence currents in each case. Recall from Chapter 12 that, in any ground connection, the neutral current is three times the zero-sequence current there, so this will always be true in faults (or unbalanced load connections) involving neutral.

In that context, we should point out again that the method of symmetrical components is not restricted to short circuits. Open-phase faults can be modeled by symmetrical components as well, although the model is a little more complex because an open-phase condition is modeled in terms of a line and thus has two ends of connection. Furthermore, *any* unbalanced case can be modeled by symmetrical components; for instance, a single-phase load connected line-to-line in a three-phase system (as some transformers are) can be modeled exactly like a line-to-line fault as long as we use the correct machine reactances.

While a detailed presentation of unbalanced short circuits is outside the scope of this book, the material presented above should enable you to have some understanding of how these cases are treated. More information and a number of examples can be found in the Glover text listed in the recommended references at the end of this chapter.

POWER SYSTEM PROTECTION

In discussing synchronous generators, we mentioned that they must be protected against faults and overcurrents. Protection strategies involve protective devices (such as breakers and fuses), instrument transformers to sense system variables (see Chapters 7 and 15), and often very complex systems to compare these variables and send trip signals to a breaker in the event they pose a threat.

Current transformers (CTs) are placed on the phase conductors at critical places in the network, and often one is placed on the neutral as well to sense neutral (residual) current. Potential transformers (PTs) or their equivalent are placed on the phases (usually phase to ground) to sense voltages. The resulting signals are sent (via wire or some form of telemetry) to some common location where mechanical relays or their electronic or digital equivalents apply logic to determine whether to send trip signals to the appropriate breakers.

Some typical relaying or protection strategies are as follows:

- **Time-overcurrent protection**. The simplest protective strategy is probably time-overcurrent protection. A 120V residential breaker, or even the simple fuse, behaves like this—the time required to clear a fault is in some way inverse to the magnitude of the current, so that faults are cleared slowly for currents that are not much greater than a threshold level but are cleared much faster for a higher current. The fact that protective devices behave (or can be made to behave) in this way allows **coordination** of protective devices, in which particular devices (typically downstream) are chosen to act first, with upstream devices serving as **backup**. A time-overcurrent device is characterized by a time-current curve, which gives the relationship between trip time and current graphically. Such a device may also have an **instantaneous** unit or function which responds quickly within some specified time for large currents above a certain level. As with motor-starting and other industrial-control schemes, the term instantaneous means *without intentional delay*.

- **Differential relaying**. Differential relaying, mentioned earlier in the book, has to do with sampling quantities on either side of a device (such as a line or transformer) and comparing them in order to decide if a fault is present between the two sampling points. If a fault in that region is detected, signals can be sent to trip the affected line or transformer out of the circuit, operating breakers at both ends if necessary. Differential protection may be applied to a bus, in which case the currents in the lines attached to the bus are compared. Of course, in a three-phase system there are three live lines for every device, complicating the number of signals to be compared and individual relay elements to use.

- **Directional-sensing relaying**. These devices can be used to help the protection scheme determine which direction the fault lies from a given point. They are also used where the direction of flow in a component is critical or when current or power flow in a reverse direction signals a possible problem. For instance, a backup generator is sometimes installed to provide service to critical parts of a facility when the utility tie is lost. The backup machine may not be able to meet the total load of the facility in the event of loss of the utility. Reverse-power relaying can be used to ensure that a breaker between the small network consisting of the backup generator and its critical load, and the rest of the system, opens upon loss of the utility tie. This would prevent the backup generator from trying to assume the entire plant load and thus tripping on overcurrent.

- **Distance relaying**. The principle of distance relaying is often used in protecting against faults on transmission lines. Because the distance relay responds to the impedance between its location and a fault, it can be set to trip only for faults within its *reach* on the line in question and to *block* tripping for faults outside of its zone.

- **Backup protection.** Like other real-world devices, relays (whether solid-state or mechanical) and their associated circuits can fail, and thus critical components are often protected with at least another level of protection which is deployed only if the primary protection fails. Backup protection may be as complicated as involving a completely separate set of relays or may be as simple as a larger upstream fuse.

IEEE Standard Device Numbers

Relay devices or their equivalent are often designated by standardized device numbers and acronyms. In fact, many "devices and functions used in electrical substations and generating plants and in installations of power utilization and conversion apparatus" are given standard designations in *IEEE PC37.2/D3.1 Draft Standard for Electrical Power System Device Function Numbers, Acronyms and Contact Designations*. Such designations have been used for many years and are presently being updated by IEEE.

In particular, system-protection engineers have used many of these standard designations for years, a few of which are as follows:

- 50—instantaneous overcurrent relay
- 32—directional power relay
- 27—undervoltage relay
- 21—distance relay
- 36—polarity or polarizing voltage device
- 37—undercurrent or underpower relay
- 51—ac inverse time overcurrent relay
- 52—ac circuit breaker

It is possible for questions about such devices to appear on the exam. Hopefully the problem will define the device verbally so that one can understand what is meant, because the list of such device identifiers is large.

RECOMMENDED REFERENCES

Glover, Sarma, and Overbye. *Power Systems Analysis and Design*. 4th ed. Thomson, 2008.

Horowitz and Phadke. *Power Systems Relaying*. Research Studies Press Ltd. and John Wiley & Sons Inc., 1992.

CHAPTER 14

Power Quality and Power Electronics

OUTLINE

ELECTRIC POWER QUALITY 190
Grounding ■ Some Terminology ■ Brief History of Power-Quality Issues ■ Harmonics ■ Total Harmonic Distortion (THD) ■ Harmonics and Phase Sequence

POWER ELECTRONICS 197
Simple Half-Wave Diode Rectifier ■ Full Wave Rectifier ■ Single-Phase Full Wave Rectifier with Capacitive Filter ■ Three-Phase Rectifiers ■ Three-Phase Rectifier with Capacitive Filter

RECOMMENDED REFERENCES 204

It is useful to review power-quality and power-electronics issues in the same chapter, because a prime motivator for power-quality concern is the preponderance of electronic equipment in the modern power system. Conversely, many electronic devices are far more *sensitive* to network disturbances than historical equipment, such as motors and incandescent lighting.

Not all power-quality problems arise from electronic equipment, although many do. Weather-related factors (particularly lightning) and abnormal occurrences, such as flashovers, can cause problems that we now classify under the broad category of power quality. Arcing devices, such as arc furnaces, as well as arc lamps and welding equipment, can cause waveform distortions. However, *most steady-state waveform distortion in low-voltage systems results from electronic switching devices*. Historically, nonlinear ferromagnetic devices, such as transformers, caused most of the observable periodic waveform distortion in the past, but this has been overshadowed in recent decades by the effects caused by the prevalence of electronic devices.

A brief review of the power-quality area, including steady-state waveform distortion, may help you answer a question or two on the exam and will enable a more meaningful discussion of power-electronic device fundamentals, which follows in this chapter.

Refer also to Chapter 15, which presents some information on harmonics, and on non-sinusoidal waves and their rms and average values.

Because the power-quality area is broad, an exam problem in the power-quality area might be anticipated to relate to a definition or concept. These types of problems are sometimes very simple and easy to answer if you know the concept. For example, you will want to know what is meant by a *harmonic*, and how to find the frequency of a particular harmonic, such as the fifth. It might also be useful to know which harmonic components can appear in the neutral of a three-phase, four-wire system (multiples of three in balanced systems; any harmonics in an unbalanced system). Perhaps a question may ask whether a particular harmonic relates to the positive-, negative-, or zero sequence (see Table 14.1). A typical numerical problem might involve the calculation of total harmonic distortion (see Example 14.1). Note that, in reality, any system that contains harmonic components above the fundamental is unbalanced, although the concepts of unbalance and harmonics are often treated separately by many authors, and the severity of the effect is dependent upon the degree of unbalance or the level of harmonic presence.

As far as electronic devices are concerned, just knowing what kind of distortion each produces may be sufficient to answer a question. For example, in three-phase rectifier circuits, six-pulse bridges produce strong 5th and 7th, 11th and 13th, 17th, 19th, etc. harmonics, while twelve-pulse bridges produce 11th and 13th, 23rd and 25th, etc. harmonics. This is easily remembered by noting that the harmonics produced by a 6-pulse rectifier lie one number above and below multiples of 6, while the harmonics produced by a 12-pulse rectifier lie one number above and below multiples of 12. It is also worth recalling that the amplitudes of the harmonics typically decrease as the order of the harmonic ascends.

Recognizing the type of waveform generated by a three-phase rectifier may be the essence of an exam problem. This chapter discusses the nature of the waveforms of several commonly used, single- and three-phase rectifier circuits.

ELECTRIC POWER QUALITY

The term **power quality** refers to how closely a given power-system quantity (usually the voltage available at the load point) meets a set of objectives. Because most power-quality discussions center on the voltage, which is of course the common feature of parallel-connected loads, a better term might be *voltage quality*, but we use the term here that is validated by conventional use.

In defining power-quality metrics, we usually assume a set of ideal objectives for the voltage at the receiving point in our system, which might include the following:

- Sinusoidal voltage at constant desired nominal rms value
- Zero percent voltage regulation
- No steady-state distortion (e.g., no harmonics above the fundamental)
- No spikes or other high-frequency distortions
- No sags, swells or other low-frequency distortions
- No outages

Of course, some of these ideal objectives overlap, and none can be guaranteed. How closely we approach these objectives is embodied in the term *power quality*.

Power-quality problems usually manifest themselves (or at least are usually observed) in low-voltage systems, and usually near loads. In these types of contexts, and particularly in the case of steady-state harmonic distortion, the problems are usually caused by loads and not by the upstream system. Weather-related and switching-related problems, in contrast, may have their roots in the transmission system, as well as in the distribution system.

Grounding

Power quality is often associated with grounding. In fact, improper or inadequate connection to ground will often result in problems in the electric power system, as well as in communication, instrumentation, and data-transmission systems. For this reason, organizations concerned with systems as varied as power, computers, and telephone have developed procedures and standards for grounding, bonding, and shielding.

Although a detailed description of grounding is outside the focus of this book, several important items should be mentioned. Codes, such as the National Electrical Code and the National Electrical Safety Code (see Chapter 16) prescribe grounding in careful terms. From the viewpoint of these codes, this is a safety issue; proper grounding is also a requirement of the successful operation of an electric power system.

Instructors in grounding seminars often make a particular point of noting that *one of the most important procedures to follow in grounding is to achieve a low-impedance, single-point ground*. In fact, this is emphasized in the codes. Electrical systems are typically grounded at the point of common connection with the utility, and at **derived sources**, such as the low-voltage secondary of a transformer. AC electrical systems at the 120/240 V user level are grounded by means of a **neutral conductor** (white wire) which is often called the **grounded** conductor. Metallic exposed parts are grounded through a safety ground (green wire), often called the **grounding** conductor. Each of these ground connections follows a different path back to the common-ground or derived-source point, where they are tied together at this point alone. This is because the two ground conductors in these types of systems have different functions: one is designed to carry neutral current, and the other is designed to carry no current under normal circumstances, but to reduce the voltages on exposed conducting surfaces to zero. Of course, these functions are often compromised during a fault, or in the case of incorrect wiring and installation. We might note that there are two types of **grounding** conductors. Grounding electrode conductors connect to ground rods, ground pipes, building steel, cold metal water lines, or other conductors in contact with earth, or employ concrete-encased grounds. Their function is to establish system ground. On the other hand, equipment grounding conductors run with the phase and neutral conductors to provide safety grounding in the event of a fault condition, as mentioned above.

Some Terminology

Probably in no other area of electric power work does common use of the terminology involved become as confusing as in the power-quality area. Such layman's terms as *glitch*, while often used by trained electric power and communication personnel, as well as less-trained individuals, have very little technical meaning, although they are usually understood to mean some kind of disturbance or deviation from what is normal and desired. Here are some generally-accepted technical definitions, which have more definite meaning:

- **Flicker** (or **voltage flicker**) is a cyclic or random variation in the voltage between the limits of 90 percent and 110 percent of nominal. The variation is slow enough to be seen (in lighting systems) or at least perceived.

- A **sag** occurs when the rms voltage is 10 to 90 percent below nominal for 0.5 cycle to 1 minute.

- A **swell** occurs when the rms voltage is 10 to 80 percent above nominal for 0.5 cycle to 1 minute.

- Abrupt, intra-cycle changes in voltage are called **surges**, **spikes**, or **impulses**.

- **Undervoltage** happens when the nominal voltage drops below 90 percent for more than one minute; **overvoltage** occurs when the nominal voltage rises above 110 percent for more than one minute.

Each of these typically has a different cause. Some of these are discussed in the next section.

Brief History of Power-Quality Issues

Probably the first related effect to receive detailed analytical and remedial attention was what became known as voltage flicker. This problem, which we might now call a voltage sag or dip, was often observed in manufacturing plants early in the 20th century as motors were brought on line without the benefit of reduced-voltage starting. It manifested itself as a change in brightness of incandescent lamps, hence the term. Office staff and others who worked under such lighting registered complaints, which were traceable to this phenomenon. Ultimately, using statistical population samples of office workers, curves were developed for perceptible and annoying thresholds in terms of percent voltage dip versus frequency of occurrence. These types of problems were ultimately solved by means such as power-factor correction, isolating lighting loads, system stiffening, and use of reduced-voltage motor starting methods.

The "second wave" of observed problems began to manifest itself around the 1950s as commercial and residential Freon cooling became widespread. The resulting preponderance of compressor motors led to voltage-drop and loss problems in electrical systems due to the lower power-factor of these loads compared with the predominate lighting load of earlier decades, coupled with the higher currents that resulted from increasing load. It was mostly solved using capacitor installations for power-factor correction. There was still relatively little waveform distortion, and the problems could be solved analytically using steady-state, fundamental-frequency techniques.

A "third wave" of problems began to surface several decades later with the advent of electronic switching devices, such as switched-mode power supplies (in computers, televisions, printers, and the like) and solid-state motor controllers. Here, a more complicated dynamic emerged. Problems involving grounding, harmonics, and resulting conductor heating in transformers and motors necessitated consideration of waveform distortion, in turn requiring new approaches. Some of these issues, which are mostly traceable to solid-state electronic devices, will be addressed in the following sections.

Harmonics

Some of the deviations from sinusoidality mentioned above, such as occasioned by switching surges and lightning strikes, are transient effects. They often appear as random disturbances, but can usually be anticipated to an extent and addressed; for example, adequate surge protection can be used to isolate sensitive equipment from lightning-induced surges. More frequently, we are concerned with steady-state distortion, in which the distortion of the wave is nearly the same from one cycle to the next. Steady-state distortion is usually quantified in terms of the concept of harmonics. This kind of distortion is often attributable to power-electronic devices, as discussed in more detail in the section "Power Electronics" later in this chapter. Chapter 15 discusses other nonlinearities that can cause steady-state distortion.

A **harmonic** is simply a sinusoidal signal (or component of a signal) whose frequency is an integral multiple of the fundamental frequency. The fundamental frequency is also known as the **first harmonic**. The **second harmonic** is then of twice the frequency as the fundamental (e.g., 120 Hz in a 60 Hz system); the **third harmonic** is of three times the frequency of the fundamental (180 Hz in a 60 Hz system), and so on. In most distorted waves, the lower-order harmonics contain the most energy, and we are rarely concerned with harmonics much above the 25th; however, instruments that analyze power-system harmonics sometimes address harmonics up to the 50th or so (the 50th harmonic of a 60 Hz signal is 3kHz, and is thus near the middle of the audio range). See Chapter 15 for some discussion and an example of the use of harmonics.

Harmonics in a power system are primarily produced by nonlinear devices, most of which are either electronic loads or (historically) devices, such as motors or transformers having ferromagnetic cores. Ferromagnetic materials have nonlinear B-H curves and thus cause either the current or voltage (or both) to be nonlinear.

The current in the primary of a transformer with the secondary circuit open (which we call **exciting current**) is typically highly nonlinear, with a prominent third harmonic (see, for example, Figure 14.1). Such distortion has been noted for well over a century; it was shown theoretically to be present even before oscilloscopes and other waveform-observing devices were developed, and could often be detected audibly as well.

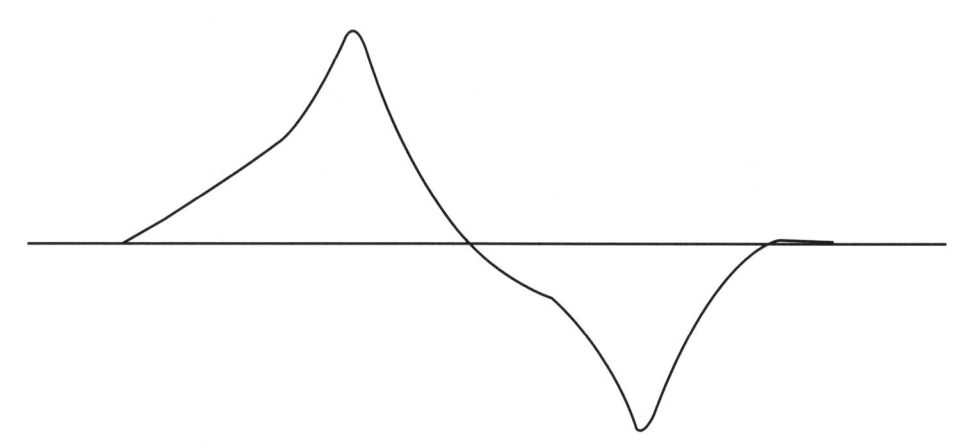

Figure 14.1 Typical transformer exciting current (containing strong fundamental and third harmonic)

However, transformer exciting current is usually a small fraction (perhaps 1 or 2 percent) of full-load current, so this effect is often masked in loaded systems. Far more serious sources of waveform distortion in today's world include electronic motor drives and switched power supplies, such as found in computers, monitors, televisions, etc. One reason that these devices distort the load is that they usually "grab" energy near the peak of the voltage wave, thus effectively flattening the wave (see Figure 14.2), where the flattened wave is compared with a sinusoid. Note that for each half cycle the absolute value of the wave increases slightly into the peak area and is thus the voltage wave is not truly "flat"; this is characteristic of such distortion. Because of their sheer numbers, office devices, such as computers, monitors, laser printers, and copiers can collectively cause considerable voltage distortion in a large office complex. This kind of distortion is somewhat self-perpetuating in the sense that because flat-topping reduces the peak energy available, electronic devices can get less energy from the flat-topped wave than from a sinusoid, and thus they reduce the wave peak even further. However, this kind of distortion is widespread and is probably not as pernicious as, for instance, spikes on the line or other high-frequency effects. In the larger power-systems context, motor drives are the prevailing culprit. Rectifiers and inverters (which are found in motor drives) are responsible for the waveform distortion in these devices. Because these are electronic devices, they can be meaningfully discussed in this chapter.

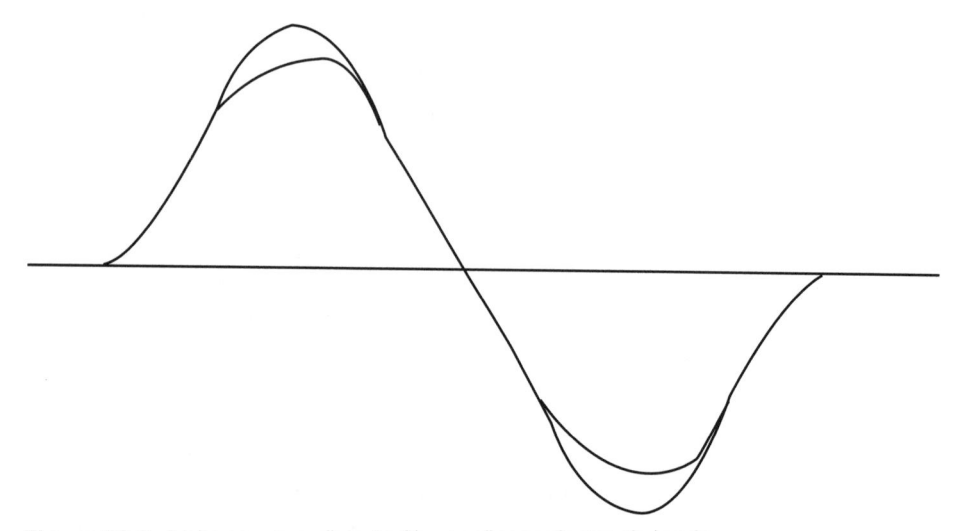

Figure 14.2 Voltage wave distorted by nonlinear electronic loads

A fact worth noting is that if a waveform is **symmetric** about its horizontal (time) axis (that is, if the negative half looks just like the positive half folded over the horizontal axis), only odd harmonics are present. This is typical of the waveforms produced by ferromagnetic devices and by most electronic devices operating normally, and can be observed in Figures 14.1 and 14.2. In practice, therefore, even harmonics are usually very small in typical steady-state power applications; their presence is often symptomatic of a failed solid-state device. Furthermore, as mentioned earlier, the lower-order harmonics predominate, so there is often a relatively high 3rd harmonic content, and possibly noticeable 5th and 7th (see below).

In a three-phase system, the third harmonic is often associated with neutral current and thus with the zero sequence (see Chapter 12 covering symmetrical components). This is because **triplen** (multiples of 3) harmonics, including the third, which are present in the phase conductors, superpose in the neutral. This has been found to be a serious problem in older, large office buildings whose neutrals were undersized in anticipation of balanced, fundamental-frequency loads. Modern single-phase nonlinear loads can result in high neutral currents in a three-phase system (typical of large building systems), which can lead to safety issues.

Harmonic pairs centering around multiples of 3, such as the (5th, 7th) and (11th, 13th) are associated prominently with three-phase rectifiers or motor drives.

Total Harmonic Distortion (THD)

As engineers, we prefer to define numerical metrics to express any deviation from an ideal (efficiency, voltage regulation, and voltage ratio are three such important metrics). One popular measure of how well a steady-state waveform approaches sinusoidality is **total harmonic distortion** (THD), defined here for voltage as

$$THD = \left(\sqrt{\sum_{i=2}^{N} V_i^2}\right) / V_1 \quad (14.1)$$

where

V_1 is the rms value of the fundamental of the wave
V_i is the rms value of harmonic i of the wave
N is the highest-order harmonic of concern (usually 50 or, often, much less).

Because harmonics tend to taper off in amplitude as the order increases, Equation 14.1 can be bounded at some convenient value of N for any given problem. In some cases, which are typically evidenced by a rather smooth wave without tight curvatures, harmonics above the fifth or so do not contribute meaningfully to THD. This becomes evident when we normalize all harmonics on a fundamental of unity amplitude, especially because the harmonic amplitudes are further reduced through the squaring process.

Note that THD is a smoothed measure and tells us nothing about the amplitudes or phases of individual harmonics. Thus, the concept of THD is akin to an averaging procedure and cannot be used to obtain a picture of the waveform. In this regard it is somewhat analogous to the concept of rms. Nonetheless, THD as a steady-state distortion measure is widely used, especially in specifying distortion limits for components or systems. Furthermore, it is obvious that, if a low THD is measured, we can be assured that this part of the system is relatively free of harmonics.

Concepts of harmonics and THD are particularly useful in considering specific power-electronic applications (as discussed in the remainder of the chapter) for which the circuits invariably produce distorted waves. Even with fairly sinusoidal voltage applied to one of those devices, the current will be distorted. If numerical values are given, it is usually simple to find THD. Example 14.1 typifies this relatively straightforward calculation.

Example 14.1

A certain electronic load supplied with a sinusoidal voltage has a somewhat distorted current wave, whose harmonic components are displayed on a harmonic analyzer as follows:

Harmonic	Amplitude (A)	Phase (degrees)
1	32.0	0
3	2.6	37
5	1.1	28

All other harmonic components are insignificant. Find the THD of this current.

Solution

THD is not a function of the phase angles of the harmonics, only the magnitudes. Using Equation 14.1,

$$THD = \left(\sqrt{(2.6^2 + 1.1^2)}\right) / 32.0 = 0.0882 = 8.82\%.$$

Harmonics and Phase Sequence

There is an interesting and useful correlation between harmonic order and phase sequence (usually expressed in terms of symmetrical components as defined in Chapter 12) for a three-phase system. Consider the following repeating table, in which + represents the positive sequence, – represents the negative sequence, and 0 represents the zero sequence.

Table 14.1 Harmonics in terms of sequence

Harmonic number	1	2	3
Sequence	+	–	0
Harmonic number	4	5	6
Sequence	+	–	0
Harmonic number	7	8	9
Sequence	+	–	0
Harmonic number	10	11	12
Sequence	+	–	0
etc.			

We can make the following observations from this table:

- The fundamental (first harmonic) is of positive sequence. This corresponds to our convention that the positive (abc) sequence is the normal directional sequence. Three-phase generators are driven in a direction to produce positive-sequence voltages by definition. For a three-phase motor, it implies that the fundamental-frequency component tends to move the shaft in the direction of the positive sequence.

- Some harmonics tend to aid this motion in a motor; for example, the 4th, 7th, and 10th.

- Harmonics such as the 2nd, 5th, 8th, and 11th contribute to an opposite rotational direction, and thus, would tend to retard the direction of a three-phase motor.

- Harmonics that are multiples of three (such as 3rd, 6th, 9th, and 12th) contribute to the zero sequence, and thus, are not identified with a rotational direction in a three-phase machine. However, we could show that currents of these sequences are additive in the neutral. These harmonics that are multiples of three are known as triplen harmonics, particularly if they are odd (recall that symmetric waveforms will not contain even harmonics, so most of the prominent harmonic components that we measure will be odd).

- Some positive-sequence harmonics are even and some are odd. This is true for the negative and zero sequences, as well.

We should emphasize that a negative-sequence signal in a three-phase machine does not necessarily imply the presence of harmonics; a fundamental-frequency, negative-sequence signal can be produced by unbalances, such as single-phase loading. Because the revolving field produced by the fundamental-frequency negative sequence revolves at the same speed as the positive-sequence field but in the opposite direction, the net effect of a fundamental-frequency negative-sequence in a synchronous machine is a field that sweeps by the rotor poles at double frequency. In practice, they may cause serious heating, which is sometimes known as *negative-sequence heating*. Harmonic components can contribute to this heating.

POWER ELECTRONICS

Power electronics refers to the use of semiconductor devices in the switch mode to process large amounts of power while converting voltages and currents to desired forms, such as dc or ac at a desired frequency. The most common applications include the following:

- **Rectifiers**: These circuits convert ac to dc. They form the front end (source side) for most power electronic applications.

- **dc-dc converters**: Extensively used in regulated power supplies, these converters change a dc voltage or current to a variable dc voltage or current.

- **Controlled Rectifiers**: These circuits also convert ac to dc, but the voltage, current and power delivered can be controlled by an electronic signal.

- **Inverters**: These convert dc to ac. In many applications, the output voltage and current are variable frequency sinusoids; however, in principle, the output waveforms can be arbitrary; for example, in power audio amplifiers.

- **Cycloconverters**: These convert an ac voltage at a given frequency to an ac voltage at a different, possibly variable, frequency.

Other desired functions can be obtained by combining these basic functions. For example, an alternative way of converting an ac voltage to another ac voltage at a different frequency is to rectify the incoming ac source and then use an inverter to obtain the desired ac output. Many of the circuits can be operated in a bidirectional way in terms of power transfer. The term **converter** is often used to describe such circuits.

Older methods of power conversion involved rotating electric machines. The advantages of power electronics include higher efficiency, smaller footprint, simpler and faster feedback control, no moving parts, and reduced maintenance.

Devices such as triacs, thyristors, and silicon controlled rectifiers (SCRs) can be driven by timed signals to switch or "turn on" at precise times. For purposes of exam review, we focus on nonswitched rectifier devices in the following discussions. It is often easy to make the step from rectifiers to switched devices by noting that a switched device (even if forward-biased) does not conduct until switched by application of an external signal, and then typically stays in the conducting mode until a voltage polarity reversal turns it off.

Simple Half-Wave Diode Rectifier

A half-wave diode rectifier with a resistive load is shown in Figure 14.3. The sinusoidal ac voltage source is connected to the resistive load through a diode.

An ideal diode has the property that it conducts in the forward direction only if the voltage at the anode (terminal A) is more positive than that at the cathode (Terminal K). In the half-wave rectifier shown, this happens during the positive half cycle of the ac voltage. In the positive half of the ac cycle, the ideal diode appears as a short and the entire source voltage appears across the load. The resulting current is proportional to the voltage:

$$i = v_{ac} / R, \quad v_{ac} > 0. \tag{14.2}$$

The resulting waveforms of the ac applied voltage, dc side (or load) voltage, and current are as shown in Figure 14.4.

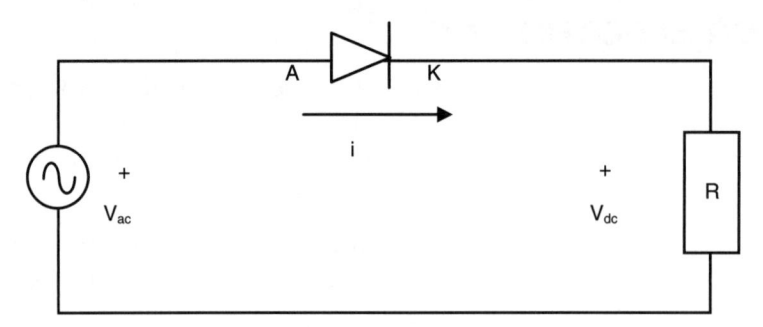

Figure 14.3 A single-phase, half-wave rectifier

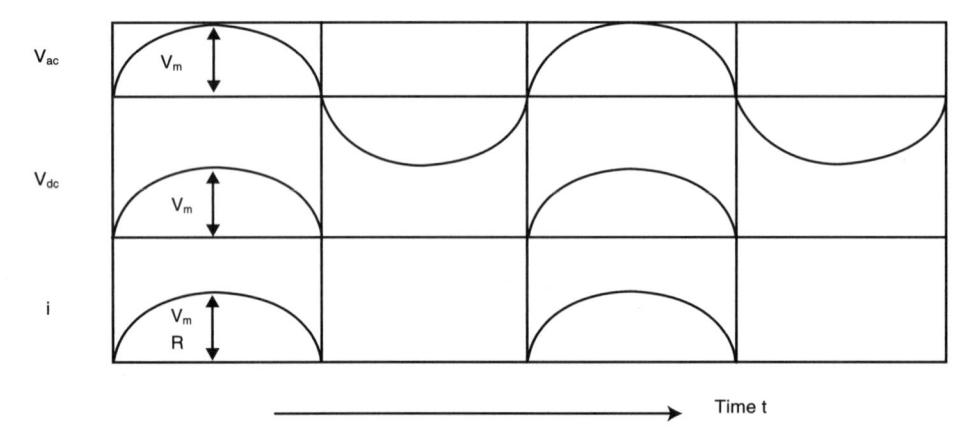

Figure 14.4 Voltage and current waveforms for a single-phase, half-wave rectifier

The voltage across the resistor is 'dc' in the sense that its time-average value is greater than zero. We say that the output voltage is **dc with ripple**.

The average or dc value of the output voltage is

$$V_{dc} = V_m/\pi \text{ volts} \tag{14.3}$$

and the power deliver to the resistor is

$$P = V_{dc}^2/R \text{ watts.} \tag{14.4}$$

Full Wave Rectifier

Figures 14.5 and 14.6 show two full-wave rectifier configurations. The circuit in Figure 14.5 requires a center-tapped transformer but only two diodes. The bridge in Figure 14.6 requires four diodes, but can be connected directly to the ac source, or to the output of a simple two-winding transformer.

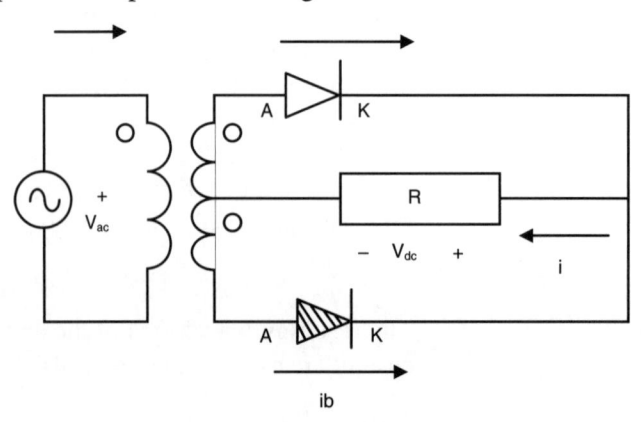

Figure 14.5 Center-tapped, single-phase, full-wave rectifier

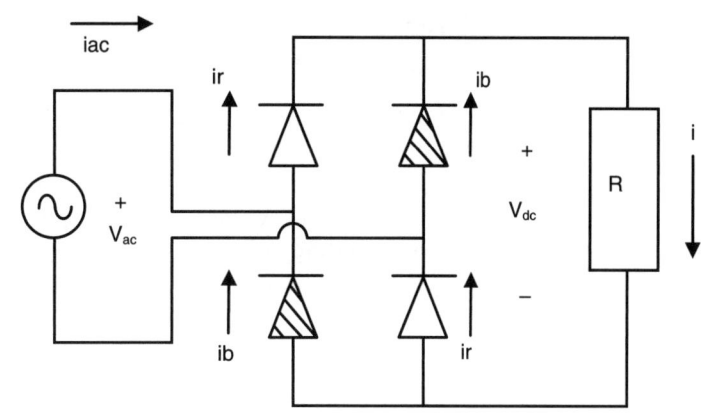

Figure 14.6 Single-phase, full-wave bridge rectifier

In both circuits, the diodes shown without cross-hatching conduct when the incoming voltage is positive, and the cross-hatched diodes conduct when the incoming voltage is negative. As shown in Figure 14.7, the output voltage is always positive, that is, it is a **rectified** version of the incoming sinusoidal voltage. Because the load is entirely resistive, the current waveform is the same as that of the voltage and, is therefore, not shown. For the full-wave rectifier, the average or dc value of the output voltage is

$$V_{dc} = 2V_m / \pi \text{ V} \tag{14.5}$$

and the power delivered to the resistor is

$$P = V_{dc}^2 / R \text{ watt.} \tag{14.6}$$

(See Figure 15.3 in the next chapter for the average and rms values of several frequently discussed waveforms, including those shown here).

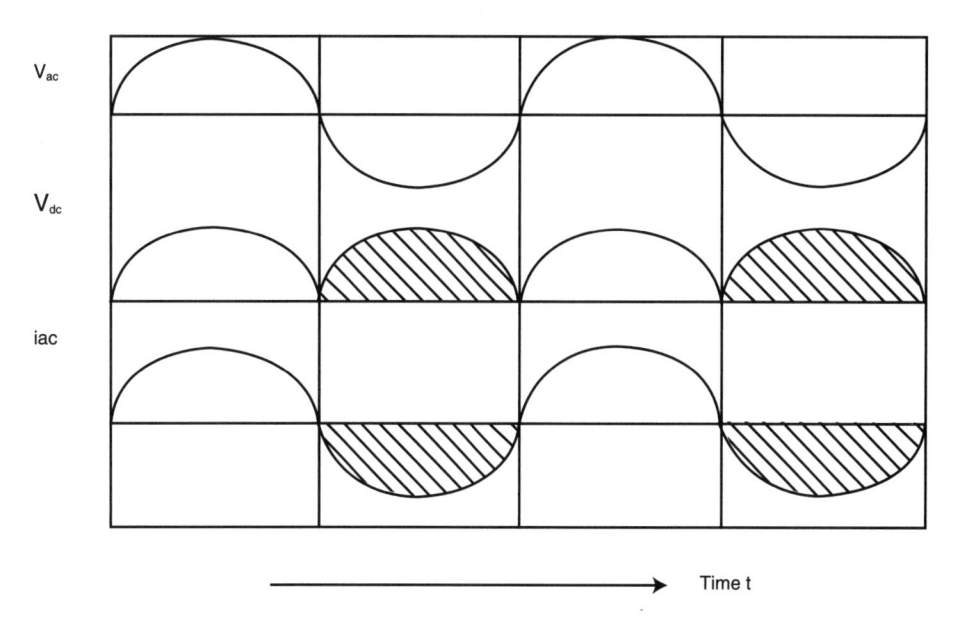

Figure 14.7 Voltage and current waveforms for single-phase, full-wave rectifiers

Single-Phase Full Wave Rectifier with Capacitive Filter

In most applications, rectifiers serve as a front end to other power processing circuits; for example, for voltage regulators in electronic equipment and inverters in variable-speed motor drives. For this reason, the **ripple** in the output voltage needs to be minimized. A series inductor can be used to reduce this ripple in the output current and thus, the voltage ripple in the resistor; a shunt (parallel) capacitor can be included to reduce voltage ripple and thus, current ripple in the resistor. The capacitor filter is the most common, and is shown in Figure 14.8.

The capacitor modifies circuit behavior. The diodes no longer conduct for the entire half cycle. Rather, each pair of diodes conducts only when the incoming voltage is larger than the capacitor voltage; this pattern is illustrated in Figure 14.9. The crosshatched diodes begin conducting at point A when the incoming voltage exceeds the output voltage. In this region, the capacitor charges from the ac source, and its voltage follows the source voltage. When the incoming voltage falls below the capacitor voltage, the diodes cease conduction. The source is effectively disconnected, and the capacitor slowly discharges into the resistor. As a consequence,

- the voltage at the output has a smaller ripple; it looks more like a dc voltage; and

- the current drawn from the supply is no longer sinusoidal. It is a sequence of narrow pulses as shown in Figure 14.9.

A larger capacitor reduces voltage ripple further, but makes the input ac current pulses narrower and of larger magnitude. *Recognizing the difference in the waveforms between these two single-phase cases (unfiltered and filtered), and similarly for the three-phase cases discussed below, might provide critical information if an exam problem provides waveforms and asks you to tell the kind of circuit that produces them.*

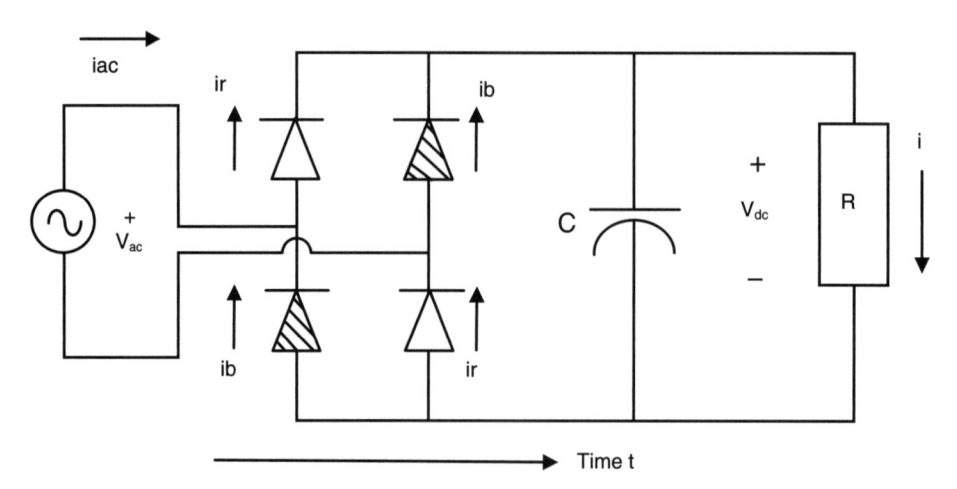

Figure 14.8 Single-phase, full-wave bridge rectifier with capacitive filter

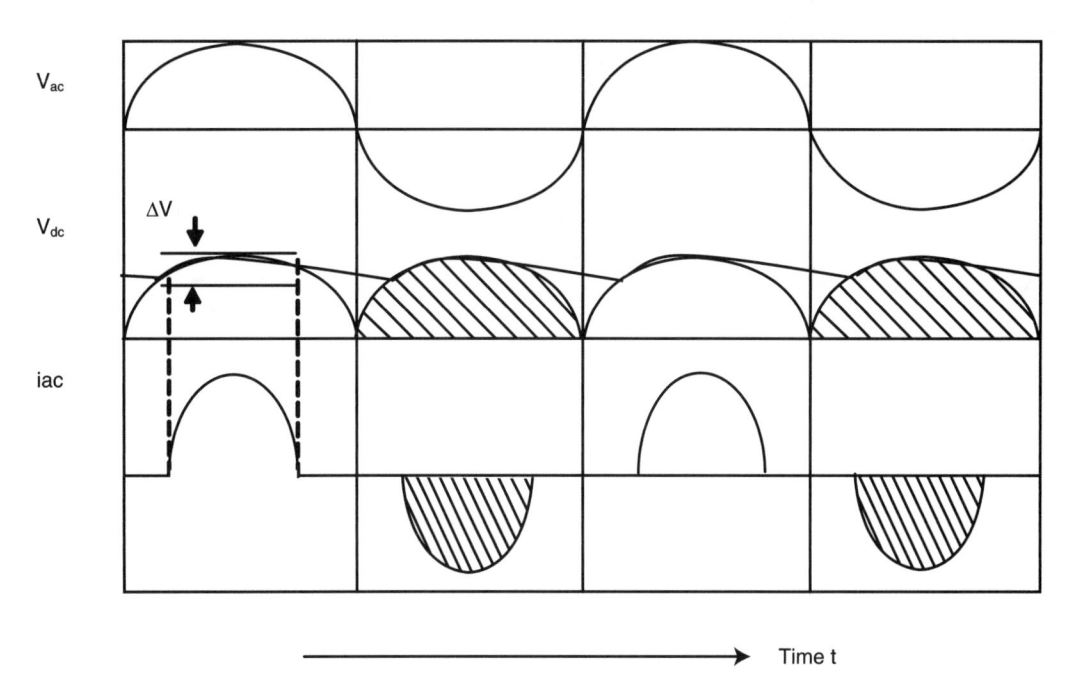

Figure 14.9 Voltage and current waveforms for single-phase, full-wave rectifiers with capacitive filter

With a large capacitor, the ripple is relatively small, and for all practical purposes, the dc output voltage is

$$V_{dc} \approx V_m \text{ V.}$$

Of additional interest in this circuit is the **peak-to-peak voltage ripple**, labeled ΔV in Figure 14.9. We can express its ratio to V as

$$\Delta V / V = 1 / (2f\,RC) \tag{14.7}$$

where R is the resistance in ohms, C is the capacitance in farads, and f is the source frequency in Hz.

Three-Phase Rectifiers

Multi-phase rectifiers are used in higher-power circuits, usually 10 kW and above. A **three phase bridge** is shown in Figure 14.10. The labels a, b, and c denote the three phases of the incoming balanced, three-phase voltage source. The six diodes are identified by capital letters, as shown. Note the ungrounded-wye connection, which is usual for such devices (and incidentally for induction motors, but not for three-phase transformers, generators, or derived sources).

As in the single-phase case, the ideal diode conducts when it is forward-biased; that is, when its anode voltage is higher than its cathode voltage. Thus, diode A, the top diode connected to phase a, will conduct when the line-to-neutral voltage V_{an} is greater than each of the line-to-neutral voltages V_{bn} and V_{cn}.

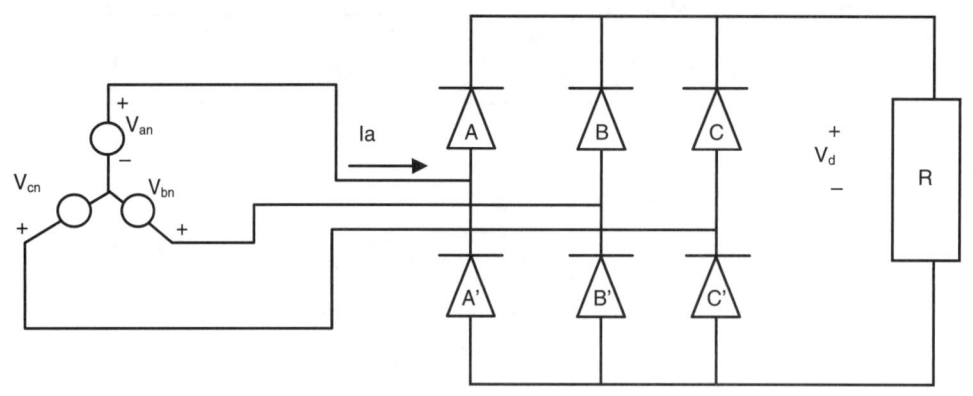

Figure 14.10 Three-phase bridge rectifier

Figure 14.11 shows the waveforms for line-to-line voltages, line-to-neutral voltages, output dc voltage V_d, and the input ac current I_a on phase a. The boxed region encloses the area where V_{an} is greater than each of the line-to-neutral voltages V_{bn} and V_{cn} when diode A conducts. To aid in understanding the operation of the circuit, the lower part of the figure shows the current paths in the circuit for each half of this part of the cycle, with non-conducting diodes omitted.

In the first half of this region, we see that V_{bn} is smaller than V_{an} and V_{cn}. Thus, diode B' conducts. Diodes A and B' provide the path for the source to supply the load during this time, and the voltage V_d across the load equals the source voltage V_{ab}. This voltage is shown in Figure 14.11 as the bold line.

In the second half of the boxed time-region, we see that V_{an} is still greater than V_{bn} and V_{cn}, but now V_{cn} is smaller than V_{an} and V_{bn}. The conduction path now is through diodes A and C', resulting in $V_d = -V_{ca}$. This yields the second 'scallop' or 'pulse' in the output voltage. There are six such pulses in V_d for each cycle of ac voltage. For this reason, the three-phase bridge is also called a **six-pulse rectifier**.

It is seen that the three-phase rectifier has a smaller magnitude of ripple than the single-phase rectifier. Further, the ripple frequency is six times the source frequency. This means that a much smaller capacitor is needed to filter the output voltage.

The dc or average output voltage is given by

$$V_d = 3\,V_m/\pi \ \text{V} \tag{14.8}$$

where V_m is the peak value of line-line voltage.

Three-Phase Rectifier with Capacitive Filter

A three-phase bridge with a capacitive filter is shown in Figure 14.12, and the corresponding waveforms are given in Figure 14.13. This circuit is very extensively used in variable speed motor drives.

As in the case of the single-phase rectifier with capacitive filter, the diodes now conduct only when the bridge output voltage exceeds the capacitor voltage. When the bridge output voltage falls below the capacitor voltage, conduction ceases. Thus, the conduction pattern (sequence) does not change, but the diodes conduct for shorter periods, with higher magnitudes of current. The output voltage approaches a constant dc voltage with a very small ripple. The ac currents on the source side are characterized by the double peaks shown. Also note the characteristic sawtooth appearance of the ripple voltage.

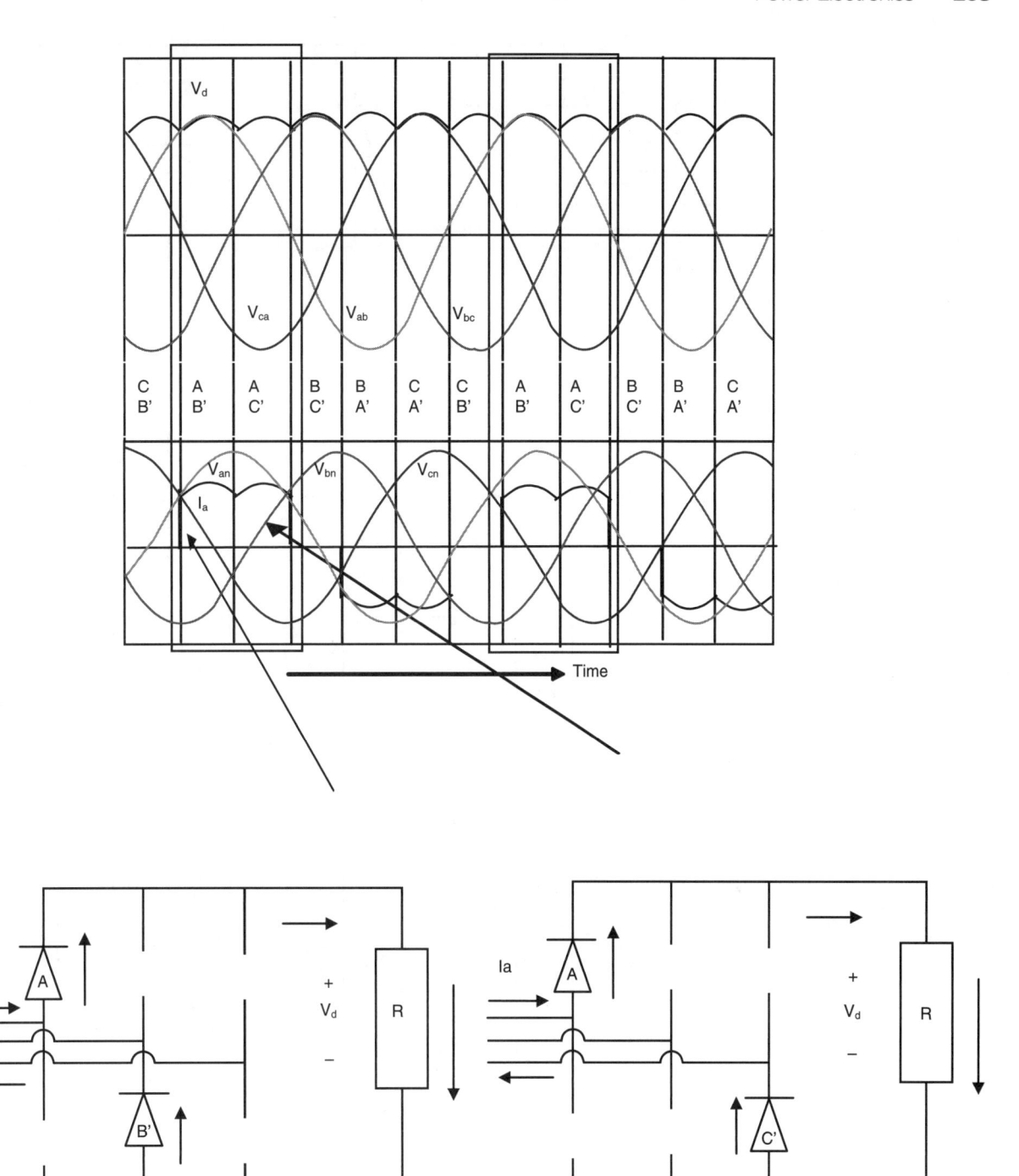

Figure 14.11 Voltage and current waveforms for a three-phase bridge rectifier (diodes not conducting in the discussed range are omitted for clarity)

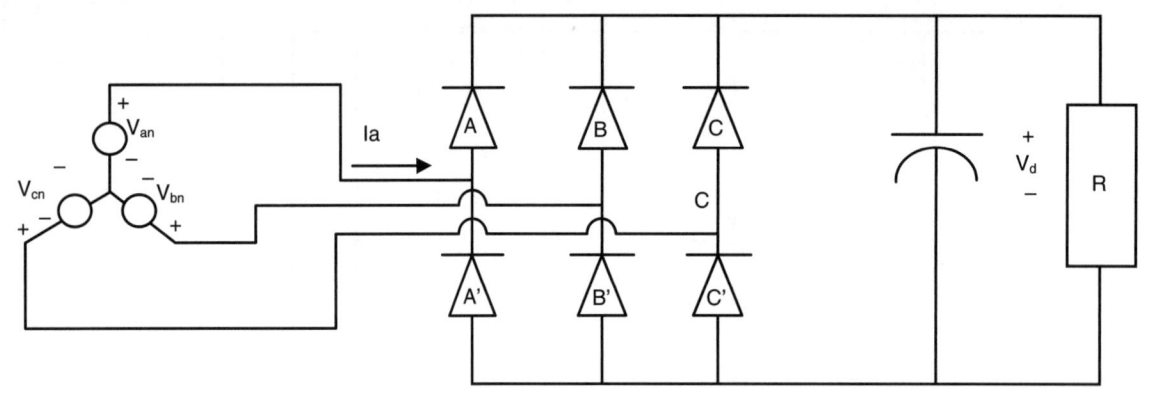

Figure 14.12 Three-phase bridge rectifier with capacitive filter

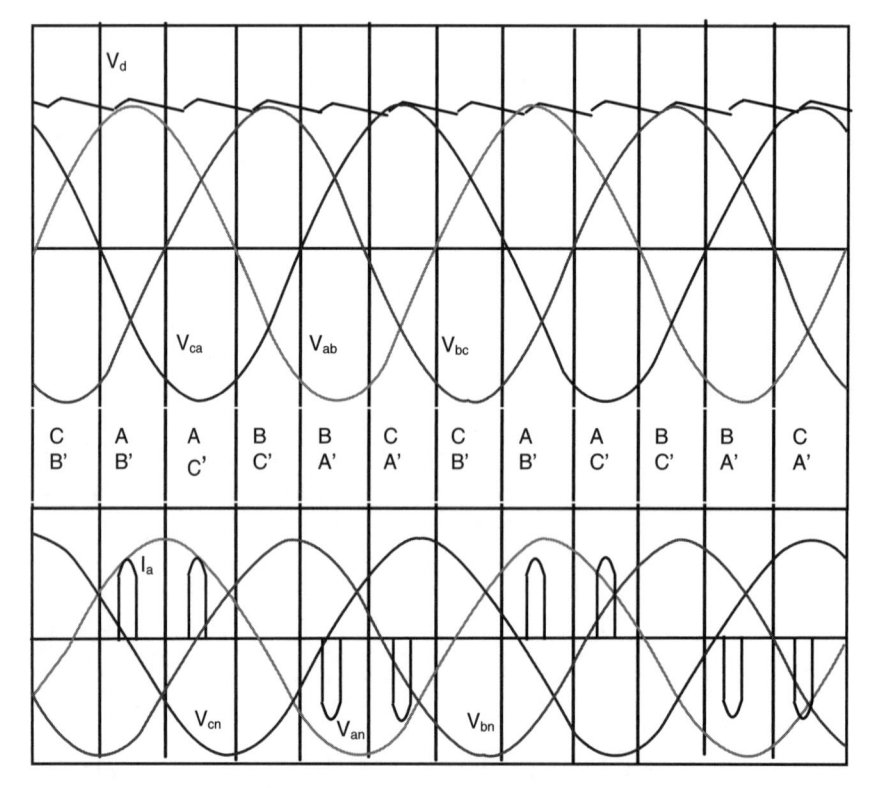

Figure 14.13 Voltage and current waveforms for a three-phase bridge rectifier with capacitive filter

RECOMMENDED REFERENCES

Dugan, Roger C., Mark F. McGranaghan,, Surya Santoso, and H. Wayne Beaty. *Electrical Power Systems Quality*. 2nd ed. McGraw-Hill, 2003.

Heydt, Gerald T. *Electric Power Quality*. Stars in a Circle Publications. 1991.

Rashid, Muhammad H. *Power Electronics Circuits, Devices, and Applications*. 2nd ed. Prentice Hall, 1988.

CHAPTER 15

Measurement, Reliability, and Lighting

OUTLINE

FUNDAMENTALS OF MEASUREMENT AND INSTRUMENTATION 206

METERS AND WAVEFORMS 206
The D'Arsonval Meter Movement ■ AC Measurements with the D'Arsonval Meter Movement ■ The Digital Multimeter (DMM) ■ Burden as Associated with an Ammeter ■ Other Waveform Considerations ■ Other AC Measuring Devices ■ Instrument Transformers ■ Watt and VAR Measurement

POWER SYSTEM RELIABILITY 217

FUNDAMENTALS OF LIGHTING (ILLUMINATION ENGINEERING) 219
Terminology of Lighting Units and Concepts ■ Guidelines for Lighting Particular Indoor Tasks

RECOMMENDED REFERENCES 222

This chapter provides some information in several "special applications" areas that the **National Council of Examiners for Engineering and Surveying** (NCEES) lists as possible areas of questioning on the PE Power exam. Including engineering economics, these topics are listed as a possible 10 percent of the exam content. Thus, it is likely that you will not see more than one or two problems in each of these areas.

Of these topic areas, measurement and instrumentation is probably the one in which the typical engineering graduate has had the most experience, because many engineering courses include laboratory work. Reliability issues are sometimes introduced implicitly in the undergraduate curriculum, but more often are found in upper-level or graduate courses. Not all students graduating in the electric power option would have covered these two topics at more than a basic level. Lighting (illumination) is not usually introduced in college courses in power systems. If you have had limited academic or practical exposure to these topics, this chapter provides some basic overview that should be helpful in the exam, including some representative practical problems.

Engineering economics, which is listed with the special application topics in the NCEES specifications, is a wide field of its own embodying numerous ideas. Chapter 17 provides detailed coverage of engineering economics.

FUNDAMENTALS OF MEASUREMENT AND INSTRUMENTATION

Instrumentation and measurement are at the heart of most laboratory work in electrical engineering. Furthermore, some basic measurements (e.g., voltage and current) are made frequently in the field. Accordingly, most electrical engineers will be familiar with these, as well as measurement of resistance and continuity testing.

Both **analog** and **digital** instruments are widely available today. The difference between them might be compared with the difference between the slide rule (or nomograph) and the digital calculator. An analog meter, regardless of the type of meter movement present, typically has a pointer (or the equivalent) that traverses a scale. To read it, you need to **interpolate** between numbered or marked values. A digital instrument, on the other hand, provides a digital readout, with the number of displayed digits usually decided by the manufacturer.

Accuracy is an issue with both types of instruments. Accuracy can be defined as the relation between the observed value and the actual value of the quantity. In an analog instrument, accuracy is a function of both the internals of the meter (including multipliers and shunts, and the engraving of the scale) and the way the instrument is read.

In using digital meters, it is tempting to believe the reading to be correct to the number of decimal places displayed. However, one should consult the instrument's manual to check the accuracy of the device.

For measuring ac values, many types of analog instruments have **nonlinear** scales, with the scale values cramped at the low end and expanded at the high end. This reduces the measurement accuracy for low values of the measured quantity. A wise precaution is to choose the range, if possible, that is easiest to read for the expected value, being careful to stay away from the extremes of the scale.

Strictly speaking, the term **precision** in an instrumentation context has to do with *repeatability*: a high-precision meter will return substantially the same reading as the same value is sampled repeatedly. Sometimes, this term is mistaken for *accuracy*.

Parallax error has to do with the angular position of the viewer with respect to the needle and scale of an analog instrument; we might call it the "sundial effect." In quality analog instruments, parallax error is sometimes reduced by placing a small mirror behind the indicator; the reader then moves the eye until the indicator casts no visible shadow.

METERS AND WAVEFORMS

The **waveform** of the quantity to be sampled can be an important issue. Some aspects of this are discussed below. You might also want to refer to Chapter 14 for additional discussions of nonsinusoidal waveforms.

With newer types of meters, such as digital multimeters (DMM), even the simplest of measurements may yield an incorrect interpretation if the waveform is different from a sinusoid. Of course, some analog-type meters presented the same kinds of problems. To use an ac meter accurately, we need to know something about the waveform of the quantity to be measured.

This was not as much of a problem in the days prior to significant waveform distortion posed by modern electronic loads. Now, it is often important to specify (and pay more for) a **true-rms-reading** instrument if we really want to know the rms value of a distorted waveform.

The D'Arsonval Meter Movement

Before the advent of digital electronics, the usual tool for measuring dc currents was the **D'Arsonval meter movement**, which basically consists of a moving coil of wire (carrying a small current) balanced so as to rotate around an iron cylinder in the presence of a strong, stationary magnetic field (see Figure 15.1). The current-carrying coil is balanced by a pair of opposing coil springs, which also serve to carry the current to and from the coil. The coil mechanism is delicately balanced, often with needle bearings and jewels, and often the scale travel is limited to about 90 degrees. Sometimes, a very long pointer is provided, which traverses one or more expanded scales. In actuality, the meter movement is really a low-torque, precision rotary solenoid. Because this movement is still widely used, it is important to understand how it translates the measured quantity into a scale reading.

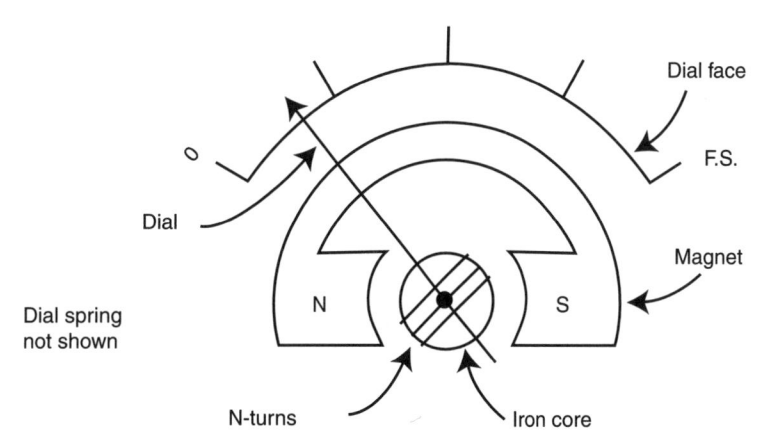

Figure 15.1. The D'Arsonval meter movement

Recall that current flow i in the presence of a magnetic field of magnetic flux density B produces a force that is proportional to conductor length l:

$$F = B\, i\, l. \tag{15.1}$$

(This simplified form of the equation actually applies to a linear, translational problem, but the concept is applicable here.)

As shown in Figure 15.1, the basic design of the meter entails arranging a permanent magnet such that an air gap is in a radial form, with wire formed in a rectangular coil and an indicating needle attached to it along with a restraining spiral spring.

The torque produced is proportional to current if all other parameters are fixed:

$$T = 2N\, B\, l\, r\, i \tag{15.2}$$
$$= K_1\, i$$

where r is radius, N is the number of turns on the coil, and K_1 is a constant. This developed torque will equal the restraining torque of the spiral springs T_s:

$$T_s = K_s\, \theta \tag{15.3}$$

where K_s is a constant.

The angular displacement of the needle is easily found to be proportional to the current:

$$\theta = K_s I. \tag{15.4}$$

By its nature, the d'Arsonval movement is best suited to dc measurements, because pulsating or alternating values would result in varying torque and thus, a tendency of the needle to attempt to track these pulsations. Thus, the needle would vibrate for ac waveforms and the reading would be meaningless. As we will show, ac signals can be conditioned to provide a meaningful response in this meter movement.

Because full scale on the D'Arsonval meter may represent a very small current (perhaps milliamperes or even microamperes), the meter coil is usually placed in parallel with a low **shunt** resistance when measuring currents. This meter may be used to measure voltage as well; this is done by connecting a high resistance (**multiplier**) in series with the coil (see Figure 15.2). This meter will read the average value of current through the coil, which for true dc would of course be constant. The resistance of the meter coil is R_m. The inductance of the coil is not an issue because the current is dc.

If the resistance of the meter coil is known, along with full-scale current, it is a simple matter to calculate the multiplier or shunt required for any desired full-scale value of V or I (so long as they are greater than the capability of the meter movement itself).

Example 15.1

As an example of the use of the D'Arsonval meter, assume that 10 mA gives full-scale reading for a particular meter and the coil resistance is 5.0 ohms. What resistance R_{ser} should be added in series so that the meter will act as a voltmeter with a 150-volt range?

Solution

The multiplier (series) resistance is found by first finding the total resistance required:

$$R_{tot} = V / I = 150 / (10 \times 10^{-3}) = 5000 \ \Omega$$
$$R_{ser} = 15000 - 5 = 4995 \ \Omega,$$

(or we could estimate $R_{ser} = 15000$ with only 0.033% error.)

Example 15.2

Find the shunt resistance required to be placed across the meter coil in order for the D'Arsonval meter movement of Example 15.1 to act as an ammeter with a full-scale deflection of 20 A.

Solution

For a total current of 20A into the circuit of Figure 15.2b, a current of 10ma must flow through the meter. For this case, the voltage across the meter movement must therefore be

$$V_m = I R = (10 \times 10^{-3}) \, 5.0 = 50 \ \text{mV},$$

so the voltage across the shunt must be the same. Because for full-scale deflection, the meter takes 10 ma, the shunt must pass 19.99 A, therefore

$$R_{shunt} = V / I = (50 \times 10^{-3}) / 19.99 = 2.501 \times 10^{-3} \, \Omega,$$

which is a very small value of resistance.

Meter shunts and multipliers are made to be very accurate, with small tolerances of error, and thus are expensive devices that need to be treated carefully.

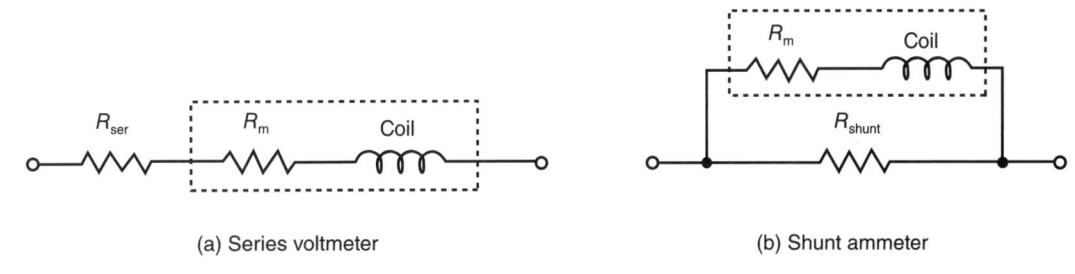

(a) Series voltmeter

(b) Shunt ammeter

Figure 15.2 A D'Arsonval voltmeter/ammeter circuit

AC Measurements with the D'Arsonval Meter Movement

Here we consider how the D'Arsonval meter might be used to measure an ac voltage or current. First, we assume the inductance of the coil is negligible. Of course, the meter alone with the series resistor as shown in Figure 15.2a would read the average voltage, which is zero for a sinusoidal applied voltage (the same would be true for the ammeter connection of Figure 15.2b). However, using a **diode** (here assumed ideal) in series with R_{ser}, the full-scale rms voltage may be found.

Figure 15.3 presents several waveforms with their average and rms values. This figure is useful in understanding how an instrument behaves in the presence of such waveforms. For example, we know that the current in a sinusoidally-excited circuit containing a diode and a resistance is a *half-wave-rectified* sinusoidal signal as shown in the second row, first column of Figure 15.3. The average value for this half-wave sinusoid, in terms of voltage, is

$$V_{avg} = V_m / \pi.$$

If this signal is applied to the previously described 150 V dc meter, the supplied ac voltage must thus have a peak (maximum) voltage of

$$V_{max} = \pi V_{avg} = \pi \, 150 = 471 \text{ V}$$

in order for the needle to move to the full-scale mark.

From Figure 15.3, the rms value of this wave must be

$$V_{rms} = V_{max} / 2 = 471 / 2 = 236 \text{ V}.$$

The scale of the meter of Example 15.1 would have to be re-engraved to reflect this if we were building a meter to accurately read the rms or average of that particular wave.

It is worth noting that some of the relationships for the entries in Figure 15.3 can be arrived at or verified simply by applying common sense. For example, the rms value of the half-wave-rectified sinusoid just discussed was stated to be

$V_m / 2$. Let's see if that makes sense in terms of the rms value of a full-wave-rectified sinusoid or a complete sinusoid.

For the full-wave-rectified sinusoid, shown in the lower left box in Figure 15.3, the rms value must simply be the combination of that of the two positive half-sinusoids. For one of these, we just noted the rms value of voltage to be stated as $V_m / 2$. Recalling that the total rms value of a periodic signal is the square root of the sum of the squares of the rms values of its components, or

$$V_{rms} = \sqrt{V^2_{1rms} + V^2_{2rms} + V^2_{3rms} + \cdots}, \quad (15.5)$$

the rms value of the full-wave-rectified sinusoid must be

$$V_{rms} = \sqrt{(V_m/2)^2 + (V_m/2)^2} = V_m / \sqrt{2}$$

as shown in the figure. Note that the full sinusoid (shown in the upper left box of Figure 15.3) must have the same rms value, because power is proportional to the square of voltage or current, and thus, the negative half of the sinusoid has the same heating effect as the positive half.

Similar logic can be carried out with regard to other waveforms shown in the figure.

The Digital Multimeter (DMM)

Digital multimeters (and digital panel meters) are widely used today. The DMM requires a little more knowledge of both the waveform and the meter itself in order to interpret its readings properly. Many such meters will measure the **true rms** voltage of an ac signal, but this should be verified in the device's specifications. However, if there is a dc component, it will have to be measured separately. The relationship between the total rms value of the waveform and the ac component and the **dc component** is given as

$$\text{rms}_{total} = [(\text{ac component rms})^2 (\text{dc component})^2]^{1/2}. \quad (15.6)$$

Burden as Associated with an Ammeter

For current measurements, when the DMM is inserted in series with the circuit, you may have to consider the voltage drop across the DMM itself; and if so, the current being measured will be in error. This voltage drop is referred to as the **burden voltage**. These burden voltages vary for different scales of a particular meter. Of course, this problem may exist for an analog instrument as well, but one can easily evaluate its effect in terms of the shunt resistance and the resistance of the meter movement.

Example 15.3

For a DMM, a typical burden voltage V_B might be 0.3 volts for a 200 mA scale and 0.9 volts on the 2,000 mA scale, when measuring full-scale values. The displayed reading as a percentage of full scale would be

$$\text{Reading} = [(100 \times \text{actual reading}) / (\text{full scale})] \times (\text{full scale burden voltage}), \quad (15.7)$$

Suppose, for this instrument, the current is being measured in a simple series circuit made up of a 10 V source and load resistance of 7.5 ohms. The expected current is, of course,

$$I = V / R = 10 / 7.5 = 1.33 \text{ A}.$$

However, suppose the displayed current is 1.254 A; this takes into account the burden voltage (0.9 volts on the 2,000 mA scale), which is

$$V_B = (1.254 / 2.000)(0.9) = (0.627)(0.9) = 0.564 \text{ V}.$$

The percentage error is

$$\% \text{ error} = (100) V_B / (V_{source} - V_B) \quad (15.8)$$
$$= (100)(0.564) / (10 - 0.564) = 5.98\%.$$

To obtain the true current, we must increase the displayed current by the 5.98% error:

$$I = (1.254)(1 + 0.0598) = 1.33 \text{ A}.$$

When making or calculating measurements, keep in mind the basic fundamentals with regard to waveforms and the limits of the measuring device.

The idea of burden is important in instrument transformers and will be discussed there, also.

Other Waveform Considerations

Recall that when determining the average value of a quantity, for example, a current, this value would be the average value of a varying current over a period of time that would deliver the same charge as a constant current over the same period. That is,

$$I_{av} = \frac{1}{T} \int_0^T i(t) dt. \quad (15.9)$$

Also, recall from Chapter 4 that the rms value is defined as being numerically equivalent to what would be required to provide the same heating effect through a resistance as a constant dc source would over the same period of time. That is, the average value of a varying power p(t) over the same time period would transfer the same energy W. Thus, recalling that P_{av} always has the same meaning as P without a subscript,

$$P_{av} T = W = \int_0^T p(t) dt \quad (15.10)$$

$$P = \frac{1}{T} \int_0^T i^2 R \, dt = I^2 R = I_{rms}^2 R. \quad (15.11)$$

Again, refer to Figure 15.3 for a few of the standard waveforms with their average and rms values given.

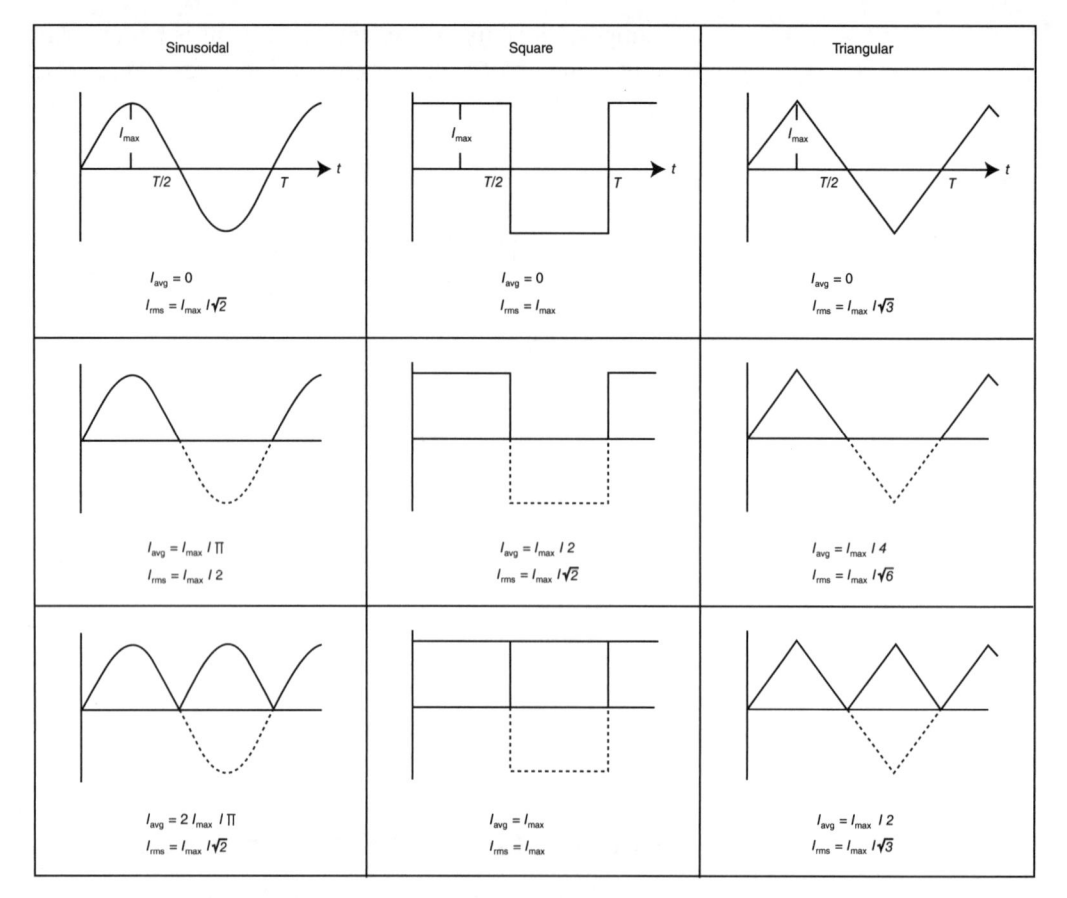

Chapter 15 Measurement, Reliability, and Lighting

Sinusoidal	Square	Triangular
$I_{avg} = 0$ $I_{rms} = I_{max}/\sqrt{2}$	$I_{avg} = 0$ $I_{rms} = I_{max}$	$I_{avg} = 0$ $I_{rms} = I_{max}/\sqrt{3}$
$I_{avg} = I_{max}/\pi$ $I_{rms} = I_{max}/2$	$I_{avg} = I_{max}/2$ $I_{rms} = I_{max}/\sqrt{2}$	$I_{avg} = I_{max}/4$ $I_{rms} = I_{max}/\sqrt{6}$
$I_{avg} = 2 I_{max}/\pi$ $I_{rms} = I_{max}/\sqrt{2}$	$I_{avg} = I_{max}$ $I_{rms} = I_{max}$	$I_{avg} = I_{max}/2$ $I_{rms} = I_{max}/\sqrt{3}$

Figure 15.3 Average and rms values for various waveforms

Example 15.4

For the voltage waveform shown in Figure 15.4, determine the average and rms values.

Solution

If a waveform is periodic, both the average and rms values are relatively easy to determine. Because the parts of this waveform are rectangular in shape, the values can be easily found by applying the meaning of Equation 15.11 discretely. For this wave, the period T is seen to be 4 seconds, and the average and rms values are given by

$$V_{av} = \text{(Net Area Over One Period)} / T$$
$$= (1 \times 1 + 3 \times 2 + 0 \times 1)/4 = 1.75 \text{ V}$$

and

$$V_{rms} = \sqrt{(\Sigma \text{ of Areas of Voltage Squared Over One Period})/T}$$
$$= \sqrt{(1^2 \times 1) + (3^2 \times 2) + (0^2 \times 1)/4} = 2.18 \text{ V}.$$

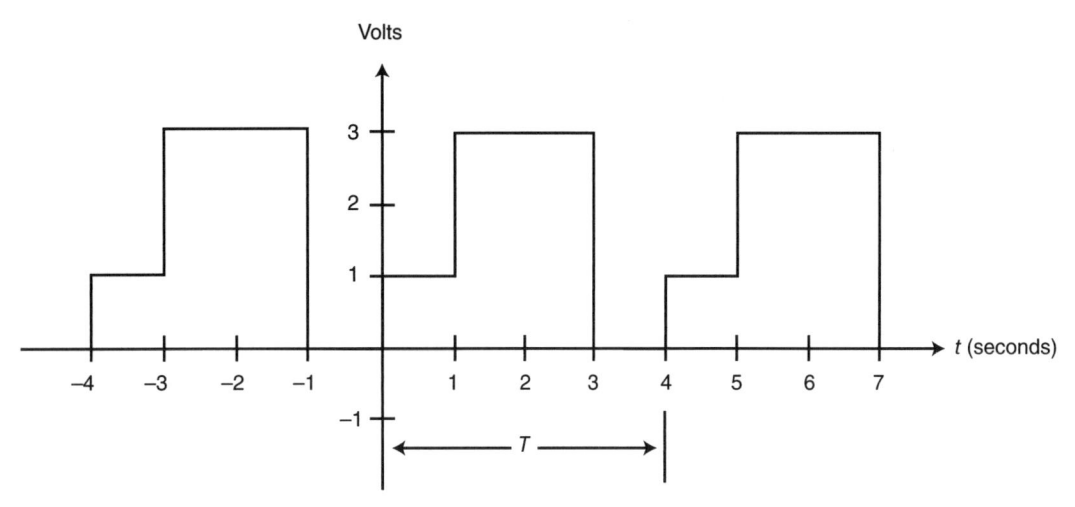

Figure 15.4 A periodic waveform

Also of interest in waveform measurements are their Fourier coefficients. The rms value of a Fourier series can be found by using Equation 15.5. It may be expressed as

$$V_{rms} = \sqrt{A_{dc}^2 + \sum (A_n/\sqrt{2})^2} \qquad (15.12)$$

where

A_{dc} is the dc term
A_n is the magnitude of the nth harmonic (where A_1 is of course the magnitude of the fundamental).

This equation may now be explained in words to mean that the rms value of the function is equal to the square root of the sum of the dc term squared, plus the squares of all the rms values of the harmonics, as we might have expected from Equation 15.5.

Example 15.5

Consider the voltage waveform in Figure 15.5a, which represents the output of a slightly saturated transformer in series with a three-volt battery. The components of the waveform are shown in Figure 15.5b. Determine the rms value of the complete voltage wave.

Solution

The equation expressing this waveform is given in terms of its dc value and two harmonics: the fundamental and third harmonic. Specifically, there is a dc value of 3 volts, a fundamental wave of $5\cos(\omega_0 t + \theta_1)$, and a third harmonic of $\cos(3\omega_0 t + \theta_3)$.

The rms value of the complete voltage wave is

$$= \sqrt{3^2 + (5/\sqrt{2})^2 + (1/\sqrt{2})^2} = \sqrt{22} = 4.69 \text{ V}.$$

The use of Fourier coefficients is also important in power-quality and power-electronics issues (see Chapter 14), because it is often very convenient when working with periodic waves to represent them by harmonics.

(a) Waveform (b) Fourier series equivalent

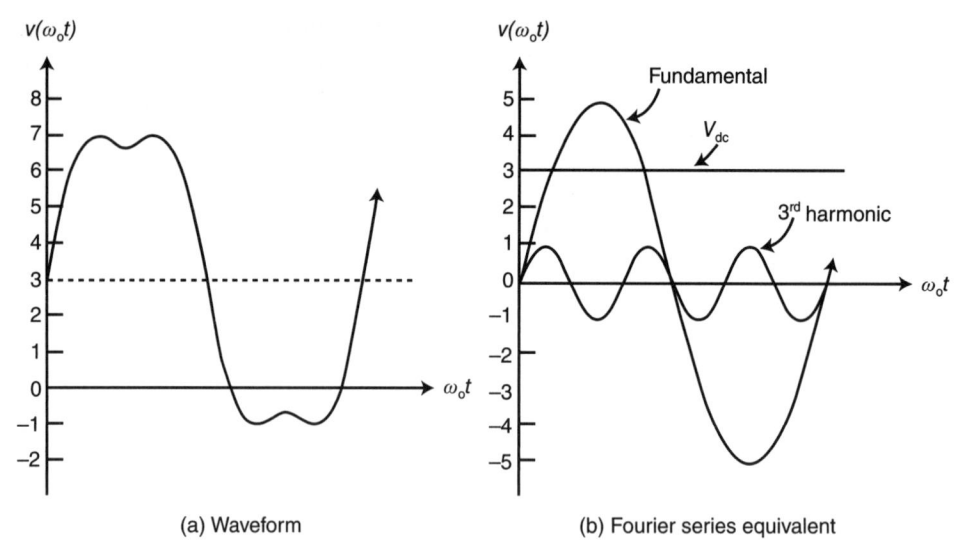

Figure 15.5 Distorted wave with dc bias

Other AC Measuring Devices

One relatively inexpensive meter movement that can be used for ac measurements is the **moving vane** movement, in which a small vane of iron moves within a stationary coil. This device has a nonlinear scale and high losses, making it an often excessive burden for voltage measurement.

Another approach is sometimes used for measurement of high-frequency currents. An ac current flows through a small heater in the presence of a **thermocouple**. The thermocouple (consisting of the junction of two small wires of dissimilar material) produces a dc voltage, which is usually applied to a d'Arsonval movement.

Many of these instruments previously described are more in the nature of portable or panel-mounted measuring devices used with relatively low (120V or less) voltages. In power-system applications, it is often necessary to obtain accurate values of considerably higher voltages and currents (too large, in fact, to be permitted near the instrument). For this purpose, an *instrument transformer* or equivalent inductive or capacitive device is used as described in the following section.

Instrument Transformers

Once we get above 120V or 480V levels, most voltages and currents are unwieldy and unsafe to measure using simple, hand-held instruments. Clip-on ammeters can be used to measure some larger currents, but the mere fact of the higher voltage level dictates safety precautions that require some kind of reduction of voltage and current. This is also necessary for system protection and relaying, because these devices work with lower currents and voltages as well, values which may be brought safely into metering enclosures and buildings.

In the United States, standard nominal values for power-systems instrumentation are 120V and 5A. Values of voltage and current in the transmission or distribution system are typically transformed to this nominal level by potential transformers (PT's) and Current transformers (CT's), respectively.

Each of these devices behaves similarly to a standard ideal transformer in theory. We should note, however, the concept of burden.

Because actual transformers are not ideal, an increasing RL load causes the load voltage to drop. This can be a serious problem in a PT, because we desire a secondary voltage that is directly proportional to the primary voltage. In order to operate in an acceptable range, the PT should have a small burden, meaning a relatively high impedance in the secondary circuit (for a CT, the opposite is true: it is best that the secondary circuit is as close to a short circuit as possible). Published curves for PT's and CT's define acceptable values of load for acceptable measurement accuracy.

A couple of examples (similar to sample PE exam problems we have seen) will illustrate. Here we will assume small burdens and, therefore, a linear operating range.

Example 15.6

A PT for a 4160V, three-phase system, which samples the phase-a voltage to ground is rated 2400/120V (the 2400V is of course the nominal line-to-ground value of the 4160V line-to-line voltage). The measured voltage on the secondary side is 122.5V. To what primary line-to-line voltage does that correspond?

Solution

Using standard ideal-transformer theory,

$$V_{pri} = \sqrt{3}\,(2400\,/\,120)\,(122.5) = 4244 \text{ V}.$$

As an additional exercise, we compute this in per-unit on a 4160V base to be

$$4244\,/\,4160 = 1.02 \text{ pu}.$$

Example 15.7

A CT has the number 100 stamped in large letters on the device. If a current of 65A flows in the primary circuit, what value of current would flow on the shorted secondary side?

Solution

The "100" indicates the rated primary current of the CT. We assume a standard rated secondary current of 5A. Thus the current transformation ratio is 100/5 or 20, and

$$I_{sec} = 65\,(5\,/\,100) = 3.25 \text{ A}.$$

Problems as simple as these have appeared on the exam, sometimes posed in the context of a relaying problem, so be alert for them. Finding a few easy problems like this will save you considerable time for the more challenging ones.

Watt and VAR Measurement

Although analog wattmeters have been in use for well over a century, the principle underlying their operation is often not as easily grasped as that of the voltmeter or ammeter. This is probably because the wattmeter must sample *both* V and I and do it correctly, with regard to polarity and connection.

Analog, mechanical wattmeters are often of the *dynamometer* type, in which the magnetic fields of a stationary coil and a moving coil interact to yield a torque that rotates the indicator against the force of a spring. The displayed value of power, to be correct, must obviously be the value of real or average power

$$P = V I \cos \varphi$$

where V and I are rms values.

You might question how the wattmeter "calculates" the cosine of the phase angle between phase voltage **V** and phase current **I**. The answer, of course, lies in the fact that the mechanical torque driving the needle is proportional to the cosine of this angle.

In any power-measuring device, whether mechanical wattmeter or electronic transducer, care must be taken to relate the polarities of the V and I samples properly. Most commercial wattmeters will have either a plus sign on one voltage terminal and a minus sign on the other, and similar markings for the current terminals, or each coil will have an unmarked terminal and one marked ±. Usually we connect the two terminals with the ± sign or the minus sign to ground or reference, or we connect both the opposite terminals to reference. That will yield an upscale reading for power from source to load if the power indeed flows in that direction.

Wattmeters are usually fitted with shunts and multipliers, as previously described, to allow proper scaling, They may be used with CT's and PT's also, in which case the proper CT or PT multiplier must be applied just as in the two previous examples.

The VARmeter is simply a wattmeter whose input signals (or at least one of them) are phase-shifted internally so that the meter gives a reading proportional to V I sin φ.

One particular connection of interest, which seems to appear occasionally in PE exam study materials, is what is sometimes called the **two-wattmeter connection** for measuring power in a three-phase circuit. This simple but ingenious circuit, known since the early days of ac power systems, is shown in Figure 15.6. Basically, it involves connecting two wattmeters (or power transducers) as shown. The total power passing from left to right in the three-phase circuit is the sum of the meter readings. The only restriction is that the sum of the currents in the three lines must be zero; that is, there must be no neutral current. It is applicable to waveforms other than the sinusoid. If one of the wattmeters reads negative when this connection is used in a balanced three-phase circuit, this simply means that the power factor of the circuit is less than 0.5, and either the V of I coil terminals of the negative-reading meter must be interchanged to obtain a reading.

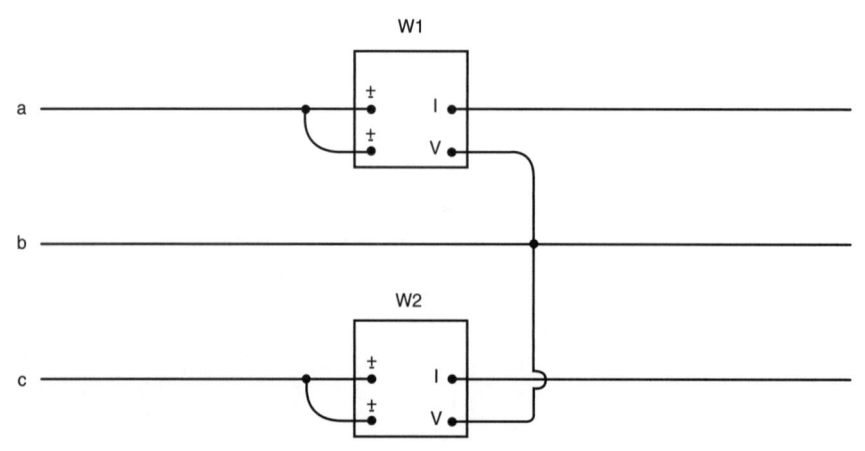

Figure 15.6 Use of two-wattmeter connection

POWER SYSTEM RELIABILITY

Reliability calculation is a very important aspect of power system engineering and design. In bulk system planning, reliability analysis is used to compare alternative planning scenarios. Distribution companies are mandated to report reliability indices calculated from historical data to regulation commissions. In facilities design for critical systems, both the normal and backup supply must be designed to provide a specific level of reliability.

Reliability is especially important with regard to electric distribution systems. Although the term is often used in its general sense, there are some particular meanings and implications that will be covered in the following paragraphs.

The basis of quantitative modeling and calculation of reliability lies in probability. **Reliability** *is defined as the ability of a system to perform its function over an intended period of time*. Depending on the context, there can be several ways to measure or quantify reliability. The simplest index is the probability that the component or system will perform as intended. The probability of failure, that is, that the component will not operate as intended at some time, is a complementary measure of **unreliability**.

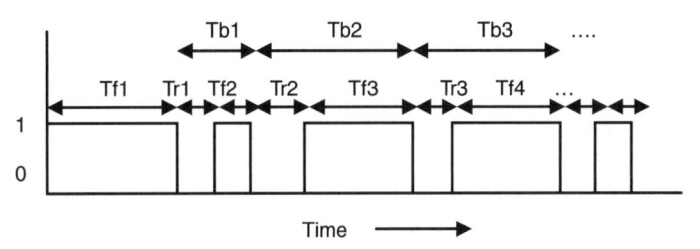

Figure 15.7 Time history of a repairable component

Figure 15.7 shows an example of the operating history of a repairable component that goes through *operate-fail-repair-operate* cycles. A value of 1 means that the component is operating as intended, or *up*; a value of 0 means that the component is not operating as intended, or is *down*. The frequency interpretation of probability allows us to write the reliability as the probability p that the component will operate as intended, as

$$p = (\text{Total time that the component is up}) / (\text{Total time}). \qquad (15.13)$$

In many instances this diagram includes a 'down' state due to scheduled or planned maintenance; it is then common to call p the **Availability** and typically treat it as synonymous with Reliability. When components are subject to maintenance or replacement, the two concepts diverge. Note that the numerical value of P provides a metric for the concept, which can be evaluated and compared.

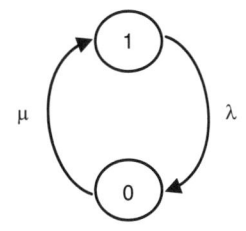

Figure 15.8 Two-state Model of a Repairable Component

In typical modeling the component transitions from the 'up' to the 'down' state because of random failures, and back to the 'up' state through repair as illustrated in the state diagram in Figure 15.8. The time to failure and time to repair (e.g., Tf1 and Tr1) are random variables that can be described by appropriate probability models that are not discussed here. A common model is the exponential distribution.

The characteristics can be described by the following commonly-used terms.

Mean time to failure MTTF: The average value of times to failure Tf1,Tf2,Tf3,…

Mean time to repair MTTR: The average value of times to repair Tr1,Tr2,Tr3,…

$$\text{Failure rate } \lambda: \lambda = 1/\text{MTTF} \tag{15.14}$$

$$\text{Repair rate } \mu: \mu = 1/\text{MTTR} \tag{15.15}$$

$$\text{Mean time between failures MTBF: MTBF} = \text{MTTF} + \text{MTTR} \tag{15.16}$$

$$\text{Frequency of failure: } f = \lambda \mu / (\lambda + \mu) \tag{15.17}$$

$$\text{Availability p: } p = \mu / (\lambda + \mu) \tag{15.18}$$

$$\text{Forced Outage rate q: } q = \lambda / (\lambda + \mu) \tag{15.19}$$

The concepts and definitions above apply to *systems* of components, as well as individual components. The first step in modeling and computing system reliability is to define the relation between component failure or status, and system failure. This process is often called *Failure Modes and Effects Analysis (FMEA)*. One way to express the results of FMEA is though reliability block-diagrams.

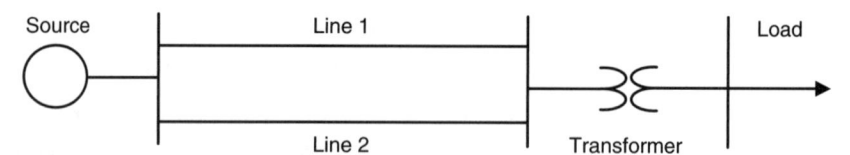

Figure 15.9 Sample Power System

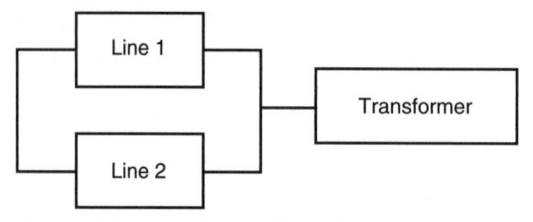

Figure 15.10 Reliability Block Diagram for System in Figure 15.9

As an example, consider the simple radial power system in Figure 15.9. The function of this system is to supply power to the load. For simplicity, assume that each line, and the transformer, has a rating larger than the load. The source is completely reliable. It follows that the load can be served as long as one of the lines and the transformer are up.

The corresponding reliability block diagram is shown in Figure 15.10. Note that its topology is exactly the same as that of the power system. The block diagram shows the logical relationship between component and system failure: Either line 1 *or* line 2 *and* the transformer must be in service to supply the load. The parallel connection of the boxes labeled Line 1 and Line 2, respectively, represents the *or* operation, while the series connection of these boxes with the box labeled Transformer represents the *and* operation.

Once the reliability diagram is constructed, straightforward formulas exist to compute system reliability. These formulas are collected in Figure 15.11 for parallel components, and Figure 15.12 for series components.

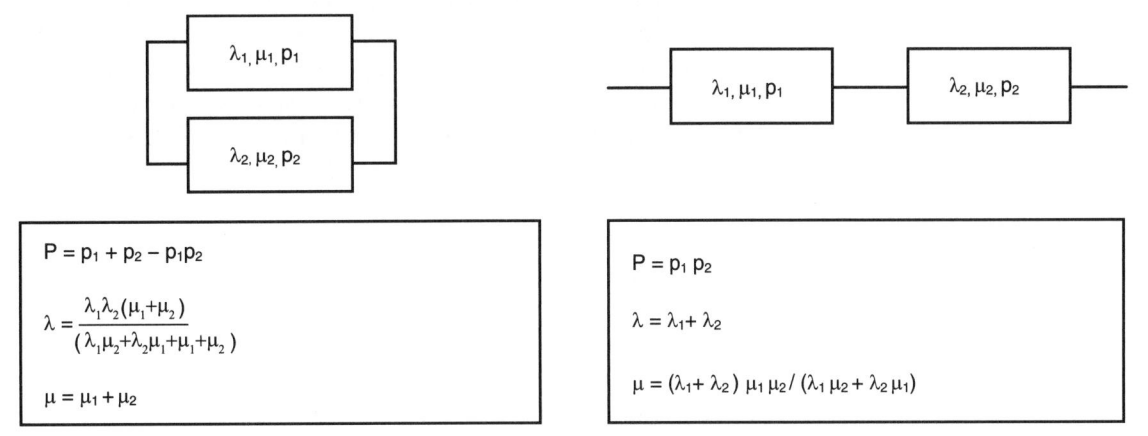

$$P = p_1 + p_2 - p_1 p_2$$

$$\lambda = \frac{\lambda_1 \lambda_2 (\mu_1 + \mu_2)}{(\lambda_1 \mu_2 + \lambda_2 \mu_1 + \mu_1 + \mu_2)}$$

$$\mu = \mu_1 + \mu_2$$

Figure 15.11 Components in Parallel

$$P = p_1 p_2$$

$$\lambda = \lambda_1 + \lambda_2$$

$$\mu = (\lambda_1 + \lambda_2)\, \mu_1 \mu_2 / (\lambda_1 \mu_2 + \lambda_2 \mu_1)$$

Figure 15.12 Components in series

Using these formulas, we can compute the reliability of the power system. In the context of this system, p represents the probability of supplying the load. We note that the quantity $1 - p$ is called the **Loss of Load Probability (LOLP)**:

$$\text{LOLP} = 1 - p \qquad (15.20)$$

Example 15.8

Calculate the availability and Loss of Load Probability (LOLP) of the system in Figure 15.9, assuming an availability of 0.99 for lines, and 0.995 for the transformer.

Solution

Let p_l denote the availability of the parallel lines, treated as one component. Using the formula in Figure 15.11,

$$p_l = 0.99 + 0.99 - 0.99\,(0.99) = 0.999900.$$

Now, treat this parallel connection of line and transformer as a series connection of two components. Using the formula in Figure 15.12, system availability is

$$p = 0.999900\,(0.995) = 0.9949005$$

and

$$\text{LOLP} = 1 - p = 0.0050995.$$

FUNDAMENTALS OF LIGHTING (ILLUMINATION ENGINEERING)

Among the earliest electric lights were **arc lamps**. Typically consisting of carbon electrodes brought together to strike an arc and then separated, they were noisy, smelly, and produced an intense, glaring light. Furthermore, they required mechanical attention, either manual or automatic, as the electrodes burned away. It was the unsuitability of these bright light sources that led investigators in the 1870s, such as Edison and Swan, to develop the incandescent lamp for indoor and other applications requiring less light, or to *subdivide the light*, as it was then termed.

Several principal lamp types are currently in use. They include incandescent (including tungsten halogen), fluorescent, and high intensity discharge (HID).

The principle of operation of the **incandescent lamp** is simply to heat a filamentary conductor to incandescence. Many types of materials (originally plant fibers) were used in early incandescent lamps by Edison and others in the 1870s. A major step forward was the use of tungsten, a metal with a very high melting point, for the filament, along with effective means of exhausting the air from the globe, and including some inert gas to suppress filament degradation.

Incandescent lamps produce a significant amount of heat, and thus, their efficacy (see below) is low. They also darken with age as the filament boils away and is deposited on the surface of the glass.

Halogen lamps, which are incandescent lamps filled with a halide gas, offer longer life and increased light output by allowing the filament to operate hotter. This also improves **color rendition**, because as a filament is energized with increasing current, it progressively moves through red, orange, yellow, and close to white light. This is why lamps operated at lowered voltages (for example, in series with a diode) offer much longer life but at the expense of color.

The color of an incandescent lamp is noticeably "warm," which makes it pleasant for many indoor applications, but its use even domestically is being challenged by compact fluorescent and other designs that last longer and exhibit greater efficacy.

Fluorescent lamps have experienced many improvements during close to a century of use. Most of us are most familiar with tubular fluorescents, such as the standard 4- and 8-foot linear tubes, or circular lamps. These lamps typically have a filament at each end, which is heated to cause a current to flow through the enclosed gas, a low-pressure mercury vapor with an inert gas, such as argon, krypton, or neon or a mixture of these. Primarily ultraviolet energy is produced, which fluoresces the phosphor coat on the interior of the tube. Although cheaper to operate than incandescent lamps, traditional fluorescents are high in ultraviolet output, can flicker noticeably, have difficulty operating in low temperatures, and require ballasts and perhaps starters. Electronic ballasts have been used in recent decades. The color of the light produced depends upon the composition of the phosphor coating, and such designations as **warm white** and **cool white** are available. In fact, manufacturers offer a variety of color renditions. "Warmer" colors (leaning toward the red end of the spectrum) are often chosen for showing meats in refrigerated displays, for example, where the use of a "cooler" color would emphasize, undesirably, the blue-green spectrum. *Rapid-start* and *instant-start* lamps are used in applications where no starter is present.

Compact fluorescents, in a variety of sizes and shapes and with standard incandescent lamp bases, are widely marketed as incandescent substitutes for residential and other low-level installations.

High intensity discharge (HID) lamps include designs such as **mercury-vapor**, **metal halide**, and **high- and low-pressure sodium** lamps. Such lamps are typically used in street lighting and other outdoor applications, garages, and other high-bay or open areas.

Metal-halide lamps are often used for sports lighting and other areas where low maintenance and good lighting quality are important. Low-pressure sodium lamps offer the highest efficacy in lumens per watt, with good color for fog penetration and surveillance, but their orange color is objectionable for many applications.

Neon lights are widely used in signs and in small, low-power indicators. Large neon lights, such as the tubes used in signage, should not be used close to areas accessible to people (including children) because of the high voltages (often on the order of 15 kV) required.

Terminology of Lighting Units and Concepts

A few additional terms relevant to illumination engineering are summarized here.

Lumen (lm): unit of light or light volume. The light output of a lamp is measured in lumens.

Candlepower (cp): unit of light intensity from a small or point source.

Footcandle (FC): unit of surface illumination. The average illumination in footcandles on a surface is the number of lumens striking the surface, divided by the square footage of the surface. A footcandle is therefore one lumen per square foot.

Efficacy (measured in lumens per watt) refers to the amount of light produced by a light source compared with the electric power consumed by the source; for example, lumens produced per watt of energy consumed. The term *efficacy* is sometimes used in place of *efficiency*, because efficiency is a ratio of similar quantities (power out versus power in) whereas the performance measure here is a ratio of two different types of quantities.

Color rendering refers to how well the light source and surrounding reflective objects render the colors of items to the typical eye. As mentioned earlier, for example, some "cooler" fluorescent lamps (which are rich in ultraviolet) are avoided in favor of "warmer" ones where raw meat is displayed, because they tend to make the foods more bluish-green. Low-pressure sodium-vapor lamps, which have high efficacy values and are used in outdoor areas, such as parking lots, give an orange cast to many objects.

Guidelines for Lighting Particular Indoor Tasks

Here are some approximate recommended guidelines for the amount of illumination (in FC) on a surface. They were determined for young, normal-sighted adults, and therefore might need to be increased for others:

Standard office work	100 FC
Bookkeeping, accounting, or other similar detailed work	150 FC
Office machine operation	150 FC
Drafting	200 FC
Conference rooms, conversation places	30 FC
Conference rooms where more detailed seeing is necessary	100 FC

These values can be used with Equation 15.21 to determine the amount of light required for a particular situation.

$$L = (FC)(A) / CU \qquad (15.21)$$

where
 L is the total net lumens required
 FC is the average desired illumination in footcandles
 A is the area of floor or other surface to be illuminated
 CU is the coefficient of utilization, which is a property of fixture employed and reflection properties of the room cavity.

Example 15.9

It is desired to light a 20 × 30 foot office space where office machinery is operated. The coefficient of utilization is estimated to be 0.8. Choose an appropriate footcandle value and figure the total number of lumens necessary, assuming a 50 percent margin above the minimum recommended level.

Solution

From the previously presented table, choose a level of 150 FC. Then

$$L = [(150)(1.5)][(20)(30)] / 0.8 = 168750 \text{ lumens}.$$

This gives an idea of how many light fixtures we need, because once the type of light source is chosen, we can go to manufacturer's data to find the number of lumens per fixture.

Note that in most straightforward and typical indoor situations like the one in this example, where it is assumed that a number of fixtures are uniformly located around the room, the height of the ceiling is not a factor.

RECOMMENDED REFERENCES

Billinton, Roy. *Power System Reliability Evaluation*. Gordon and Breach, 1976.

Gonen, Turan. *Electric Power Distribution System Engineering*. McGraw-Hill, 1976.

Useful guides concerning lighting selection and design are readily available from manufacturers of lighting products.

CHAPTER 16

Codes, Standards, and Safety

OUTLINE

CODE-RELATED EXAM QUESTIONS 224

AN OVERVIEW OF ELECTRICAL CODES AND STANDARDS 224
Differences Between the NEC and the NESC ■ Some Electric Code History

THE NATIONAL ELECTRICAL CODE 226
Chapter 3—Wiring Methods and Materials ■ Chapter 4—Equipment for General Use

THE NATIONAL ELECTRICAL SAFETY CODE 229

ELECTRICAL SAFETY 230
Step and Touch Potentials ■ Reasonable Precaution ■ Insulation by Isolation

ELECTRICAL SHOCK AND OTHER INJURY 231
Macroshocks ■ Microshocks ■ Shock Protection and Prevention

RECOMMENDED REFERENCES 236

The **National Council of Examiners for Engineering and Surveying** (NCEES) indicates that the power option of the PE exam in electrical engineering may include questions relating to the National Electrical Code (NEC) and the National Electrical Safety Code (NESC), and that this material may comprise as much as 12.5 percent of the exam. The NCEES website indicates that "Code information required to solve questions will be consistent with the last edition of the code issued before the year of the exam." Accordingly, this review provides a brief synopsis of the latest issues of the two standards as of 2009, the date of publication of this book. You should be sure to consult appropriate versions of the code for the year in which you are taking the PE exam. Few changes are typically made in the NESC or NEC from one issue to the next, especially in items such as conductor characteristics, but it is possible that a problem may be included on the exam just to verify that you are aware of the most current code.

Because of the length and scope of the NEC and NESC, this review must focus on a brief introduction to each and point to some important sections. These include sections required for the solution of problems seen in several exam review resources, in which case we point out the important tables or concepts. We recommend that you review these codes and familiarize yourself with their contents. Preparing a set of tabs and inserting them at the appropriate places may help in quick access of important material, and will have the added advantage of forcing you to become more familiar with the documents.

In addition to providing an overview of the NEC and NESC, this chapter focuses on electrical safety issues, including electric shock. But before getting into any of these details, we will begin with some insight into what types of code questions you can expect on the exam.

CODE-RELATED EXAM QUESTIONS

Because the code documents are large (several hundred pages) and packed with intensely detailed statements and tables, you may wonder what kinds of questions you are likely to find on the PE exam. Code-related questions often involve one of the following:

- Look-up questions, in which you need to find the correct value from a table (typical of this is finding permissible ampacity of a conductor).

- Problems in which you can use common sense or a general understanding of the situation to help you arrive at an answer.

The latter type of problem is often stated without any numerical data. *If you can rule out one or two of the responses, you will have significantly increased your probability of a correct answer.*

In any case, be sure you carefully note the parameters of the problem (conductors in magnetic or nonmagnetic duct, copper or aluminum conductors, etc.) before selecting an answer. In these types of problems, both identifying the relevant table and reading it correctly are critical. Some problems may include answer choices that test your care in using the correct table. For example, remember that in a single-phase circuit, there are two conductors, so the total impedance of the circuit is twice the impedance of one conductor.

We have seen a problem or two in various exam review resources that relates to situations that the codes in question *do not* cover. Becoming familiar with code content will help you answer these types of questions.

Because many of the potential exam questions on codes and standards can be answered only with reference to the code in question, *we recommend that you obtain and familiarize yourself with a copy of the NEC and the NESC.* In the absence of these, the pocket-sized quick-guides or handbooks may be useful if they present tables for frequently-used conductors, for example, but they can't take the place of the actual code volume.

AN OVERVIEW OF ELECTRICAL CODES AND STANDARDS

Before exploring details of the NEC and NESC, it is useful to discuss the topic of codes and standards in general. Sometimes, just knowing some facts about one of these codes, such as the kinds of material and situations they cover, may enable you to answer an exam problem.

The NEC and NESC are sometimes known as *consensus standards*. This means that they have been assembled, modified, and updated by committees representing a variety of interested organizations and agencies that have a stake in the issues. Like many engineering standards, they are updated by consensus. Usually, any person or organization may propose new material or modifications for the next update cycle.

It is important to recognize what these documents are and what they are not. *The underlying purpose of the NEC and the NESC (and many other codes) is safety*. They are not design handbooks (although they present useful information for installation designers), nor do they in general indicate how to analyze engineering problems. They will not tell you, for example, how to build a transformer, how to design a generation facility, or how much current can be carried by a particular conductor type before the insulation or the wire melts. The NEC will tell you, however, the allowable safe current that a particular wire may carry. The NESC will tell you the minimum safe clearance for a transmission crossing a highway or railroad.

It is sometimes said that a code or standard "reads like a law book." In fact, many jurisdictions adopt the NEC in particular as law, sometimes with some modifications. Electricians and other workers are often required to pass tests on these standards, because their work must conform to them. Typically, electrical and most other engineering work must be inspected before being approved for use. Again, the guiding purpose is to protect the public (which is assumed to not understand the nature of hazards and threats to health and safety to the extent required of knowledgeable and trained persons). Legal entities recognize the greater responsibility placed on trained persons (such as electricians and engineers) to act in the public interest. In the case of an accident, the outcome of litigation often hinges on how closely the relevant codes were followed in the design, construction, and maintenance of the system in question.

There are many codes and standards in existence, and a large subset of them relates to electrical work. Some well-known additional examples include the IEEE "color book" series, or the large C57 standard dealing with transformers.

A code or standard often is denoted by its name, some unique designation that typically follows it through successive issues, and the date of the present issue. For example, the latest edition of the National Electrical Safety Code is designated C2-2007.

A code or standard typically includes an often detailed table of contents, an introduction, statement of purpose, and possibly a glossary of definitions related to the particular specialty it covers. It is good to be familiar with this content, because parts of exam problems might be answered with reference to statements of purpose or definitions. Furthermore, reading the introductory material can inform you of the scope and intent of the standard.

Codes are living documents, with committees continually soliciting and examining suggestions for change and acting on these suggestions as appropriate. In addition, such committees may issue rulings on specific cases submitted to them. Sometimes, these rulings lead to changes or insertions in future editions of the code.

Differences Between the NEC and the NESC

Although there are a few exceptions, *in general, the NEC relates to premises wiring (both within buildings, and external wiring) and the NESC relates to outdoor equipment, such as transmission and distribution lines and substations.*

The NEC, for example, provides guidelines on the number and spacing of outlets on a branch circuit and will tell you the minimum gauge of wire required for such circuits. It provides safety guidelines for installation of equipment in special locations, such as lights in swimming pools, and for particular situations, such as theaters and healthcare facilities.

The NESC, in contrast, concerns topics, such as electric supply stations, clearances for wires above roadways and railroads, and required "grades of construction" (relating to factors such as ice loading). Both codes address some common topics, such as surge arresters and grounding methods.

Some Electric Code History

Before examining the contents of the current NEC and NESC, a little history lesson is in order.

Within two decades of the invention of the incandescent lamp, an increase in electrically-caused fires prompted insurers and others to raise the question of adequate standards for electrical equipment and its installation and operation. Installers commonly created their own guidelines as they worked, and the lack of standards and knowledge, as well as instances of shoddy work and corner-cutting, led to fires, injury, and death. Electrical handbooks of the 1890s often cite case histories of these kinds of occurrences and attempt to provide guidelines for safer installation and operation, but the lack of an enforceable standard is evident.

By the mid 1890s, several electrical codes had been written and put into use, often conflicting with each other. But shortly before the turn of the century, more that 1000 persons were involved in reviewing the electrical codes of the US and Europe, resulting in the issue of the first edition of what we now call the NEC. The *secretariat*, or managing body, of the NEC is now, quite appropriately, the National Fire Protection Association (NFPA).

The NESC is a younger document, but its history parallels that of the NEC; development began in 1913 at the U.S. National Bureau of Standards. It is now under the auspices of the Institute of Electrical and Electronics Engineers (IEEE).

THE NATIONAL ELECTRICAL CODE

As mentioned earlier, the NEC is published by the National Fire Protection Association and is designated as NFPA 70. Also published by NFPA is the *NEC 2008 Handbook*, a 1400-page detailed reference containing the complete text of the NEC, plus numerous discussions and illustrations. Recent editions of the *Handbook* contain color illustrations, making it particularly useful. We highly recommend this book as a guide to study and use of the NEC.

Topics in the NEC are designated by *chapters* and *articles*. The Code begins with Article 90 (Introduction). This is followed by Chapter 1, which includes Article 100 (definitions) and Article 110, which addresses requirements for electrical installations. Chapter 2 includes articles numbered 2xx, Chapter 3 includes articles numbered 3xx, and so on. *Chapters 3 and 4, in particular, contain material that seems likely to be seen on the exam, so we discuss parts of these chapters specifically below*. However, that doesn't mean that other parts of the NEC should not be studied for the exam. Table 16.1 summarizes all of the chapters and *annexes* (found at the end of the document) in the NEC by topic.

Chapter 3—Wiring Methods and Materials

There are some important tables in Article 310 of Chapter 3. For example, Table 310.16 gives allowable ampacities of a number of conductor types rated 0-2000 V in a raceway or cable, or directly buried in earth, while Table 310.17 provides this information for cables in "free air." Note that there are several columns for different temperature ratings. Because so much important data is contained in the tables in this article, it might be wise to examine all of these tables carefully and to study their applications.

A typical code-related problem might be to find the ampacity of a given conductor, perhaps in terms of motor rating (see next section).

Chapter 4—Equipment for General Use

Code-related questions concerning motors seem to appear on the exam frequently, so it would be wise to become especially familiar with this material. Article 430 addresses "motors, motor circuits and controllers."

Tables in Section 430 provide required overload protectors and locked-rotor currents indicated by code letters. Just knowing where to find this information quickly can be useful.

For example, Section 430.74 discusses the inadvertent starting of a motor by ground fault in an improperly-connected control circuit. You should examine some motor-starting circuits and use common sense in answering questions about them on the exam.

Also, **code letters** that indicate kVA/HP under locked-rotor conditions are given in Table 430.7(B).

You might need to use information in this section and then refer back to the appropriate ampacity table of Chapter 3 as part of a problem. For example, Article 430.22 states that conductors supplying "a single motor used in a continuous duty application shall have an ampacity of not less than 125 percent of the motor's full-load current rating…" It is easy to miss this kind of important statement if you are not familiar with the Code.

Table 16.1 Content Summary for the NEC

Chapter/Annex	Topical Coverage
Chapter 2—Wiring and Protection	Grounded conductors and grounding, branch circuits and feeders, and some protection issues including ground fault circuit interrupter (GFCI) and feeder ground-fault protection.
Chapter 3—Wiring Methods and Materials	Conductors and cable types of different voltage ratings and guidelines for their use. The chapter covers most wiring installations. Ampacity (current-carrying capacity) tables are given for conductors in various situations in Article 310. Methods for pulling and supporting conductors are cited.
Chapter 4—Equipment for General Use	Flexible cords and cables, fixture wires, and wiring devices, such as switches and receptacles. Guidelines for switchboards, panelboards, light fixtures, transformers, batteries and other devices are discussed. The chapter also covers motors and applications for refrigeration and heating equipment.

Table 16.1 Content Summary for the NEC (continued)

Chapter/Annex	Topical Coverage
Chapter 5—Special Occupancies	Guidelines for classified hazardous locations, commercial auto repair facilities, aircraft hangars, health-care facilities, theaters, television studios, carnivals and fairs, recreational vehicle parks, and other non-traditional locations. A detailed table lists hazardous location categories. Some particular considerations for health-care facilities are given.
Chapter 6—Special Equipment	Electric signs, welders, dumbwaiters, chair lifts, electroplating, pipe organs, photovoltaic facilities, and other special applications. Of particular note is the section on swimming pools and their associated equipment, and other facilities that require particular safety precautions because of their proximity to water.
Chapter 7—Special Conditions	Emergency and standby systems, some signaling and low-voltage equipment, as well as optical-fiber systems, fire alarm systems, and similar topics.
Chapter 8—Communication Systems	Commercial radio and television equipment, community signal distributing systems, and network-powered broadband communication systems.
Chapter 9—Tables	Tables for fill of conduits, tubing bend radii, wire dimensions, conductor electrical properties, and others. *We recommend becoming familiar with what tables are included here.*
Annex A—Product Safety Standards	Applicable standards, such as IEEE and UL for a variety of devices.
Annex B—Application Information for Ampacity Calculation	Tables for ampacities of various conductor types.
Annex C—Conduit and Tubing Fill Tables for Conductors and Fixture Wires of the Same Size	Primarily gives the maximum allowable number of conductors in different types of conduits. A number of tables are included.
Annex D—Examples	Examples for determining the kVA requirements for typical dwellings, store buildings, range loads, motor circuits, etc.
Annex E—Types of Construction	Provides fire resistance ratings for types of walls and other construction features.
Annex F	Discusses availability and reliability for critical power systems
Annex G	Provides some compliance requirements for SCADA systems
Annex H—Administration and Enforcement	Provides a "model set of rules for adoption by jurisdictions," including topics such as recordkeeping, issuing of permits, inspectors' qualifications, and appeals.

THE NATIONAL ELECTRICAL SAFETY CODE

Table 16.2 summarizes the major sections of the NESC, with some important contents indicated. There are some gaps in the section numbers—this is intentional; not all section numbers are used in the current edition of the code. In addition to the sections, keep in mind that the appendices are an important part of the NESC.

Table 16.2 Content Summary for the NESC

Section/Part	Topical Coverage
Section 1	Introduction: purpose, scope, and applications of the code.
Section 2	Definition of special terms
Section 3	References
Section 9	Grounding methods: information on allowable grounding methods for both "electric supply" and "communications" facilities. Grounding questions are likely to be answered here.
Part 1 (Sections 10–19)	Electric supply stations: safety clearances, ventilation and illumination, exits, guarding live parts, flammable and dangerous materials, surge protection, etc.
Part 2 (Sections 20–29)	Safety rules for the maintenance of overhead electric supply and communication lines: Voltage classifications, minimum clearances (an important concept that may lead to questions having numerical answers), required grades of construction, mechanical strength requirements, and of course, accessibility and guarding.
Part 3 (Sections 30–39)	Underground equipment: similar issues as Part 2 (above).
Part 4	*Operation* of electric supply and communication lines and equipment: precautions, inspection, minimum approach distance to live parts, work on energized equipment, and procedures for de-energizing equipment for worker protection.
Appendices A, B, and C	Calculation of clearances between lines and between lines and structures, rights of way, etc. It is likely that numerical questions on clearances will require an understanding of this material.
Appendix D	Brief treatment of overvoltage factor; included in the NESC printing for informational purposes only.
Appendix E	Bibliography of mostly other standards relating to the NESC.

ELECTRICAL SAFETY

As discussed, the primary purposes of the NEC and the NESC have been the protection of humans and property from hazardous conditions. In this section, we review a few general safety topics that may help to place the discussions of the NEC and NESC into perspective.

Step and Touch Potentials

Recall that earth is a conductor of electricity. Its conductivity varies depending upon type of earth, amount of moisture, salinity and other chemicals, and other factors. Furthermore, the potential gradients at or near the earth's surface are likely to be larger in areas near the points at which large currents enter or leave the earth. Especially when large short-circuit currents or lightning-induced surges flow through the earth, a potential difference can be developed along the surface of the ground. This potential gradient may be quite large even between points a few feet apart.

Step potential refers to electrical potential possible between a pair of points on the ground that may be separated the approximate distance of a standing person's feet. If one is standing with the legs several feet apart in a substation with moist earth, For example, a dangerous current might be conducted through the body from one foot to another during the brief time that a ground fault is present. Substation *ground mats* (conducting metallic grids buried under the earth, gravel, and other fill) are designed to reduce this potential. However, a prudent course of action when standing near electrical equipment is to keep one's feet close together to minimize step potential. This is particularly important when the ground and one's feet are wet.

Touch potential, on the other hand, relates to possible conduction through the body due to touching a piece of equipment. The path may be through one or the other leg to ground, or through both arms if one is touching two different conducting objects simultaneously. It is wise to avoid touching equipment whenever possible while standing on conducting material, even if the equipment seems securely grounded. It is also prudent to avoid touching two different metallic objects at once in such situations, even if they appear to be grounded. The presence of a ground conductor or braid firmly connected to a chassis and disappearing into the earth is no guarantee of effective grounding; the author is reminded of utility observations in which grounding braids were found to be nearly disintegrated a few inches below the earth's surface in one swampy area of the South.

In particular, maintenance people who must be near such equipment on a regular basis and who come into particular, intentional contact with metallic parts as part of their work should pay careful attention to these issues.

Reasonable Precaution

In designing and constructing electrical facilities, such as substations and open-wire lines, it is incumbent upon technical persons to protect the public from anticipated hazards. However, perfect protection against all hazards cannot be guaranteed. For example, a high chain-link fence topped with "concertina wire" can be installed around a substation and the property locked. Signage can be posted in visible places warning of the potential hazards. The property can be kept locked and security-lighted. The very presence of these warning devices is likely to act as a deterrent for most persons who might otherwise be tempted to wander onto the property. However, they do nothing to stop the person with a chain cutter who is determined to cut through the fence to steal copper wire or aluminum busbar.

Here again the code comes into play, in this case defining the reasonable precautions that must be taken to ensure public safety. The NESC prescribes warning and guarding including those cited above (signage, fences of at least minimum height topped with sharp wire, locks, etc.) In the case of litigation as the result of an accident, the utility or other entity will want to show that it carried out at least the minimum protection requirements as defined in the NESC.

In many cases, **adequate inspection** is also a legal duty of those operating dangerous electrical equipment. The author is aware of a case in which a distribution conductor on a pole in a remote mountain area pulled off an insulator, sagging to the extent that it was a safety hazard. Fortunately, inspection of this remote line revealed the hazard, although possibly months after the wire left the crossarm and not until hikers noticed the arcing sounds between conductor and bushes and reported it. Frequent inspection is essential.

Insulation by Isolation

You have probably noticed that transmission and distribution conductors, busbars in substations, and other current-carrying parts are often not covered with insulation as are cables. There are several reasons for this. First, insulation for the voltage levels required on such equipment would necessarily be expensive, and probably thick and weighty. Aging insulation materials often develop cracks that allow moisture to enter, negating the electrical insulation effect they are intended to provide. Thermal conduction from conductor heating might be impaired, as well.

More importantly, insulation is achieved in such cases, at least ideally, by isolating the conductors at a sufficient height to render them safe for most applications. This is sometimes called **insulation by isolation**, and the required distances above earth and other structures are specified in the NESC.

The author is aware of cases in which such isolation has been compromised through the erection of, for example, an amateur-radio antenna in close proximity to a transmission line. Another possible compromise of safety, which is not as obvious, involves high-pressure farm watering systems, particularly the moving or rotating kind, for which a jet of water might become a dangerous conductor between an overhead high-voltage line and equipment below. The utility company must inspect regularly in order to be aware of such potential safety compromises.

ELECTRICAL SHOCK AND OTHER INJURY

Ever since Benjamin Franklin (and possibly one or more of his contemporaries) demonstrated that lightning is electricity, we have been aware of some of the hazards of electric currents and discharges. The widespread use of lightning rods to protect property, including such diverse entities as cathedral spires and ships at sea, from fire or other injury, dates from approximately the same period.

Since early times, injuries and deaths due to lightning were certainly observed but were not well understood. Probably the effects upon structures were much better appreciated: bricks and stone blown out by expanding steam, and fires due to lightning strikes to ship masts, have been tangible evidence for centuries of the power of this force. The possibility of conducting lightning into structures through the introduction of telegraph wires in the 1840s, and telephone conductors in the

1880s, opened new and necessary areas for lightning protection and safety. Furthermore, with the purposeful production of electricity and its intentional disposition into buildings in the late nineteenth century for motors and lighting, the need for electrical safety took additional directions.

When considering electrical effects to the human body, we often think of **electrocution** (death resulting from current passing through the body) and **burns**. However, indirect causes and effects are important, as well. For example, a mild electrical shock can cause a muscular reaction and resulting injury from a tool or sharp surface. Persons using aluminum ladders are particularly susceptible to sudden and unexpected injury should they accidentally make contact between the grounded ladder and a live conducting item, such as roof flashing or a conduit or pipe energized due to faulty insulation. In any case, prudent safety practices, based upon knowledge of what electricity can do, are always essential, even if one is not intentionally working on electrical equipment.

This section provides some medical information concerning the effects of electrical currents on the human body. While detailed exam questions on this material are not anticipated, it may help familiarize you with the issue, and underscore the motivation for safety codes. You will notice that much of this discussion relates to observations made in connection with the use of medical equipment under controlled conditions; however, the extensions to other situations (such as would be more likely encountered by engineers) are obvious.

Whether we are dealing with a person making repairs to the exterior of a building or a patient in a hospital bed, most shocks today occur when the subject accidentally becomes part of a circuit between a 60 Hz power source and ground. As in other areas of analysis presented in this book, it is helpful to visualize the electric circuit involved. The shock circuit is often quite circuitous and not obvious, which probably has a great deal to do with why such circuits seem to accidentally appear.

The physiological damage that results from shock takes the form of resistive heating and electrochemical burning of the tissue, and the stimulation of excitable nerve and muscle tissue.

Macroshocks

When shock current enters and exits the body through the skin, generally only a small fraction of the current flows through the heart. In order for currents through the heart to reach dangerous levels, the externally applied currents must typically be rather large, and are thus referred to as **macroshocks**. The following macroshock effects are the result of 60 Hz currents applied to the hands. Note that each effect can occur over a range of current values due to different current densities within different subjects.

The **threshold of perception** occurs when enough current flows to excite the nerve endings of the skin. In contrast to the current levels found even in domestic appliances, for example, it does not take much current for this to be noticeable. The range of threshold of perception current is from about 0.5 milliamps (mA) to about 7 mA.

Involuntary muscle contractions result from higher current levels. The **maximum "let go" current** is defined as the highest current level from which a subject can voluntarily withdraw. The range of maximum "let go" current also varies among subjects and is from about 7 mA to about 100 mA.

Involuntary contraction and paralysis of the respiratory muscles result as current levels are increased. The range of *minimum* respiratory arrest current is from about 10 mA to about 22 mA.

At even higher current levels, if the density of current through only part of the heart is sufficient to excite cardiac muscle, then the normal sequence of synchronizing electrical events within the heart can be disrupted. Once the heart has been desynchronized, or **fibrillated**, it stops pumping. Simply ceasing the flow of the shock current will not start the heart pumping again. The range of the threshold of fibrillation current is from about 75 mA to about 400 mA.

Still higher current levels are capable of simultaneously depolarizing the entire heart muscle instead of just part of it. This concurrent relaxation, or defibrillation, results in the return of the heart's normal rhythmic activity. The range of the maximum fibrillation or minimum defibrillation current is about 1 A to about 6 A.

Yet higher current levels result in sustained contraction of the entire heart muscle. Normal heart activity ensues when the shock current is ceased. The range of the minimum complete myocardial contraction current is about 1 A to about 6 A.

Above the 1 A to 6 A current range, burns due to resistive heating usually result. Burns occur particularly at the skin contact points because of the high resistance of skin as compared to the lower resistance of internal tissues. At these high current levels, nervous tissue loses functional excitability, and skeletal muscles undergo severe contraction.

As mentioned above, macroshock hazards can exist in a wide variety of situations. In the biomedical environment, For example, they generally come from equipment with faulty hot and/or ground power line connections. A 1Ω resistance ground connection between equipment chassis ground and earth ground will conduct virtually all of the current to earth ground from a hot power line conductor with faulty insulation that is contacting chassis ground. Virtually no current flows to earth ground through a high resistance person touching the chassis. However, were the chassis ground to earth ground connection to fail as an open circuit, a macroshock hazard would be present. A person simultaneously touching the hot chassis and any other earth grounded object would then conduct all of the current from the faulty hot power line conductor to earth ground.

Microshocks

When shock current enters and exits the body *directly through the heart*, much lower current levels are considered dangerous. These internally applied currents are thus referred to as **microshocks**. Hospital patients who have direct electrical connections to the heart, For example, are termed *electrically susceptible* patients. Much of the following material relates to such situations.

Data for fibrillation caused by shock currents passing through intracardiac catheters indicate that the range of the minimum fibrillation current is from about 80 µA (microamps) to about 600 µA. The safety current limit for microshocks is 10 µA.

Microshock hazards in the biomedical environment generally come from leakage currents that invariably flow between insulated conductors at different potentials. Most of the leakage current flows through capacitive coupling between conductors, though some resistive leakage current flows through insulation, dirt, and moisture.

The leakage currents pertinent to microshocks are those leakage currents that ultimately flow between conductors with direct connections to the patient and line-powered conductors, and between conductors with direct connections to the patient and chassis ground conductors. A microshock hazard is present for a patient who has a heart catheter connected to a piece of equipment with excessive capacitive coupling between the conductive fluid in the catheter and the power supply lines of the equipment. Were the patient to touch the earth grounded hospital bed frame, a microshock circuit would be completed.

Microshock hazards in the biomedical environment can also come from current flowing through a patient connected between instruments with a voltage difference between the different chassis ground potentials.

A microshock hazard is present for a patient with an ECG monitor plugged into one outlet, and a heart catheter blood pressure monitor plugged into a different outlet. Any ground fault current flowing through the ground line between the two outlets would place the chassis of the two instruments at different potentials.

Capacitive coupling between the ECG leads and ECG chassis ground, and between the catheter and the book pressure chassis ground, would complete a circuit through which a microshock current could flow.

Figure 16.1 shows a schematic of a typical microshock circuit.

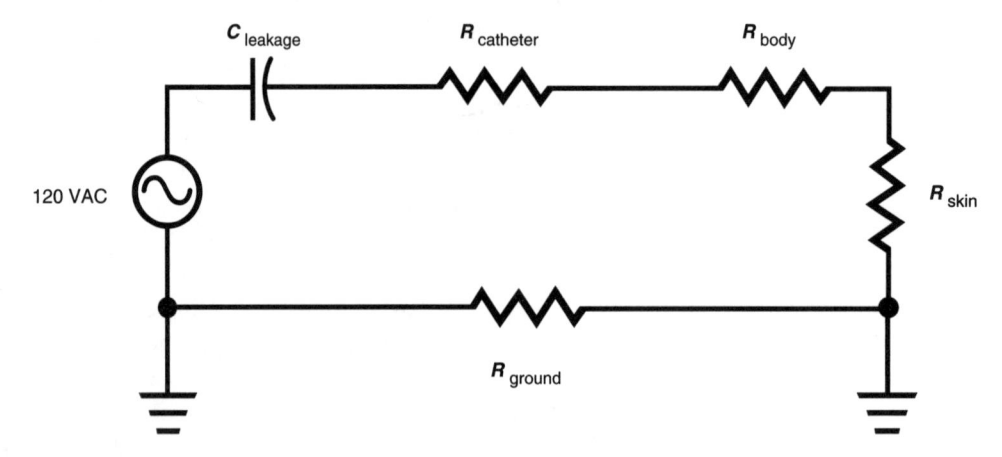

Figure 16.1 A typical microshock circuit

Example 16.1

A patient in the cardiac care unit is lying in a bed with its frame grounded to the hospital's electrical ground network. The series resistance of a hospital's electrical ground network is usually negligible (less than 1Ω). The patient's hand is resting on the bed frame. The series resistance of a patient's skin is about 100 kΩ. The internal series resistance between a patient's hand and heart is about 300Ω.

A saline-filled catheter is connected between the patient's heart and a 120 V ac line-powered blood pressure monitor. The series resistance of the saline between a patient's heart and the monitor's pressure transducer is about 40 Ω. The blood pressure monitor has leaky insulation in its power supply. The capacitance between the hot lead of the line cord and the leads of the pressure transducer is about 0.063 μF.

Is the patient described above in danger of fibrillation?

Solution

At 60 Hz, the series impedance of the leakage capacitor is

$$Z_c = -j/[(377)(0.063 \times 10^{-6})] = -j42.1 \text{ k}\Omega.$$

The total series resistance is

$$R = 100000 + 300 + 40 = 100.34 \text{ k}\Omega.$$

And thus the magnitude of the impedance is

$$Z = (100.34^2 + 42.1^2)^{0.5} = 108.8 \text{ k}\Omega.$$

The microshock current is therefore

$$I \approx 120 / 108.800 = 1.1 \text{ mA}.$$

The patient is in great danger of fibrillation as the safety current limit is 10 µA.

Shock Protection and Prevention

The typical macroshock circuit path runs (1) from the utility company's hot power line, (2) through a person to earth ground, and (3) through earth ground to the grounding point of the utility company's neutral power line. Installing a 1:1 input-to-output ratio *isolation transformer* into the power line path is one method of protecting against macroshocks. The transformer's primary coil is connected to the utility company's hot and neutral power lines. The transformer's secondary coil provides the isolated hot and neutral power lines. The macroshock circuit path has been broken by the insertion of the isolation transformer, because there is now no electrical path between the utility company's power line and the isolated hot power line.

The microshock circuit path, unfortunately, has not been broken by the insertion of the isolation transformer, because there are capacitive and resistive leakage current paths between the input and output coils of the transformer. Though useful for protecting against macroshocks, the leakage currents through isolation transformers are too high for the transformers to be used alone to protect against microshocks.

Installing ground fault circuit interrupters (GFCIs) in series with the power lines is another method of protecting against macroshocks. Any imbalance over a few milliamps between the hot and neutral power lines indicates that current is flowing to ground through an undesired path, and the interrupter cuts off power to its output.

However, GFCIs should not be used to protect critical life-support equipment. In this case, the tripping off of the power is probably more life-threatening to a patient than a ground fault. Also, the minimum trip current of GFCIs is generally too high for the interrupters to be used alone to protect against microshocks.

Microshock protection is achieved through complete electrical isolation of the patient. Ideally, all electrical signals are passed through linear isolation amplifiers and isolated power supplies. Electrical power is provided through isolation transformers. Catheters for electrified equipment (which have their transducers located at the patient's body) and the electrical connections to the transducer are, as above, isolated. Electrified equipment uses low-voltage power supplies and is constructed with low leakage insulation. Double-insulated construction is also used to prevent any contact with the grounded chassis of equipment.

Equipotential grounding systems are also important to microshock protection. By grounding all equipment within the patient environment to a single grounding point, potential differences between equipment chassis grounds due to current flow through ground loops can be avoided. Maintenance of all the shock elimination methods described above is accomplished through the regular testing of equipment for low leakage currents, low resistance of interconnections, and the integrity of insulation.

RECOMMENDED REFERENCES

National Electrical Code. National Fire Protection Association (NFPA).

National Electrical Safety Code. IEEE.

Ray, C.D., *Medical Engineering*. Year Book Medical Publishers, 1974.

Van Doren, Thomas, *Grounding and Shielding* (Short course notes, University of Missouri at Rolla, 2009)

CHAPTER 17

Engineering Economics

OUTLINE

CASH FLOW 239

TIME VALUE OF MONEY 240
Simple Interest

EQUIVALENCE 241

COMPOUND INTEREST 241
Symbols and Functional Notation ■ Single-Payment Formulas ■ Uniform Payment Series Formulas ■ Uniform Gradient

NOMINAL AND EFFECTIVE INTEREST 248
Non-Annual Compounding ■ Continuous Compounding

SOLVING ENGINEERING ECONOMICS PROBLEMS 250
Criteria

PRESENT WORTH 251
Appropriate Problems ■ Infinite Life and Capitalized Cost

FUTURE WORTH OR VALUE 253

ANNUAL COST 254
Criteria ■ Application of Annual Cost Analysis

RATE OF RETURN ANALYSIS 256
Two Alternatives ■ Three or More Alternatives

BENEFIT-COST ANALYSIS 260

BREAKEVEN ANALYSIS 261

OPTIMIZATION 262
Minima-Maxima ■ Economic Problem—Best Alternative ■ Economic Order Quantity

VALUATION AND DEPRECIATION 264
Notation ■ Straight-Line Depreciation ■ Sum-of-Years'-Digits Depreciation ■ Declining-Balance Depreciation ■ Sinking-Fund Depreciation ■ Modified Accelerated Cost Recovery System Depreciation

TAX CONSEQUENCES 269

INFLATION 270
Effect of Inflation on Rate of Return

RISK ANALYSIS 272
Probability ■ Risk ■ Expected Value

REFERENCE 274

INTEREST TABLES 275

This is a review of the field known variously as *engineering economics*, *engineering economy*, or *engineering economic analysis*. Since engineering economics is straightforward and logical, even people who have not had a formal course should be able to gain sufficient knowledge from this chapter to successfully solve most engineering economics problems.

Most problems require basic mathematics, including raising a number to a power, which are easily done on a financial or engineering calculator.

Because this material is widely used by financial persons, we will often use mathematical notation (such as the "functional notation" of Table 17.1) that is more common to those professionals. This should pose no problem to the engineer.

Economics problems involving more complex situations (such as comparing alternatives) usually involve performing several simple steps in sequence. For some examples of this, see the section Solving More Advanced Economics Problems.

There are 35 example problems scattered throughout the chapter. These examples are an integral part of the review and should be examined as you come to them.

The field of engineering economics uses mathematical and economics techniques to systematically analyze situations that pose alternative courses of action. The initial step in engineering economics problems is to resolve a situation, or each alternative in a given situation, into its favorable and unfavorable consequences or factors. These are then measured in some common unit—usually money. Factors that cannot readily be equated to money are called intangible or irreducible factors. Such factors are considered in conjunction with the monetary analysis when making the final decision on proposed courses of action.

A set of tables is presented at the end of the chapter. Historically, such tables were used to solve problems of the type found in this chapter. However, use of the relevant equations (as presented here) is straightforward for engineers using a scientific calculator. Business calculators that have these equations already programmed are also readily available. Because the correct use of these business calculators is greatly dependent upon knowledge of the particular calculator, we encourage you to work the example problems and compare the results if you use one of these business calculators. With any calculator, it is a good idea (as we have pointed out earlier) to become very conversant with the machine through problem solving well before the test.

Note that the calculator solution of these problems may yield slightly different results than use of the tables.

CASH FLOW

A cash flow table shows the "money consequences" of a situation and its timing. For example, a simple problem might be to list the year-by-year consequences of purchasing and owning a used car:

Year	Cash Flow	
Beginning of first year 0	−$4500	Car purchased "now" for $4500 cash. (The minus sign indicates a disbursement.)
End of year 1	−350	
End of year 2	−350	
End of year 3	−350	Maintenance costs are $350 per year.
	−350	
End of year 4	−350	This car is sold at the end of the fourth year for $2000. (The plus sign represents the receipt of money.)
	+2000	

This same cash flow may be represented graphically, as shown in Figure 17.1. The upward arrow represents a receipt of money, and the downward arrows represent disbursements. The horizontal axis represents the passage of time

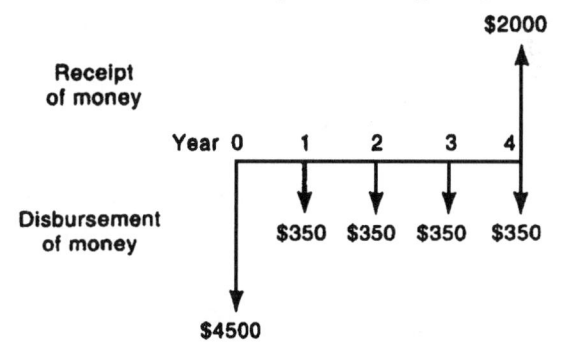

Figure 17.1 Cash flow diagram

Example 17.1

In January 2003, a firm purchased a used copier for $500. Repairs cost nothing in 2003 or 2004. Repairs are $85 in 2005, $130 in 2006, and $140 in 2007. The machine is sold in 2007 for $300. Complete the cash flow table.

Solution

Unless otherwise stated, the customary assumption is a beginning-of-year purchase, followed by end-of-year receipts or disbursements, and an end-of-year resale or salvage value. Thus the copier repairs and the copier sale are assumed to occur at the end of the year. Letting a minus sign represent a disbursement of money and a plus sign a receipt of money, we are able to set up the cash flow table:

Year	Cash Flow
Beginning of 2003	−$500
End of 2003	0
End of 2004	0
End of 2005	−85
End of 2006	−130
End of 2007	+160

Notice that at the end of 2007, the cash flow table shows +160, which is the net sum of −140 and +300. If we define year 0 as the beginning of 2003, the cash flow table becomes:

Year	Cash Flow
0	−$500
1	0
2	0
3	−85
4	−130
5	+160

From this cash flow table, the definitions of year 0 and year 1 become clear. Year 0 is defined as the *beginning* of year 1. Year 1 is the *end* of year 1, and so forth.

TIME VALUE OF MONEY

When the money consequences of an alternative occur in a short period of time—say, less than one year—we might simply add up the various sums of money and obtain the net result. But we cannot treat money this way over longer periods of time. This is because money today does not have the same value as money at some future time.

Consider this question: Which would you prefer, $100 today or the assurance of receiving $100 a year from now? Clearly, you would prefer the $100 today. If you had the money today, rather than a year from now, you could use it for the year. And if you had no use for it, you could lend it to someone who would pay interest for the privilege of using your money for the year.

Simple Interest

Simple interest is interest that is computed on the original sum. Thus if one were to lend a present sum P to someone at a simple annual interest rate i, the future amount F due at the end of n years would be

$$F = P + Pin.$$

Example 17.2

How much will you receive back from a $500 loan to a friend for three years at 10 percent simple annual interest?

Solution

$$F = P + Pin = 500 + 500 \times 0.10 \times 3 = \$650$$

In Example 17.2 one observes that the amount owed, based on 10 percent simple interest at the end of one year, is $500 + 500 \times 0.10 \times 1 = \$550$. But at simple interest there is no interest charged on the $50 interest, even though it is not paid until the end of the third year. Thus simple interest is not realistic and is seldom used. *Compound interest* charges interest on the principal owed plus the interest earned to date. This produces a charge of interest on interest, or compound interest. Engineering economics uses compound interest computations.

EQUIVALENCE

In the preceding section we saw that money at different points in time (for example, $100 today or $100 one year hence) may be equal in the sense that they both are $100, but $100 a year hence is *not* an acceptable substitute for $100 today. When we have acceptable substitutes, we say they are *equivalent* to each other. Thus at 8 percent interest, $108 a year hence is equivalent to $100 today.

Example 17.3

At a 10 percent per year (compound) interest rate, $500 now is *equivalent* to how much three years hence?

Solution

A value of $500 now will increase by 10 percent in each of the three years.

$$\begin{aligned}\text{Now} &= \$500.00\\ \text{End of 1st year} &= 500 + 10\%(500) = 550.00\\ \text{End of 2nd year} &= 550 + 10\%(550) = 605.00\\ \text{End of 3rd year} &= 605 + 10\%(605) = 665.50\end{aligned}$$

Thus $500 now is *equivalent* to $665.50 at the end of three years. Note that interest is charged each year on the original $500 plus the unpaid interest. This compound interest computation gives an answer that is $15.50 higher than the simple-interest computation in Example 17.2.

Equivalence is an essential factor in engineering economics. Suppose we wish to select the better of two alternatives. First, we must compute their cash flows. For example,

Year	Alternative A	Alternative B
0	−$2000	−$2800
1	+800	+1100
2	+800	+1100
3	+800	+1100

The larger investment in alternative *B* results in larger subsequent benefits, but we have no direct way of knowing whether it is better than alternative *A*. So we do not know which to select. To make a decision, we must resolve the alternatives into *equivalent* sums so that they may be compared accurately.

COMPOUND INTEREST

To facilitate equivalence computations, a series of compound interest factors will be derived here, and their use will be illustrated in examples.

Symbols and Functional Notation

i = effective interest rate per interest period. In equations, the interest rate is stated as a decimal (that is, 8 percent interest is 0.08).

n = number of interest periods. Usually, the interest period is one year, in which case n would be number of years.

P = a present sum of money.

F = a future sum of money. The future sum F is an amount n interest periods from the present that is equivalent to P at interest rate i.

A = an end-of-period cash receipt or disbursement (annuity) in a uniform series continuing for n periods. The entire series is equivalent to P or F at interest rate i.

G = uniform period-by-period increase in cash flows; the uniform gradient.

r = nominal annual interest rate.

Table 17.1 Periodic compounding: Functional notation and formulas

Factor	Given	To Find	Functional Notation	Formula
Single payment				
Compound amount factor	P	F	$(F/P, i\%, n)$	$F = P(1+i)^n$
Present worth factor	F	P	$(P/F, i\%, n)$	$P = F(1+i)^{-n}$
Uniform payment series				
Sinking fund factor	F	A	$(A/F, i\%, n)$	$A = F\left[\dfrac{i}{(1+i)^n - 1}\right]$
Capital recovery factor	P	A	$(A/P, i\%, n)$	$A = P\left[\dfrac{i(1+i)^n}{(1+i)^n - 1}\right]$
Compound amount factor	A	F	$(F/A, i\%, n)$	$F = A\left[\dfrac{(1+i)^n - 1}{i}\right]$
Present worth factor	A	P	$(P/A, i\%, n)$	$P = A\left[\dfrac{(1+i)^n - 1}{i(1+i)^n}\right]$
Uniform gradient				
Gradient present worth	G	P	$(P/G, i\%, n)$	$P = G\left[\dfrac{(1+i)^n - 1}{i^2(1+i)^n} - \dfrac{n}{i(1+i)^n}\right]$
Gradient future worth	G	F	$(F/G, i\%, n)$	$F = G\left[\dfrac{(1+i)^n - 1}{i^2} - \dfrac{n}{i}\right]$
Gradient uniform series	G	A	$(A/G, i\%, n)$	$A = G\left[\dfrac{1}{i} - \dfrac{n}{(1+i)^n - 1}\right]$

From Table 17.1 we can see that the functional notation scheme is based on writing (to find/given, i, n). Thus, if we wished to find the future sum F, given a uniform series of receipts A, the proper compound interest factor to use would be $(F/A, i, n)$.

Single-Payment Formulas

Suppose a present sum of money P is invested for one year at interest rate i. At the end of the year, the initial investment P is received together with interest equal to Pi, or a total amount $P + Pi$. Factoring P, the sum at the end of one year is

$P(1 + i)$. If the investment is allowed to remain for subsequent years, the progression is as follows:

Amount at Beginning of the Period	+	Interest for the Period	=	Amount at End of the Period
1st year, P	+	Pi	=	$P(1+i)$
2nd year, $P(1+i)$	+	$Pi(1+i)$	=	$P(1+i)^2$
3rd year, $P(1+i)^2$	+	$Pi(1+i)^2$	=	$P(1+i)^3$
nth year, $P(1+i)^{n-1}$	+	$Pi(1+i)^{n-1}$	=	$P(1+i)^n$

The present sum P increases in n periods to $P(1 + i)^n$. This gives a relation between a present sum P and its equivalent future sum F:

$$\text{Future sum} = (\text{present sum})(1 + i)^n$$
$$F = P(1 + i)^n.$$

This is the *single-payment compound amount formula*. In functional notation it is written

$$F = P(F/P, i, n).$$

The relationship may be rewritten as

$$\text{Present sum} = (\text{Future sum})(1 + i)^{-n}$$
$$P = F(1 + i)^{-n}.$$

This is the *single-payment present worth formula*. It is written

$$P = F(P/F, i, n).$$

Example 17.4

At a 10 percent per year interest rate, $500 now is *equivalent* to how much three years hence?

Solution

This problem was solved in Example 17.3. Now it can be solved using a single-payment formula. $P = \$500$, $n = 3$ years, $i = 10$ percent, and $F =$ unknown:

$$F = P(1 + i)^n = 500(1 + 0.10)^3 = \$665.50.$$

This problem also may be solved using a compound interest table:

$$F = P(F/P, i, n) = 500(F/P, 10\%, 3).$$

From the 10 percent compound interest table, read $(F/P, 10\%, 3) = 1.331$.

$$F = 500(F/P, 10\%, 3) = 500(1.331) = \$665.50.$$

Example 17.5

To raise money for a new business, small startup company asks you to lend it some money. The entrepreneur offers to pay you $3000 at the end of four years. How much should you give the company now if you want to realize 12 percent interest per year?

Solution

P = unknown, F = $3000, n = 4 years, and i = 12 percent:

$$P = F(1 + i)^{-n} = 3000(1 + 0.12)^{-4} = \$1906.55.$$

Alternative computation using a compound interest table:

$$P = F(P/F, i, n) = 3000(P/F, 12\%, 4) = 3000(0.6355) = \$1906.50.$$

Note that the solution based on the compound interest table is slightly different from the exact solution using a hand-held calculator. In engineering economics, the compound interest tables are considered to be sufficiently accurate.

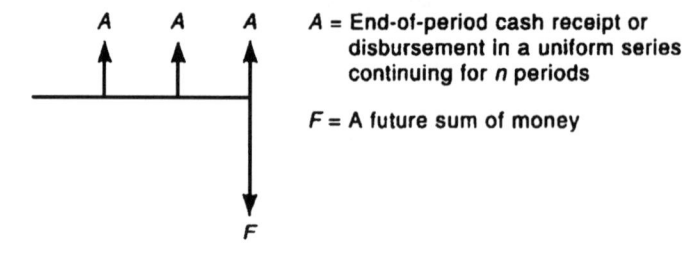

A = End-of-period cash receipt or disbursement in a uniform series continuing for n periods

F = A future sum of money

Figure 17.2 Cash flow diagram—uniform payment series

Uniform Payment Series Formulas

Consider the situation shown in Figure 17.2. Using the single-payment compound amount factor, we can write an equation for F in terms of A:

$$F = A + A(1 + i) + A(1 + i)^2. \qquad \text{(i)}$$

In this situation, with $n = 3$, Eq. (i) may be written in a more general form:

$$F = A + A(1 + i) + A(1 + i)^{n-1} \qquad \text{(ii)}$$

Multiply Eq. (ii) by $(1 + i)$ $\quad (1 + i)F = A(1 + i) + A(1 + i)^{n-1} + A(1 + i)^n \qquad \text{(iii)}$

Subtract Eq. (ii) yields: $\quad iF = -A + A(1 + i)^n.$

This produces the *uniform series compound amount formula*:

$$F = A\left[\frac{(1+i)^n - 1}{i}\right]$$

Solving this equation for A produces the *uniform series sinking fund formula*:

$$A = F\left[\frac{i}{(1+i)^n - 1}\right]$$

Since $F = P(1 + i)^n$, we can substitute this expression for F in the equation and obtain the *uniform series capital recovery formula*:

$$A = P\left[\frac{i(1+i)^n}{(1+i)^n - 1}\right]$$

Solving the equation for P produces the *uniform series present worth formula*:

$$P = A\left[\frac{(1+i)^n - 1}{i(1+i)^n}\right]$$

In functional notation, the uniform series factors are:

Compound amount (*F/A, i, n*)

Sinking fund (*A/F, i, n*)

Capital recovery (*A/P, i, n*)

Present worth (*P/A, i, n*)

Example 17.6

If $100 is deposited at the end of each year in a savings account that pays 6 percent interest per year, how much will be in the account at the end of five years?

Solution

$A = \$100$, F = unknown, $n = 5$ years, and $i = 6$ percent:
$$F = A(F/A, i, n) = 100(F/A, 6\%, 5) = 100(5.637) = \$563.70$$

Example 17.7

A fund established to produce a desired amount at the end of a given period, by means of a series of payments throughout the period, is called a *sinking fund*. A sinking fund is to be established to accumulate money to replace a $10,000 machine. If the machine is to be replaced at the end of 12 years, how much should be deposited in the sinking fund each year? Assume the fund earns 10 percent annual interest.

Solution

Annual sinking fund deposit $A = 10,000(A/F, 10\%, 12)$
$$= 10,000(0.0468) = \$468.$$

Example 17.8

An individual is considering the purchase of a used automobile. The total price is $6200. With $1240 as a down payment, and the balance paid in 48 equal monthly payments with interest at 1 percent per month, compute the monthly payment. The payments are due at the end of each month.

Solution

The amount to be repaid by the 48 monthly payments is the cost of the automobile *minus* the $1240 downpayment.

$P = \$4960$, A = unknown, $n = 48$ monthly payments, and $i = 1$ percent per month:
$$A = P(A/P, 1\%, 48) = 4960(0.0263) = \$130.45$$

Example 17.9

A couple sell their home. In addition to cash, they take a mortgage on the house. The mortgage will be paid off by monthly payments of $450 for 50 months. The couple decides to sell the mortgage to a local bank. The bank will buy the mortgage, but it requires a 1 percent per month interest rate on its investment. How much will the bank pay for the mortgage?

Solution

$A = \$450$, $n = 50$ months, $i = 1$ percent per month, and P = unknown:

$$P = A(P/A, i, n) = 450(P/A, 1\%, 50) = 450(39.196) = \$17{,}638.20$$

Uniform Gradient

At times, one will encounter a situation where the cash flow series is not a constant amount A; instead, it is an increasing series. The cash flow shown in Figure 17.3 may be resolved into two components (Figure 17.4). We can compute the value of P^* as equal to P' plus P, and we already have the equation for P': $P' = A(P/A, i, n)$.

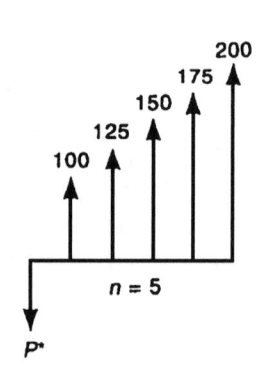

Figure 17.3 Cash flow diagram—uniform gradient

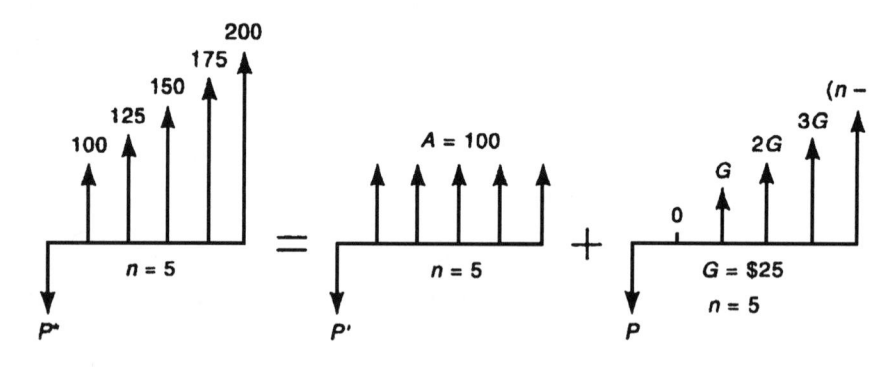

Figure 17.4 Uniform gradient diagram resolved

The value for P in the right-hand diagram is

$$P = G\left[\frac{(1+i)^n - 1}{i^2(1+i)^n} - \frac{n}{i(1+i)^n}\right]$$

This is the *uniform gradient present worth formula*. In functional notation, the relationship is $P = G(P/G, i, n)$.

Example 17.10

The maintenance on a machine is expected to be $155 at the end of the first year, and it is expected to increase $35 each year for the following seven years (Exhibit 1). What sum of money should be set aside now to pay the maintenance for the eight-year period? Assume 6 percent interest.

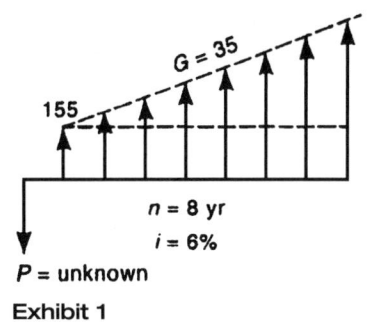

Exhibit 1

Solution

$$P = 155(P/A, 6\%, 8) + 35(P/G, 6\%, 8)$$
$$= 155(6.210) + 35(19.841) = \$1656.99.$$

In the gradient series, if—instead of the present sum, P—an equivalent uniform series A is desired, the problem might appear as shown in Figure 17.5. The relationship between A' and G in the right-hand diagram is

$$A' = G\left[\frac{1}{i} - \frac{n}{(1+i)^n - 1}\right].$$

In functional notation, the uniform gradient (to) uniform series factor is: $A' = G(A/G, i, n)$.

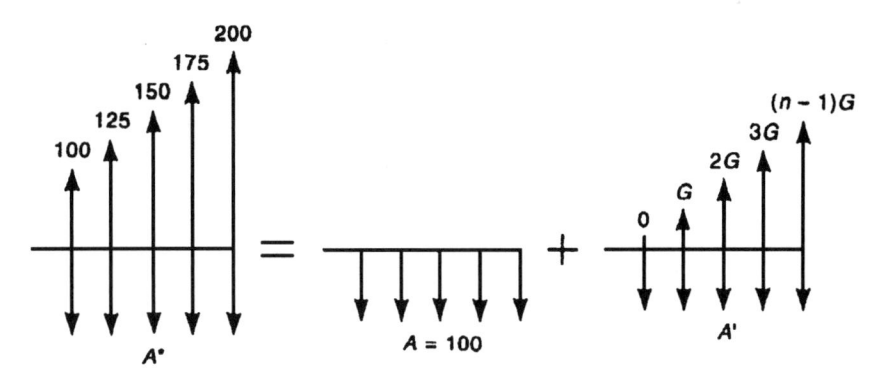

Figure 17.5 Uniform series, uniform gradient cash flow diagram

The uniform gradient uniform series factor may be read from the compound interest tables directly, or computed as

$$(A/G, i, n) = \frac{1 - n(A/F, i, n)}{i}.$$

Note carefully the diagrams for the uniform gradient factors. The first term in the uniform gradient is zero and the last term is $(n-1)G$. But we use n in the equations and function notation. The derivations (not shown here) were done on this basis, and the uniform gradient compound interest tables are computed this way.

Example 17.11

For the situation in Example 17.10, we wish now to know the uniform annual maintenance cost. Compute an equivalent A for the maintenance costs.

Solution

Refer to Exhibit 2. The equivalent uniform annual maintenance cost is

$$A = 155 + 35(A/G, 6\%, 8) = 155 + 35(3.195) = \$266.83.$$

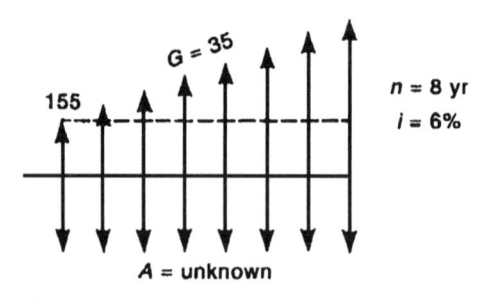

Exhibit 2

Standard compound interest tables give values for eight interest factors: two single payments, four uniform payment series, and two uniform gradients. The tables do *not* give the uniform gradient future worth factor, $(F/G, i, n)$. If it is needed, it may be computed from two tabulated factors:

$$(F/G, i, n) = (P/G, i, n)(F/P, i, n).$$

For example, if $i = 10$ percent and $n = 12$ years, then $(F/G, 10\%, 12) = (P/G, 10\%, 12)(F/P, 10\%, 12) = (29.901)(3.138) = 93.83$.

A second method of computing the uniform gradient future worth factor is

$$(F/G, i, n) = \frac{(F/G, i, n) - n}{i}.$$

Using this equation for $i = 10$ percent and $n = 12$ years, $(F/G, 10\%, 12) = [(F/A, 10\%, 12) - 12]/0.10 = (21.384 - 12)/0.10 = 93.84$.

NOMINAL AND EFFECTIVE INTEREST

Nominal interest is the annual interest rate without considering the effect of any compounding. *Effective interest* is the annual interest rate taking into account the effect of any compounding during the year.

Non-Annual Compounding

Frequently an interest rate is described as an annual rate, even though the interest period may be something other than one year. A bank may pay 1 percent interest on the amount in a savings account every three months. The *nominal* interest rate in this situation is $4 \times 1\% = 4\%$. But if you deposited $1000 in such an account, would you have $104\%(1000) = \$1040$ in the account at the end of one year? The answer is no, you would have more. The amount in the account would increase as follows.

Amount in Account

Beginning of year:	1000.00
End of three months:	1000.00 + 1%(1000.00) = 1010.00
End of six months:	1010.00 + 1%(1010.00) = 1020.10
End of nine months:	1020.10 + 1%(1020.10) = 1030.30
End of one year:	1030.30 + 1%(1030.30) = 1040.60

At the end of one year, the interest of $40.60, divided by the original $1000, gives a rate of 4.06 percent. This is the *effective* interest rate.

$$\text{Effective interest rate per year:} \quad i_{\text{eff}} = (1 + r/m)^m - 1$$

where r = nominal annual interest rate
m = number of compound periods per year
r/m = effective interest rate per period.

Example 17.12

A bank charges 1.5 percent interest per month on the unpaid balance for purchases made on its credit card. What nominal interest rate is it charging? What is the effective interest rate?

Solution

The nominal interest rate is simply the annual interest ignoring compounding, or $12(1.5\%) = 18\%$.

$$\text{Effective interest rate} = (1 + 0.015)^{12} - 1 = 0.1956 = 19.56\%$$

Continuous Compounding

When m, the number of compound periods per year, becomes very large and approaches infinity, the duration of the interest period decreases from Δt to dt. For this condition of *continuous compounding*, the effective interest rate per year is

$$i_{\text{eff}} = e^r - 1$$

where r = nominal annual interest rate.

Table 17.2 Continuous compounding: Functional notation and formulas

Factor	Given	To Find	Functional Notation	Formula
Single payment				
Compound amount factor	P	F	(F/P, r %, n)	$F = P[e^{rn}]$
Present worth factor	F	P	(P/F, r %, n)	$P = F[e^{-rn}]$
Uniform payment series				
Sinking fund factor	F	A	(A/F, r %, n)	$A = F\left[\dfrac{e^r - 1}{e^{rn} - 1}\right]$
Capital recovery factor	P	A	(A/P, r %, n)	$A = P\left[\dfrac{e^r - 1}{1 - e^{-rn}}\right]$

Factor	Given	To Find	Functional Notation	Formula
Compound amount factor	A	F	(F/A, r %, n)	$F = A\left[\dfrac{e^{rn}-1}{e^{r}-1}\right]$
Present worth factor	A	P	(P/A, r %, n)	$P = A\left[\dfrac{1-e^{-rn}}{e^{r}-1}\right]$

r = nominal annual interest rate, n = number of years.

Example 17.13

Five hundred dollars is deposited each year into a savings bank account that pays 5 percent nominal interest, compounded continuously. How much will be in the account at the end of five years?

Solution

$A = \$500$, $r = 0.05$, $n = 5$ years.

$$F = A(F/A, r\%, n) = A\left[\dfrac{e^{rn}-1}{e^{r}-1}\right] = 500\left[\dfrac{e^{0.05(5)}-1}{e^{0.05}-1}\right] = \$2769.84$$

Example 17.14

If the bank in Example 17.12 changes its policy and charges 1.5 percent per month, compounded continuously, what nominal and what effective interest rate is it charging?

Solution

Nominal annual interest rate, $r = 12 \times 1.5\% = 18\%$

Effective interest rate per year, $i_{\text{eff}} = e^{0.18} - 1 = 0.1972 = 19.72\%$

SOLVING ENGINEERING ECONOMICS PROBLEMS

The techniques presented so far illustrate how to convert single amounts of money, and uniform or gradient series of money, into some equivalent sum at another point in time. These compound interest computations are an essential part of engineering economics problems.

The typical situation is that we have a number of alternatives; the question is, which alternative should we select? The customary method of solution is to express each alternative in some common form and then choose the best, taking both the monetary and intangible factors into account. In most computations an interest rate must be used. It is often called the minimum attractive rate of return (MARR), to indicate that this is the smallest interest rate, or rate of return, at which one is willing to invest money.

Criteria

Engineering economics problems inevitably fall into one of three categories:

1. *Fixed input.* The amount of money or other input resources is fixed.
   Example: A project engineer has a budget of $450,000 to overhaul a plant.

2. *Fixed output.* There is a fixed task or other output to be accomplished. *Example*: A mechanical contractor has been awarded a fixed-price contract to air-condition a building.

3. *Neither input nor output fixed.* This is the general situation, where neither the amount of money (or other inputs) nor the amount of benefits (or other outputs) is fixed. *Example*: A consulting engineering firm has more work available than it can handle. It is considering paying the staff to work evenings to increase the amount of design work it can perform.

There are five major methods of comparing alternatives: present worth, future worth, annual cost, rate of return, and benefit-cost analysis. These are presented in the sections that follow.

PRESENT WORTH

Present worth analysis converts all of the money consequences of an alternative into an equivalent present sum. The criteria are:

Category	Present Worth Criterion
Fixed input	Maximize the present worth of benefits or other outputs
Fixed output	Minimize the present worth of costs or other inputs
Neither input nor output fixed	Maximize present worth of benefits minus present worth of costs, or maximize net present worth

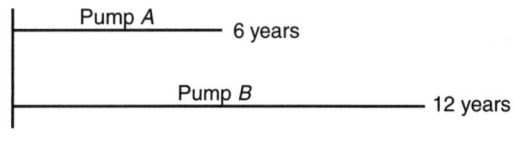

Figure 17.6 Improper present worth comparison

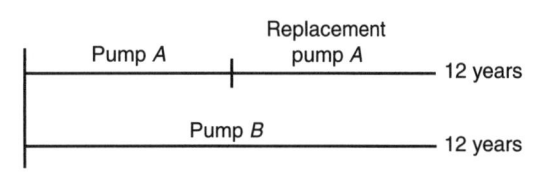

Figure 17.7 Proper present worth comparison

Appropriate Problems

Present worth analysis is most frequently used to determine the present value of future money receipts and disbursements. We might want to know, for example, the present worth of an income-producing property, such as an oil well. This should provide an estimate of the price at which the property could be bought or sold.

An important restriction in the use of present worth calculation is that there must be a common analysis period for comparing alternatives. It would be incorrect, for example, to compare the present worth (PW) of cost of pump *A*, expected to last 6 years, with the PW of cost of pump *B*, expected to last 12 years

(Figure 17.6). In situations like this, the solution is either to use some other analysis technique (generally, the annual cost method is suitable in these situations) or to restructure the problem so that there is a common analysis period.

In this example, a customary assumption would be that a pump is needed for 12 years and that pump A will be replaced by an identical pump A at the end of 6 years. This gives a 12-year common analysis period (Figure 17.7). This approach is easy to use when the different lives of the alternatives have a practical least-common-multiple life. When this is not true (for example, the life of *J* equals 7 years and the life of *K* equals 11 years), some assumptions must be made to select a suitable common analysis period, or the present worth method should not be used.

Example 17.15

Machine *X* has an initial cost of $10,000, an annual maintenance cost of $500 per year, and no salvage value at the end of its 4-year useful life. Machine *Y* costs $20,000, and the first year there is no maintenance cost. Maintenance is $100 the second year, and it increases $100 per year thereafter. The machine has an anticipated $5000 salvage value at the end of its 12-year useful life. If the minimum attractive rate of return (MARR) is 8 percent, which machine should be selected?

Solution

The analysis period is not stated in the problem. Therefore, we select the least common multiple of the lives, or 12 years, as the analysis period.

Present worth of cost of 12 years of machine *X*:

$$PW = 10{,}000 + 10{,}000(P/F, 8\%, 4) + 10{,}000(P/F, 8\%, 8) + 500(P/A, 8\%, 12)$$
$$= 10{,}000 + 10{,}000(0.7350) + 10{,}000(0.5403) + 500(7.536) = \$26{,}521.$$

Present worth of cost of 12 years of machine *Y*:

$$PW = 20{,}000 + 100(P/G, 8\%, 12) - 5000(P/F, 8\%, 12)$$
$$= 20{,}000 + 100(34.634) - 5000(0.3971) = \$21{,}478.$$

Choose machine *Y*, with its smaller PW of cost.

Example 17.16

Two alternatives have the following cash flows:

	Alternative	
Year	A	B
0	−$2000	−$2800
1	+800	+1100
2	+800	+1100
3	+800	+1100

At a 4 percent interest rate, which alternative should be selected?

Solution

The net present worth of each alternative is computed:

Net present worth (NPW) = PW of benefit − PW of cost
NPW_A = 800(P/A, 4%, 3) − 2000 = 800(2.775) − 2000 = $220.00
NPW_B = 1100(P/A, 4%, 3) − 2800 = 1100(2.775) − 2800 = $252.50.

To maximize NPW, choose alternative B.

Infinite Life and Capitalized Cost

In the special situation where the analysis period is infinite ($n = \infty$), an analysis of the present worth of cost is called *capitalized cost*. There are a few public projects where the analysis period is infinity. Other examples are permanent endowments and cemetery perpetual care.

When n equals infinity, a present sum P will accrue interest of Pi for every future interest period. For the principal sum P to continue undiminished (an essential requirement for n equal to infinity), the end-of-period sum A that can be disbursed is Pi (Figure 17.8). When $n = \infty$, the fundamental relationship is

$$A = Pi.$$

Some form of this equation is used whenever there is a problem involving an infinite analysis period.

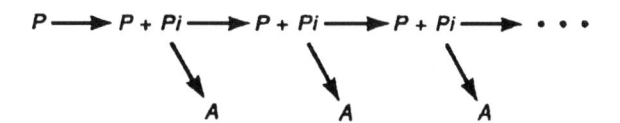

Figure 17.8 Infinite life, capitalized cost diagram

Example 17.17

In his will, a man wishes to establish a perpetual trust to provide for the maintenance of a small local park. If the annual maintenance is $7500 per year and the trust account can earn 5 percent interest, how much money must be set aside in the trust?

Solution

When $n = \infty$, $A = Pi$ or $P = A/i$. The capitalized cost is $P = A/i$ = $7500/0.05 = $150,000.

FUTURE WORTH OR VALUE

In present worth analysis, the comparison is made in terms of the equivalent present costs and benefits. But the analysis need not be made in terms of the present—it can be made in terms of a past, present, or future time. Although the numerical calculations may look different, the decision is unaffected by the selected point in time. Often we do want to know what the future situation will be if we take some particular course of action now. An analysis based on some future point in time is called *future worth analysis*.

Category	Future Worth Criterion
Fixed input	Maximize the future worth of benefits or other outputs
Fixed output	Minimize the future worth of costs or other inputs
Neither input nor output fixed	Maximize future worth of benefits minus future worth of costs, or maximize net future worth

Example 17.18

Two alternatives have the following cash flows:

	Alternative	
Year	A	B
0	−$2000	−$2800
1	+800	+1100
2	+800	+1100
3	+800	+1100

At a 4 percent interest rate, which alternative should be selected?

Solution

In Example 17.16, this problem was solved by present worth analysis at year 0. Here it will be solved by future worth analysis at the end of year 3.

Net future worth (NFW) = FW of benefits − FW of cost

$$NFW_A = 800(F/A, 4\%, 3) - 2000(F/P, 4\%, 3)$$
$$= 800(3.122) - 2000(1.125) = +\$247.60.$$

$$NFW_B = 1100(F/A, 4\%, 3) - 2800(F/P, 4\%, 3)$$
$$= 1100(3.122) - 2800(1.125) = +\$284.20.$$

To maximize NFW, choose alternative B.

ANNUAL COST

The annual cost method is more accurately described as the method of equivalent uniform annual cost (EUAC). Where the computation is of benefits, it is called the method of equivalent uniform annual benefits (EUAB).

Criteria

For each of the three possible categories of problems, there is an annual cost criterion for economic efficiency.

Category	Annual Cost Criterion
Fixed input	Maximize the equivalent uniform annual benefits (EUAB)
Fixed output	Minimize the equivalent uniform annual cost (EUAC)
Neither input nor output fixed	Maximize EUAB − EUAC

Application of Annual Cost Analysis

In the section on present worth, we pointed out that the present worth method requires a common analysis period for all alternatives. This restriction does not apply in all annual cost calculations, but it is important to understand the circumstances that justify comparing alternatives with different service lives.

Frequently, an analysis is done to provide for a more-or-less continuing requirement. For example, one might need to pump water from a well on a continuing basis. Regardless of whether each of two pumps has a useful service life of 6 years or 12 years, we would select the alternative whose annual cost is a minimum. And this still would be the case if the pumps' useful lives were the more troublesome 7 and 11 years. Thus, if we can assume a continuing need for an item, an annual cost comparison among alternatives of differing service lives is valid. The underlying assumption in these situations is that the shorter-lived alternative can be replaced with an identical item with identical costs, when it has reached the end of its useful life. This means that the EUAC of the initial alternative is equal to the EUAC for the continuing series of replacements.

On the other hand, if there is a specific requirement to pump water for 10 years, then each pump must be evaluated to see what costs will be incurred during the analysis period and what salvage value, if any, may be recovered at the end of the analysis period. The annual cost comparison needs to consider the actual circumstances of the situation.

Examination problems are often readily solved using the annual cost method. And the underlying "continuing requirement" is usually present, so an annual cost comparison of unequal-lived alternatives is an appropriate method of analysis.

Example 17.19

Consider the following alternatives:

	A	B
First cost	$5000	$10,000
Annual maintenance	500	200
End-of-useful-life salvage value	600	1000
Useful life	5 years	15 years

Based on an 8 percent interest rate, which alternative should be selected?

Solution

Assuming both alternatives perform the same task and there is a continuing requirement, the goal is to minimize EUAC.

Alternative A:

$$\text{EUAC} = 5000(A/P, 8\%, 5) + 500 - 600(A/F, 8\%, 5)$$
$$= 5000(0.2505) + 500 - 600(0.1705) = \$1650.$$

Alternative B:

$$\text{EUAC} = 10{,}000(A/P, 8\%, 15) + 200 - 1000(A/F, 8\%, 15)$$
$$= 10{,}000(0.1168) + 200 - 1000(0.0368) = \$1331.$$

To minimize EUAC, select alternative B.

RATE OF RETURN ANALYSIS

A typical situation is a cash flow representing the costs and benefits. The rate of return may be defined as the interest rate where PW of cost = PW of benefits, EUAC = EUAB, or PW of cost – PW of benefits = 0.

Example 17.20

Compute the rate of return for the investment represented by the following cash flow table.

Year:	0	1	2	3	4	5
Cash flow:	–$595	+250	+200	+150	+100	+50

Solution

This declining uniform gradient series may be separated into two cash flows (Exhibit 3) for which compound interest factors are available.

Note that the gradient series factors are based on an *increasing* gradient. Here the declining cash flow is solved by subtracting an increasing uniform gradient, as indicated in the figure.

PW of cost – PW of benefits = 0

$$595 - [250(P/A, i, 5)] - 50(P/G, i, 5) = 0$$

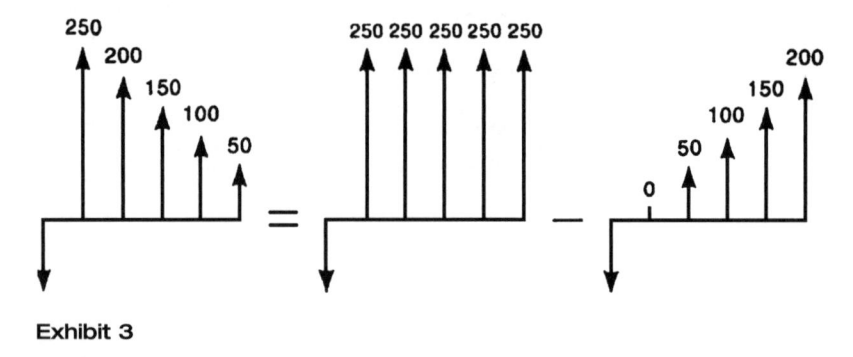

Exhibit 3

Try $i = 10\%$:

$$595 - [250(3.791) - 50(6.862)] = -9.65$$

Try $i = 12\%$:

$$595 - [250(3.605) - 50(6.397)] = +13.60$$

The rate of return is between 10 percent and 12 percent. It may be computed more accurately by linear interpolation:

$$\text{Rate of return} = 10\% + (2\%)\left(\frac{9.65 - 0}{13.60 + 9.65}\right) = 10.83\%.$$

Two Alternatives

Compute the incremental rate of return on the cash flow representing the difference between the two alternatives. Since we want to look at increments of *investment*, the cash flow for the difference between the alternatives is computed by taking the higher initial-cost alternative minus the lower initial-cost alternative. If the incremental rate of return is greater than or equal to the predetermined minimum attractive rate of return (MARR), choose the higher-cost alternative; otherwise, choose the lower-cost alternative.

Example 17.21

Two alternatives have the following cash flows:

Year	Alternative A	Alternative B
0	−$2000	−$2800
1	+800	+1100
2	+800	+1100
3	+800	+1100

If 4 percent is considered the minimum attractive rate of return (MARR), which alternative should be selected?

Solution

These two alternatives were previously examined in Examples 17.16 and 17.18 by present worth and future worth analysis. This time, the alternatives will be resolved using a rate-of-return analysis.

Note that the problem statement specifies a 4 percent MARR, whereas Examples 17.16 and 17.18 referred to a 4 percent interest rate. These are really two different ways of saying the same thing: The minimum acceptable time value of money is 4 percent.

First, tabulate the cash flow that represents the increment of investment between the alternatives. This is done by taking the higher initial-cost alternative minus the lower initial-cost alternative:

Year	Alternative A	Alternative B	Difference Between Alternatives B − A
0	−$2000	−$2800	−$800
1	+800	+1100	+300
2	+800	+1100	+300
3	+800	+1100	+300

Then compute the rate of return on the increment of investment represented by the difference between the alternatives:

$$\text{PW of cost} = \text{PW of benefits}$$
$$800 = 300(P/A, i, 3)$$
$$(P/A, i, 3) = 800/300 = 2.67$$
$$i = 6.1\%.$$

Since the incremental rate of return exceeds the 4 percent MARR, the increment of investment is desirable. Choose the higher-cost alternative B.

Before leaving this example, one should note something that relates to the rates of return on alternative A and on alternative B. These rates of return, if calculated, are:

	Rate of Return
Alternative A	9.7%
Alternative B	8.7%

The correct answer to this problem has been shown to be alternative B, even though alternative A has a higher rate of return. The higher-cost alternative may be thought of as the lower-cost alternative plus the increment of investment between them. Viewed this way, the higher-cost alternative B is equal to the desirable lower-cost alternative A plus the difference between the alternatives.

The important conclusion is that computing the rate of return for each alternative does *not* provide the basis for choosing between alternatives. Instead, incremental analysis is required.

Example 17.22

Consider the following:

	Alternative	
Year	A	B
0	−$200.0	−$131.0
1	+77.6	+48.1
2	+77.6	+48.1
3	+77.6	+48.1

If the MARR is 10 percent, which alternative should be selected?

Solution

To examine the increment of investment between the alternatives, we will examine the higher initial-cost alternative minus the lower initial-cost alternative, or A − B.

	Alternative		Increment
Year	A	B	A − B
0	−$200.0	−$131.0	−$69.0
1	+77.6	+48.1	+29.5
2	+77.6	+48.1	+29.5
3	+77.6	+48.1	+29.5

Solve for the incremental rate of return:

$$\text{PW of cost} = \text{PW of benefits}$$
$$69.0 = 29.5(P/A, i, 3)$$
$$(P/A, i, 3) = 69.0/29.5 = 2.339.$$

From compound interest tables, the incremental rate of return is between 12 percent and 18 percent. This is a desirable increment of investment; hence we select the higher-initial-cost alternative A.

Three or More Alternatives

When there are three or more mutually exclusive alternatives, proceed with the same logic presented for two alternatives. The components of incremental analysis are listed below.

Step 1. Compute the rate of return for each alternative. Reject any alternative where the rate of return is less than the desired MARR. (This step is not essential, but helps to immediately identify unacceptable alternatives.)

Step 2. Rank the remaining alternatives in order of increasing initial cost.

Step 3. Examine the increment of investment between the two lowest-cost alternatives as described for the two-alternative problem. Select the better of the two alternatives and reject the other one.

Step 4. Take the preferred alternative from step 3. Consider the next higher initial-cost alternative, and proceed with another two-alternative comparison.

Step 5. Continue until all alternatives have been examined and the best of the multiple alternatives has been identified.

Example 17.23

Consider the following:

	Alternative	
Year	A	B
0	−$200.0	−$131.0
1	+77.6	+48.1
2	+77.6	+48.1
3	+77.6	+48.1

If the MARR is 10 percent, which alternative, if any, should be selected?

Solution

One should carefully note that this is a *three-alternative* problem, where the alternatives are A, B, and *Do nothing*. In this solution we will skip step 1. Reorganize the problem by placing the alternatives in order of increasing initial cost:

	Alternative		
Year	Do Nothing	B	A
0	0	−$131.0	−$200.0
1	0	+48.1	+77.6
2	0	+48.1	+77.6
3	0	+48.1	+77.6

Examine the *B – Do nothing* increment of investment:

Year	B – Do Nothing
0	–$131.0 – 0 = – $131.0
1	+ 48.1 – 0 = + 48.1
2	+ 48.1 – 0 = + 48.1
3	+ 48.1 – 0 = + 48.1

Solve for the incremental rate of return:

$$\text{PW of cost} = \text{PW of benefits}$$
$$131.0 = 48.1(P/A, i, 3)$$
$$(P/A, i, 3) = 131.0/48.1 = 2.723.$$

From compound interest tables, the incremental rate of return is about 5 percent. Since the incremental rate of return is less than 10 percent, the *B – Do nothing* increment is not desirable. Reject alternative *B*.

Year	A – Do Nothing
0	–$200.0 – 0 = –$200.0
1	+77.6 – 0 = +77.6
2	+77.6 – 0 = +77.6
3	+77.6 – 0 = +77.6

Next, consider the increment of investment between the two remaining alternatives. Solve for the incremental rate of return:

$$\text{PW of cost} = \text{PW of benefits}$$
$$200.0 = 77.6(P/A, i, 3)$$
$$(P/A, i, 3) = 200.0/77.6 = 2.577.$$

The incremental rate of return is 8 percent, less than the desired 10 percent. Reject the increment and select the remaining alternative: *Do nothing*.

If you have not already done so, you should go back to Example 17.22 and see how the slightly changed wording of the problem has radically altered it. Example 17.22 required a choice between two undesirable alternatives. This example adds the *Do nothing* alternative, which is superior to *A* and *B*.

BENEFIT-COST ANALYSIS

Generally, in public works and governmental economic analyses, the dominant method of analysis is the *benefit-cost ratio*. It is simply the ratio of benefits divided by costs, taking into account the time value of money.

$$B/C = \frac{\text{PW of benefits}}{\text{PW of cost}} = \frac{\text{Equivalent uniform annual benefits}}{\text{Equivalent uniform annual cost}}$$

For a given interest rate, a B/C ratio ≥1 reflects an acceptable project. The B/C analysis method is parallel to rate-of-return analysis. The same kind of incremental analysis is required.

Example 17.24

Solve Example 17.22 by benefit-cost analysis.

Solution

Year	Alternative A	Alternative B	Increment A – B
0	–$200.0	–$131.0	–$69.0
1	+77.6	+48.1	+29.5
2	+77.6	+48.1	+29.5
3	+77.6	+48.1	+29.5

The benefit-cost ratio for the $A - B$ increment is

$$B/C = \frac{\text{PW of benefits}}{\text{PW of cost}} = \frac{29.5(P/A, 10\%, 3)}{69.0} = \frac{73.37}{69.0} = 1.06.$$

Since the B/C ratio exceeds 1, the increment of investment is desirable. Select the higher-cost alternative A.

BREAKEVEN ANALYSIS

In business, "breakeven" is defined as the point where income just covers costs. In engineering economics, the breakeven point is defined as the point where two alternatives are equivalent.

Example 17.25

A city is considering a new $50,000 snowplow. The new machine will operate at a savings of $600 per day compared with the present equipment. Assume that the MARR is 12 percent, and the machine's life is 10 years with zero resale value at that time. How many days per year must the machine be used to justify the investment?

Solution

This breakeven problem may be readily solved by annual cost computations. We will set the equivalent uniform annual cost (EUAC) of the snowplow equal to its annual benefit and solve for the required annual utilization. Let X = breakeven point = days of operation per year.

$$\text{EUAC} = \text{EUAB}$$
$$50,000(A/P, 12\%, 10) = 600X$$
$$X = 50,000(0.1770)/600 = 14.8 \text{ days/year}.$$

OPTIMIZATION

Optimization is the determination of the best or most favorable situation.

Minima-Maxima

In problems where the situation can be represented by a function, the customary approach is to set the first derivative of the function to zero and solve for the root(s) of this equation. If the second derivative is *positive*, the function is a minimum for the critical value; if it is *negative*, the function is a maximum.

Example 17.26

A consulting engineering firm estimates that their net profit is given by the equation

$$P(x) = -0.03x^3 + 36x + 500 \quad x \geq 0$$

where x = number of employees and $P(x)$ = net profit. What is the optimal number of employees?

Solution

$$P'(x) = -0.09x^2 + 36 = 0 \quad P''(x) = -0.18x$$
$$x^2 = 36/0.09 = 400$$
$$x = 20 \text{ employees.}$$
$$P''(20) = -0.18(20) = -3.6.$$

Since $P''(20) < 0$, the net profit is maximized for 20 employees.

Economic Problem—Best Alternative

Since engineering economics problems seek to identify the best or most favorable situation, they are by definition optimization problems. Most use compound interest computations in their solution, but some do not. Consider the following example.

Example 17.27

A firm must decide which of three alternatives to adopt to expand its capacity. It wants a minimum annual profit of 20 percent of the initial cost of each increment of investment. Any money not invested in capacity expansion can be invested elsewhere for an annual yield of 20 percent of the initial cost.

Alternative	Initial Cost	Annual Profit	Profit Rate
A	$100,000	$30,000	30%
B	300,000	66,000	22
C	500,000	80,000	16

Which alternative should be selected?

Solution

Since alternative *C* fails to produce the 20 percent minimum annual profit, it is rejected. To decide between alternatives *A* and *B*, examine the profit rate for the *B* − *A* increment.

Alternative	Initial Cost	Annual Profit	Incremental Cost	Incremental Profit	Incremental Profit Rate
A	$100,000	$30,000			
			$200,000	$36,000	18%
B	300,000	66,000			

The *B* − *A* incremental profit rate is less than the minimum 20 percent, so alternative *B* should be rejected. Thus the best investment of $300,000, for example, would be alternative *A* (annual profit = $30,000) plus $200,000 invested elsewhere at 20 percent (annual profit = $40,000). This combination would yield a $70,000 annual profit, which is better than the alternative *B* profit of $66,000. Select *A*.

Economic Order Quantity

One special case of optimization occurs when an item is used continuously and is periodically purchased. Thus the inventory of the item fluctuates from zero (just prior to the receipt of the purchased quantity) to the purchased quantity (just after receipt). The simplest model for the economic order quantity (EOQ) is

$$\text{EOQ} = \sqrt{\frac{2BD}{E}}$$

where
  B = ordering cost, $/order
  D = demand per period, units
  E = inventory holding cost, $/unit/period
EOC = economic order quantity, units.

Example 17.28

A company uses 8000 wheels per year in its manufacture of golf carts. The wheels cost $15 each and are purchased from an outside supplier. The money invested in the inventory costs 10 percent per year, and the warehousing cost amounts to an additional 2 percent per year. It costs $150 to process each purchase order. When an order is placed, how many wheels should be ordered?

Solution

$$\text{EOQ} = \sqrt{\frac{2 \times \$150 \times 8000}{(10\% + 2\%)(15.00)}} = 1155 \text{ wheels}$$

VALUATION AND DEPRECIATION

Depreciation of capital equipment is an important component of many after-tax economic analyses. For this reason, one must understand the fundamentals of depreciation accounting.

Notation

BV = book value
C = cost of the property ("basis")
D_j = depreciation in year j
S_n = salvage value in year n

Depreciation is the systematic allocation of the cost of a capital asset over its useful life. *Book value* is the original cost of an asset, minus the accumulated depreciation of the asset.

$$BV = C - \Sigma(D_j)$$

In computing a schedule of depreciation charges, four items are considered.

1. Cost of the property, C (called the *basis* in tax law).

2. Type of property. Property is classified as either *tangible* (such as machinery) or *intangible* (such as a franchise or a copyright), and as either *real property* (real estate) or *personal property* (everything that is not real property).

3. Depreciable life in years, n.

4. Salvage value of the property at the end of its depreciable (useful) life, S_n.

Straight-Line Depreciation

The depreciation charge in any year is

$$D_j = \frac{C - S_n}{n}.$$

An alternative computation is

$$\text{Depreciation change in any year, } D_j = \frac{C - \text{depreciation taken to beginning of year } j - S_n}{\text{Remaining useful life at beginning of year } j}.$$

Sum-of-Years'-Digits Depreciation

$$\text{Depreciation change in any year, } D_j = \frac{\text{Remaining depreciable life at beginning of year}}{\text{Sum of year's digits for total useful life}} \times (C - S_n).$$

Declining-Balance Depreciation

Double declining-balance depreciation charge in any year, $D_j = \dfrac{2C}{m}\left(1 - \dfrac{2}{n}\right)^{j-1}$.

Total depreciation at the end of n years, $C = \left[1 - \left(1 - \dfrac{2}{n}\right)^n\right]$.

Book value at the end of j years, $BV_j = C\left(1 - \dfrac{2}{n}\right)^j$.

For 150 percent declining-balance depreciation, replace the 2 in the three equations above with 1.5.

Sinking-Fund Depreciation

Depreciation charge in any year, $D_j = (C - S_n)(A/F, i\%, n)(F/P, i\%, j - 1)$.

Modified Accelerated Cost Recovery System Depreciation

The modified accelerated cost recovery system (MACRS) depreciation method generally applies to property placed in service after 1986. To compute the MACRS depreciation for an item, one must know the following:

- Cost (basis) of the item.

- Property class. All tangible property is classified in one of six classes (3, 5, 7, 10, 15, and 20 years), which is the life over which it is depreciated (see Table 17.3). Residential real estate and nonresidential real estate are in two separate real property classes of 27.5 years and 39 years, respectively.

- Depreciation computation.

Table 17.3 MACRS classes of depreciable property

Property Class	Personal Property (All Property Except Real Estate)
3-year property	Special handling devices for food and beverage manufacture Special tools for the manufacture of finished plastic products, fabricated metal products, and motor vehicles Property with an asset depreciation range (ADR) midpoint life of 4 years or less
5-year property	Automobiles* and trucks Aircraft (of non–air-transport companies) Equipment used in research and experimentation Computers Petroleum drilling equipment Property with an ADR midpoint life of more than 4 years and less than 10 years
7-year property	All other property not assigned to another class Office furniture, fixtures, and equipment Property with an ADR midpoint life of 10 years or more, and less than 16 years

Property Class	Real Property (Real Estate)
10-year property	Assets used in petroleum refining and preparation of certain food products Vessels and water transportation equipment Property with an ADR midpoint life of 16 years or more, and less than 20 years
15-year property	Telephone distribution plants Municipal sewage treatment plants Property with an ADR midpoint life of 20 years or more, and less than 25 years
20-year property	Municipal sewers Property with an ADR midpoint life of 25 years or more
27.5 years	Residential rental property (does not include hotels and motels)
39 years	Nonresidential real property

*The depreciation deduction for automobiles is limited to $2860 in the first tax year and is reduced in subsequent years.

- Use double-declining-balance depreciation for 3-, 5-, 7-, and 10-year property classes with conversion to straight-line depreciation in the year that increases the deduction.

- Use 150%-declining-balance depreciation for 15- and 20-year property classes with conversion to straight-line depreciation in the year that increases the deduction.

- In MACRS, the salvage value is assumed to be zero.

Half-Year Convention

Except for real property, a half-year convention is used. Under this convention all property is considered to be placed in service in the middle of the tax year, and a half-year of depreciation is allowed in the first year. For each of the remaining years, one is allowed a full year of depreciation. If the property is disposed of

Table 17.4 MACRS depreciation for personal property—half-year convention

If the Recovery Year Is	The Applicable Percentage for the Class of Property Is			
	3-Year Class	5-Year Class	7-Year Class	10-Year Class
1	33.33	20.00	14.29	10.00
2	44.45	32.00	24.49	18.00
3	14.81†	19.20	17.49	14.40
4	7.41	11.52†	12.49	11.52
5		11.52	8.93†	9.22
6		5.76	8.92	7.37
7			8.93	6.55†
8			4.46	6.55
9				6.56
10				6.55
11				3.28

†Use straight-line depreciation for the year marked and all subsequent years.

prior to the end of the recovery period (property class life), a half-year of depreciation is allowed in that year. If the property is held for the entire recovery period, a half-year of depreciation is allowed for the year following the end of the recovery period (see Table 17.4). Owing to the half-year convention, a general form of the double-declining-balance computation must be used to compute the year-by-year depreciation.

DDB depreciation in any year, $D_j = \dfrac{2}{n}(C - \text{depreciation in years prior to } j)$.

Example 17.29

A $5000 computer has an anticipated $500 salvage value at the end of its five-year depreciable life. Compute the depreciation schedule for the machinery by (a) sum-of-years'-digits depreciation and (b) MACRS depreciation. Do the MACRS computation by hand, and then compare the results with the values from Table 17.4.

Solution

(a) Sum-of-years'-digits depreciation:

$$D_j = \dfrac{n-j+1}{\dfrac{n}{2}(n+1)}(C - S_n)$$

$$D_1 = \dfrac{5-1+1}{\dfrac{5}{2}(5+1)}(5000 - 500) = \quad 1500$$

$$D_2 = \dfrac{5-2+1}{\dfrac{5}{2}(5+1)}(5000 - 500) = \quad 1200$$

$$D_3 = \dfrac{5-3+1}{\dfrac{5}{2}(5+1)}(5000 - 500) = \quad 900$$

$$D_4 = \dfrac{5-4+1}{\dfrac{5}{2}(5+1)}(5000 - 500) = \quad 600$$

$$D_5 = \dfrac{5-5+1}{\dfrac{5}{2}(5+1)}(5000 - 500) = \quad \underline{300}$$

$$\$4500$$

(b) MACRS depreciation. Double-declining-balance with conversion to straight-line. Five-year property class. Half-year convention. Salvage value S_n is assumed to be zero for MACRS. Using the general DDB computation,

Year

1 (half-year) $D_1 = \dfrac{1}{2} \times \dfrac{2}{5}(5000 - 0)$ = $1,000

2 $D_2 = \dfrac{2}{5}(5000 - 1000)$ = 1600

3 $D_3 = \dfrac{2}{5}(5000 - 2600)$ = 960

4 $D_4 = \dfrac{2}{5}(5000 - 3560)$ = 576

5 $D_5 = \dfrac{2}{5}(5000 - 4136)$ = 346

6 (half-year) $D_6 = \dfrac{1}{2} \times \dfrac{2}{5}(5000 - 4482)$ = 104

$4586

The computation must now be modified to convert to straight-line depreciation at the point where the straight-line depreciation will be larger. Using the alternative straight-line computation,

$$D_5 = \dfrac{5000 - 4136 - 0}{1.5 \text{ years remaining}} = \$576.$$

This is more than the $346 computed using DDB, hence switch to straight-line for year 5 and beyond.

$$D_6 (\text{half-year}) = \dfrac{1}{2}(576) = \$288$$

Answers:

	Depreciation	
Year	SOYD	MACRS
1	$1500	$1000
2	1200	1600
3	900	960
4	600	576
5	300	576
6	0	288
	$4500	$5000

The computed MACRS depreciation is identical to the result obtained from Table 17.4.

TAX CONSEQUENCES

Income taxes represent another of the various kinds of disbursements encountered in an economic analysis. The starting point in an after-tax computation is the before-tax cash flow. Generally, the before-tax cash flow contains three types of entries:

1. Disbursements of money to purchase capital assets. These expenditures create no direct tax consequence, for they are the exchange of one asset (money) for another (capital equipment).

2. Periodic receipts and/or disbursements representing operating income and/or expenses. These increase or decrease the year-by-year tax liability of the firm.

3. Receipts of money from the sale of capital assets, usually in the form of a salvage value when the equipment is removed. The tax consequences depend on the relationship between the book value (cost − depreciation taken) of the asset and its salvage value.

Situation	Tax Consequence
Salvage value > Book value	Capital gain on differences
Salvage value = Book value	No tax consequence
Salvage value < Book value	Capital loss on difference

After determining the before-tax cash flow, compute the depreciation schedule for any capital assets. Next, compute taxable income, the taxable component of the before-tax cash flow minus the depreciation. The income tax is the taxable income times the appropriate tax rate. Finally, the after-tax cash flow is the before-tax cash flow adjusted for income taxes.

To organize these data, it is customary to arrange them in the form of a cash flow table, as follows:

Year	Before-Tax Cash Flow	Depreciation	Taxable Income	Income Taxes	After-Tax Cash Flow
0	•				•
1	•	•	•	•	•

Example 17.30

A corporation expects to receive $32,000 each year for 15 years from the sale of a product. There will be an initial investment of $150,000. Manufacturing and sales expenses will be $8067 per year. Assume straight-line depreciation, a 15-year useful life, and no salvage value. Use a 46 percent income tax rate. Determine the projected after-tax rate of return.

Solution

Straight-line depreciation, $D_j = \dfrac{C - S_n}{n} = \dfrac{\$150,000 - 0}{15} = \$10,000$ per year

Year	Before-Tax Cash Flow	Depreciation	Taxable Income	Income Taxes	After-Tax Cash Flow
0	−150,000				−150,000
1	+23,933	10,000	13,933	−6,409	+17,524
2	+23,933	10,000	13,933	−6,409	+17,524
•	•	•	•	•	•
•	•	•	•	•	•
•	•	•	•	•	•
15	+23,933	10,000	13,933	−6,409	+17,524

Take the after-tax cash flow and compute the rate of return at which the PW of cost equals the PW of benefits.

$$150{,}000 = 17{,}524 \, (P/A, i\%, 15)$$

$$(P/A, i\%, 15) = \frac{150{,}000}{17{,}524} = 8.559$$

From the compound interest tables, the after-tax rate of return is $i = 8\%$.

INFLATION

Inflation is characterized by rising prices for goods and services, whereas deflation involves a fall in prices. An inflationary trend makes future dollars have less purchasing power than present dollars. This helps long-term borrowers of money, for they may repay a loan of present dollars in the future with dollars of reduced buying power. The help to borrowers is at the expense of lenders. Deflation has the opposite effect. Money borrowed at one point in time, followed by a deflationary period, subjects the borrower to loan repayment with dollars of greater purchasing power than those borrowed. This is to the lenders' advantage at the expense of borrowers.

Price changes occur in a variety of ways. One method of stating a price change is as a uniform rate of price change per year.

f = General inflation rate per interest period
i = Effective interest rate per interest period

The following situation will illustrate the computations. A mortgage is to be repaid in three equal payments of $5000 at the end of years 1, 2, and 3. If the annual inflation rate, f, is 8% during this period, and a 12% annual interest rate (i) is desired, what is the maximum amount the investor would be willing to pay for the mortgage?

The computation is a two-step process. First, the three future payments must be converted to dollars with the same purchasing power as today's (year 0) dollars.

Year	Actual Cash Flow	Multiplied by		Cash Flow Adjusted to Today's (yr. 0) Dollars
0	—	—		—
1	+5000	× $(1 + 0.08)^{-1}$	=	+4630
2	+5000	× $(1 + 0.08)^{-2}$	=	+4286
3	+5000	× $(1 + 0.08)^{-3}$	=	+3969

The general form of the adjusting multiplier is

$$(1 + f)^{-n} = (P/F, f, n).$$

Now that the problem has been converted to dollars of the same purchasing power (today's dollars, in this example), we can proceed to compute the present worth of the future payments.

Year	Adjusted Cash Flow		Multiplied by		Present Worth
0	—		—		—
1	+4630	×	$(1 + 0.12)^{-1}$	=	+4134
2	+4286	×	$(1 + 0.12)^{-2}$	=	+3417
3	+3969	×	$(1 + 0.12)^{-3}$	=	+2825
					$10,376

The general form of the discounting multiplier is

$$(1 + i)^{-n} = (P/F, i\%, n).$$

Alternative Solution

Instead of doing the inflation and interest rate computations separately, one can compute a combined equivalent interest rate, d.

$$d = (1 + f)(1 + i) - 1 = i + f + i(f)$$

For this cash flow, $d = 0.12 + 0.08 + 0.12(0.08) = 0.2096$. Since we do not have 20.96 percent interest tables, the problem has to be calculated using present worth equations.

$$PW = 5000(1 + 0.2096)^{-1} + 5000(1 + 0.2096)^{-2} + 5000(1 + 0.2096)^{-3}$$
$$= 4134 + 3417 + 2825 = \$10,376.$$

Example 17.31

One economist has predicted that there will be 7 percent per year inflation of prices during the next 10 years. If this proves to be correct, an item that presently sells for $10 would sell for what price 10 years hence?

Solution

$$f = 7\%, P = \$10$$
$$F = ?, n = 10 \text{ years}$$

Here the computation is to find the future worth F, rather than the present worth, P.

$$F = P(1 + f)^{10} = 10(1 + 0.07)^{10} = \$19.67$$

Effect of Inflation on Rate of Return

The effect of inflation on the computed rate of return for an investment depends on how future benefits respond to the inflation. If benefits produce constant dollars, which are not increased by inflation, the effect of inflation is to reduce the before-tax rate of return on the investment. If, on the other hand, the dollar benefits increase to keep up with the inflation, the before-tax rate of return will not be adversely affected by the inflation.

This is not true when an after-tax analysis is made. Even if the future benefits increase to match the inflation rate, the allowable depreciation schedule does not increase. The result will be increased taxable income and income tax payments. This reduces the available after-tax benefits and, therefore, the after-tax rate of return.

Example 17.32

A man bought a 5 percent tax-free municipal bond. It cost $1000 and will pay $50 interest each year for 20 years. The bond will mature at the end of 20 years and return the original $1000. If there is 2% annual inflation during this period, what rate of return will the investor receive after considering the effect of inflation?

Solution

$$d = 0.05, \; i = \text{unknown}, \; j = 0.02$$
$$d = i + j + i(j)$$
$$0.05 = i + 0.02 + 0.02i$$
$$1.02i = 0.03, \; i = 0.294 = 2.94\%.$$

RISK ANALYSIS

Probability

Probability can be considered to be the long-run relative frequency of occurrence of an outcome. For example, there are two possible outcomes from flipping a coin (a head or a tail). If a coin is flipped over and over, we can expect in the long run that half the time heads will appear and half the time tails will appear. We would say the probability of flipping a head is 0.50 and of flipping a tail is 0.50. Since the probabilities are defined so that the sum of probabilities for all possible outcomes is 1, the situation is

$$\text{Probability of flipping a head} = 0.50$$
$$\text{Probability of flipping a tail} = \underline{0.50}$$
$$\text{Sum of all possible outcomes} = 1.00.$$

Example 17.33

If one were to roll one die (that is, one-half of a pair of dice), what is the probability that either a 1 or a 6 would result?

Solution

Since a die is a perfect six-sided cube, the probability of any side appearing is 1/6.

$$\text{Probability of rolling a } 1 = P(1) = 1/6$$
$$2 = P(2) = 1/6$$
$$3 = P(3) = 1/6$$
$$4 = P(4) = 1/6$$
$$5 = P(5) = 1/6$$
$$6 = P(6) = 1/6$$

Sum of all possible outcomes = 6/6 = 1. The probability of rolling a 1 or a 6 = 1/6 + 1/6 = 1/3.

In the preceding examples, the probability of each outcome was the same. This need not be the case.

Example **17.34**

In the game of blackjack, a *perfect hand* is a 10 or a face card plus an ace. What is the probability of being dealt a 10 or a face card from a newly shuffled deck of 52 cards? What is the probability of being dealt an ace in this same situation?

Solution

The three outcomes examined are to be dealt either (a) 10 or a face card, (b) an ace, or (c) some other card. Every card in the deck represents one of these three possible outcomes. To solve this problem, of course, we need to know the composition of a deck of cards. There are 4 aces; 16 10s, jacks, queens, and kings; and 32 other cards.

Probability of being dealt a 10 or a face card = 16/52 = 0.31
Probability of being dealt an ace = 4/52 = 0.08
Probability of being dealt some other card = 32/52 = 0.61
1.00

Risk

The term *risk* has a special meaning in statistics. It is defined as a situation where there are two or more possible outcomes and the probability associated with each outcome is known. In each of the two previous examples there is a risk situation. We could not know in advance what playing card would be dealt or what number would be rolled by the die. However, since the various probabilities could be computed, our definition of risk has been satisfied. Probability and risk calculations are not restricted to gambling games, but are important in many other areas as well. For example, in a particular engineering course, a student has computed the probability for each of the letter grades he might receive as follows:

Grade	Grade Point	Probability P(Grade)
A	4.0	0.10
B	3.0	0.30
C	2.0	0.25
D	1.0	0.20
F	0	0.15
		1.00

From the table we see that the grade with the highest probability is a B. This, therefore, is the most likely grade. We also see that there is a substantial probability that some grade other than a B will be received. And the probabilities indicate that if a B is not received, the grade will probably be something less than a B. But in saying that the most likely grade is a B, other outcomes are ignored. In the next section we will show that a composite statistic may be computed using all the data.

Expected Value

In the last example the most likely grade of B in an engineering class had a probability of 0.30. That is not a very high probability. In some other course, say a math class, we might estimate a probability of 0.65 of obtaining a B, again making the B the most likely grade. Then, while a B is most likely in both classes, it is more certain in the math class.

We can compute a weighted mean to give a better understanding of the total situation as represented by various possible outcomes. When the probabilities are used as the weighting factors, the result is called the *expected value* and is written

$$\text{Expected value} = \text{Outcome}_A \times P(A) + \text{Outcome}_B \times P(B) + \ldots$$

Example 17.35

An engineer wishes to determine the risk of fire loss for her $200,000 home. From a fire rating bureau she obtains the following data:

Outcome	Probability
No fire loss	0.986 in any year
$10,000 fire loss	0.010
40,000 fire loss	0.003
200,000 fire loss	0.001

Compute the expected fire loss in any year.

Solution

Expected fire loss = 10,000(0.010) + 40,000(0.003) + 200,000(0.001) = $420

REFERENCE

Newnan, Donald G. *Engineering Economic Analysis*. 5th ed. Engineering Press, San Jose, CA, 1995.

INTEREST TABLES
Compound interest factors

½% ½%

	Single Payment		Uniform Payment Series				Uniform Gradient		
	Compound Amount Factor	Present Worth Factor	Sinking Fund Factor	Capital Recovery Factor	Compound Amount Factor	Present Worth Factor	Gradient Uniform Series	Gradient Present Worth	
n	Find F Given P F/P	Find P Given F P/F	Find A Given F A/F	Find A Given P A/P	Find F Given A F/A	Find P Given A P/A	Find A Given G A/G	Find P Given G P/G	n
1	1.005	.9950	1.0000	1.0050	1.000	0.995	0	0	1
2	1.010	.9901	.4988	.5038	2.005	1.985	0.499	0.991	2
3	1.015	.9851	.3317	.3367	3.015	2.970	0.996	2.959	3
4	1.020	.9802	.2481	.2531	4.030	3.951	1.494	5.903	4
5	1.025	.9754	.1980	.2030	5.050	4.926	1.990	9.803	5
6	1.030	.9705	.1646	.1696	6.076	5.896	2.486	14.660	6
7	1.036	.9657	.1407	.1457	7.106	6.862	2.980	20.448	7
8	1.041	.9609	.1228	.1278	8.141	7.823	3.474	27.178	8
9	1.046	.9561	.1089	.1139	9.182	8.779	3.967	34.825	9
10	1.051	.9513	.0978	.1028	10.228	9.730	4.459	43.389	10
11	1.056	.9466	.0887	.0937	11.279	10.677	4.950	52.855	11
12	1.062	.9419	.0811	.0861	12.336	11.619	5.441	63.218	12
13	1.067	.9372	.0746	.0796	13.397	12.556	5.931	74.465	13
14	1.072	.9326	.0691	.0741	14.464	13.489	6.419	86.590	14
15	1.078	.9279	.0644	.0694	15.537	14.417	6.907	99.574	15
16	1.083	.9233	.0602	.0652	16.614	15.340	7.394	113.427	16
17	1.088	.9187	.0565	.0615	17.697	16.259	7.880	128.125	17
18	1.094	.9141	.0532	.0582	18.786	17.173	8.366	143.668	18
19	1.099	.9096	.0503	.0553	19.880	18.082	8.850	160.037	19
20	1.105	9051	.0477	.0527	20.979	18.987	9.334	177.237	20
21	1.110	.9006	.0453	.0503	22.084	19.888	9.817	195.245	21
22	1.116	.8961	.0431	.0481	23.194	20.784	10.300	214.070	22
23	1.122	.8916	.0411	.0461	24.310	21.676	10.781	233.680	23
24	1.127	.8872	.0393	.0443	25.432	22.563	11.261	254.088	24
25	1.133	.8828	.0377	.0427	26.559	23.446	11.741	275.273	25
26	1.138	.8784	.0361	.0411	27.692	24.324	12.220	297.233	26
27	1.144	.8740	.0347	.0397	28.830	25.198	12.698	319.955	27
28	1.150	.8697	.0334	.0384	29.975	26.068	13.175	343.439	28
29	1.156	.8653	.0321	.0371	31.124	26.933	13.651	367.672	29
30	1.161	.8610	.0310	.0360	32.280	27.794	14.127	392.640	30
36	1.197	.8356	.0254	.0304	39.336	32.871	16.962	557.564	36
40	1.221	.8191	.0226	.0276	44.159	36.172	18.836	681.341	40
48	1.270	.7871	.0185	.0235	54.098	42.580	22.544	959.928	48
50	1.283	.7793	.0177	.0227	56.645	44.143	23.463	1 035.70	50
52	1.296	.7716	.0169	.0219	59.218	45.690	24.378	1 113.82	52
60	1.349	.7414	.0143	.0193	69.770	51.726	28.007	1 448.65	60
70	1.418	.7053	.0120	.0170	83.566	58.939	32.468	1 913.65	70
72	1.432	.6983	.0116	.0166	86.409	60.340	33.351	2 012.35	72
80	1.490	.6710	.0102	.0152	98.068	65.802	36.848	2 424.65	80
84	1.520	.6577	.00961	.0146	104.074	68.453	38.576	2 640.67	84
90	1.567	.6383	.00883	.0138	113.311	72.331	41.145	2 976.08	90
96	1.614	.6195	.00814	.0131	122.829	76.095	43.685	3 324.19	96
100	1.647	.6073	.00773	.0127	129.334	78.543	45.361	3 562.80	100
104	1.680	.5953	.00735	.0124	135.970	80.942	47.025	3 806.29	104
120	1.819	.5496	.00610	.0111	163.880	90.074	53.551	4 823.52	120
240	3.310	.3021	.00216	.00716	462.041	139.581	96.113	13 415.56	240
360	6.023	.1660	.00100	.00600	1 004.5	166.792	128.324	21 403.32	360
480	10.957	.0913	.00050	.00550	1 991.5	181.748	151.795	27 588.37	480

Compound interest factors

1%   **1%**

	Single Payment		Uniform Payment Series				Uniform Gradient		
	Compound Amount Factor	Present Worth Factor	Sinking Fund Factor	Capital Recovery Factor	Compound Amount Factor	Present Worth Factor	Gradient Uniform Series	Gradient Present Worth	
n	Find F Given P F/P	Find P Given F P/F	Find A Given F A/F	Find A Given P A/P	Find F Given A F/A	Find P Given A P/A	Find A Given G A/G	Find P Given G P/G	n
1	1.010	.9901	1.0000	1.0100	1.000	0.990	0	0	1
2	1.020	.9803	.4975	.5075	2.010	1.970	0.498	0.980	2
3	1.030	.9706	.3300	.3400	3.030	2.941	0.993	2.921	3
4	1.041	.9610	.2463	.2563	4.060	3.902	1.488	5.804	4
5	1.051	.9515	.1960	.2060	5.101	4.853	1.980	9.610	5
6	1.062	.9420	.1625	.1725	6.152	5.795	2.471	14.320	6
7	1.072	.9327	.1386	.1486	7.214	6.728	2.960	19.917	7
8	1.083	.9235	.1207	.1307	8.286	7.652	3.448	26.381	8
9	1.094	.9143	.1067	.1167	9.369	8.566	3.934	33.695	9
10	1.105	.9053	.0956	.1056	10.462	9.471	4.418	41.843	10
11	1.116	.8963	.0865	.0965	11.567	10.368	4.900	50.806	11
12	1.127	.8874	.0788	.0888	12.682	11.255	5.381	60.568	12
13	1.138	.8787	.0724	.0824	13.809	12.134	5.861	71.112	13
14	1.149	.8700	.0669	.0769	14.947	13.004	6.338	82.422	14
15	1.161	.8613	.0621	.0721	16.097	13.865	6.814	94.481	15
16	1.173	.8528	.0579	.0679	17.258	14.718	7.289	107.273	16
17	1.184	.8444	.0543	.0643	18.430	15.562	7.761	120.783	17
18	1.196	.8360	.0510	.0610	19.615	16.398	8.232	134.995	18
19	1.208	.8277	.0481	.0581	20.811	17.226	8.702	149.895	19
20	1.220	.8195	.0454	.0554	22.019	18.046	9.169	165.465	20
21	1.232	.8114	.0430	.0530	23.239	18.857	9.635	181.694	21
22	1.245	.8034	.0409	.0509	24.472	19.660	10.100	198.565	22
23	1.257	.7954	.0389	.0489	25.716	20.456	10.563	216.065	23
24	1.270	.7876	.0371	.0471	26.973	21.243	11.024	234.179	24
25	1.282	.7798	.0354	.0454	28.243	22.023	11.483	252.892	25
26	1.295	.7720	.0339	.0439	29.526	22.795	11.941	272.195	26
27	1.308	.7644	.0324	.0424	30.821	23.560	12.397	292.069	27
28	1.321	.7568	.0311	.0411	32.129	24.316	12.852	312.504	28
29	1.335	.7493	.0299	.0399	33.450	25.066	13.304	333.486	29
30	1.348	.7419	.0287	.0387	34.785	25.808	13.756	355.001	30
36	1.431	.6989	.0232	.0332	43.077	30.107	16.428	494.620	36
40	1.489	.6717	.0205	.0305	48.886	32.835	18.178	596.854	40
48	1.612	.6203	.0163	.0263	61.223	37.974	21.598	820.144	48
50	1.645	.6080	.0155	.0255	64.463	39.196	22.436	879.417	50
52	1.678	.5961	.0148	.0248	67.769	40.394	23.269	939.916	52
60	1.817	.5504	.0122	.0222	81.670	44.955	26.533	1 192.80	60
70	2.007	.4983	.00993	.0199	100.676	50.168	30.470	1 528.64	70
72	2.047	.4885	.00955	.0196	104.710	51.150	31.239	1 597.86	72
80	2.217	.4511	.00822	.0182	121.671	54.888	34.249	1 879.87	80
84	2.307	.4335	.00765	.0177	130.672	56.648	35.717	2 023.31	84
90	2.449	.4084	.00690	.0169	144.863	59.161	37.872	2 240.56	90
96	2.599	.3847	.00625	.0163	159.927	61.528	39.973	2 459.42	96
100	2.705	.3697	.00587	.0159	170.481	63.029	41.343	2 605.77	100
104	2.815	.3553	.00551	.0155	181.464	64.471	42.688	2 752.17	104
120	3.300	.3030	.00435	.0143	230.039	69.701	47.835	3 334.11	120
240	10.893	.0918	.00101	.0110	989.254	90.819	75.739	6 878.59	240
360	35.950	.0278	.00029	.0103	3 495.0	97.218	89.699	8 720.43	360
480	118.648	.00843	.00008	.0101	11 764.8	99.157	95.920	9 511.15	480

Compound interest factors

1½%　　　　　　　　　　　　　　　　　　　　　　　　　　　　　　　　　　　　　　　1½%

	Single Payment		Uniform Payment Series				Uniform Gradient		
	Compound Amount Factor	Present Worth Factor	Sinking Fund Factor	Capital Recovery Factor	Compound Amount Factor	Present Worth Factor	Gradient Uniform Series	Gradient Present Worth	
n	Find F Given P F/P	Find P Given F P/F	Find A Given F A/F	Find A Given P A/P	Find F Given A F/A	Find P Given A P/A	Find A Given G A/G	Find P Given G P/G	n
1	1.015	.9852	1.0000	1.0150	1.000	0.985	0	0	1
2	1.030	.9707	.4963	.5113	2.015	1.956	0.496	0.970	2
3	1.046	.9563	.3284	.3434	3.045	2.912	0.990	2.883	3
4	1.061	.9422	.2444	.2594	4.091	3.854	1.481	5.709	4
5	1.077	.9283	.1941	.2091	5.152	4.783	1.970	9.422	5
6	1.093	.9145	.1605	.1755	6.230	5.697	2.456	13.994	6
7	1.110	.9010	.1366	.1516	7.323	6.598	2.940	19.400	7
8	1.126	.8877	.1186	.1336	8.433	7.486	3.422	25.614	8
9	1.143	.8746	.1046	.1196	9.559	8.360	3.901	32.610	9
10	1.161	.8617	.0934	.1084	10.703	9.222	4.377	40.365	10
11	1.178	.8489	.0843	.0993	11.863	10.071	4.851	48.855	11
12	1.196	.8364	.0767	.0917	13.041	10.907	5.322	58.054	12
13	1.214	.8240	.0702	.0852	14.237	11.731	5.791	67.943	13
14	1.232	.8118	.0647	.0797	15.450	12.543	6.258	78.496	14
15	1.250	.7999	.0599	.0749	16.682	13.343	6.722	89.694	15
16	1.269	.7880	.0558	.0708	17.932	14.131	7.184	101.514	16
17	1.288	.7764	.0521	.0671	19.201	14.908	7.643	113.937	17
18	1.307	.7649	.0488	.0638	20.489	15.673	8.100	126.940	18
19	1.327	.7536	.0459	.0609	21.797	16.426	8.554	140.505	19
20	1.347	.7425	.0432	.0582	23.124	17.169	9.005	154.611	20
21	1.367	.7315	.0409	.0559	24.470	17.900	9.455	169.241	21
22	1.388	.7207	.0387	.0537	25.837	18.621	9.902	184.375	22
23	1.408	.7100	.0367	.0517	27.225	19.331	10.346	199.996	23
24	1.430	.6995	.0349	.0499	28.633	20.030	10.788	216.085	24
25	1.451	.6892	.0333	.0483	30.063	20.720	11.227	232.626	25
26	1.473	.6790	.0317	.0467	31.514	21.399	11.664	249.601	26
27	1.495	.6690	.0303	.0453	32.987	22.068	12.099	266.995	27
28	1.517	.6591	.0290	.0440	34.481	22.727	12.531	284.790	28
29	1.540	.6494	.0278	.0428	35.999	23.376	12.961	302.972	29
30	1.563	.6398	.0266	.0416	37.539	24.016	13.388	321.525	30
36	1.709	.5851	.0212	.0362	47.276	27.661	15.901	439.823	36
40	1.814	.5513	.0184	.0334	54.268	29.916	17.528	524.349	40
48	2.043	.4894	.0144	.0294	69.565	34.042	20.666	703.537	48
50	2.105	.4750	.0136	.0286	73.682	35.000	21.428	749.955	50
52	2.169	.4611	.0128	.0278	77.925	35.929	22.179	796.868	52
60	2.443	.4093	.0104	.0254	96.214	39.380	25.093	988.157	60
70	2.835	.3527	.00817	.0232	122.363	43.155	28.529	1 231.15	70
72	2.921	.3423	.00781	.0228	128.076	43.845	29.189	1 279.78	72
80	3.291	.3039	.00655	.0215	152.710	46.407	31.742	1 473.06	80
84	3.493	.2863	.00602	.0210	166.172	47.579	32.967	1 568.50	84
90	3.819	.2619	.00532	.0203	187.929	49.210	34.740	1 709.53	90
96	4.176	.2395	.00472	.0197	211.719	50.702	36.438	1 847.46	96
100	4.432	.2256	.00437	.0194	228.802	51.625	37.529	1 937.43	100
104	4.704	.2126	.00405	.0190	246.932	52.494	38.589	2 025.69	104
120	5.969	.1675	.00302	.0180	331.286	55.498	42.518	2 359.69	120
240	35.632	.0281	.00043	.0154	2 308.8	64.796	59.737	3 870.68	240
360	212.700	.00470	.00007	.0151	14 113.3	66.353	64.966	4 310.71	360
480	1 269.7	.00079	.00001	.0150	84 577.8	66.614	66.288	4 415.74	480

Compound interest factors

2% 2%

	Single Payment		Uniform Payment Series				Uniform Gradient		
	Compound Amount Factor	Present Worth Factor	Sinking Fund Factor	Capital Recovery Factor	Compound Amount Factor	Present Worth Factor	Gradient Uniform Series	Gradient Present Worth	
	Find F Given P F/P	Find P Given F P/F	Find A Given F A/F	Find A Given P A/P	Find F Given A F/A	Find P Given A P/A	Find A Given G A/G	Find P Given G P/G	
n									n
1	1.020	.9804	1.0000	1.0200	1.000	0.980	0	0	1
2	1.040	.9612	.4951	.5151	2.020	1.942	0.495	0.961	2
3	1.061	.9423	.3268	.3468	3.060	2.884	0.987	2.846	3
4	1.082	.9238	.2426	.2626	4.122	3.808	1.475	5.617	4
5	1.104	.9057	.1922	.2122	5.204	4.713	1.960	9.240	5
6	1.126	.8880	.1585	.1785	6.308	5.601	2.442	13.679	6
7	1.149	.8706	.1345	.1545	7.434	6.472	2.921	18.903	7
8	1.172	.8535	.1165	.1365	8.583	7.325	3.396	24.877	8
9	1.195	.8368	.1025	.1225	9.755	8.162	3.868	31.571	9
10	1.219	.8203	.0913	.1113	10.950	8.983	4.337	38.954	10
11	1.243	.8043	.0822	.1022	12.169	9.787	4.802	46.996	11
12	1.268	.7885	.0746	.0946	13.412	10.575	5.264	55.669	12
13	1.294	.7730	.0681	.0881	14.680	11.348	5.723	64.946	13
14	1.319	.7579	.0626	.0826	15.974	12.106	6.178	74.798	14
15	1.346	.7430	.0578	.0778	17.293	12.849	6.631	85.200	15
16	1.373	.7284	.0537	.0737	18.639	13.578	7.080	96.127	16
17	1.400	.7142	.0500	.0700	20.012	14.292	7.526	107.553	17
18	1.428	.7002	.0467	.0667	21.412	14.992	7.968	119.456	18
19	1.457	.6864	.0438	.0638	22.840	15.678	8.407	131.812	19
20	1.486	.6730	.0412	.0612	24.297	16.351	8.843	144.598	20
21	1.516	.6598	.0388	.0588	25.783	17.011	9.276	157.793	21
22	1.546	.6468	.0366	.0566	27.299	17.658	9.705	171.377	22
23	1.577	.6342	.0347	.0547	28.845	18.292	10.132	185.328	23
24	1.608	.6217	.0329	.0529	30.422	18.914	10.555	199.628	24
25	1.641	.6095	.0312	.0512	32.030	19.523	10.974	214.256	25
26	1.673	.5976	.0297	.0497	33.671	20.121	11.391	229.169	26
27	1.707	.5859	.0283	.0483	35.344	20.707	11.804	244.428	27
28	1.741	.5744	.0270	.0470	37.051	21.281	12.214	259.936	28
29	1.776	.5631	.0258	.0458	38.792	21.844	12.621	275.703	29
30	1.811	.5521	.0247	.0447	40.568	22.396	13.025	291.713	30
36	2.040	.4902	.0192	.0392	51.994	25.489	15.381	392.036	36
40	2.208	.4529	.0166	.0366	60.402	27.355	16.888	461.989	40
48	2.587	.3865	.0126	.0326	79.353	30.673	19.755	605.961	48
50	2.692	.3715	.0118	.0318	84.579	31.424	20.442	642.355	50
52	2.800	.3571	.0111	.0311	90.016	32.145	21.116	678.779	52
60	3.281	.3048	.00877	.0288	114.051	34.761	23.696	823.692	60
70	4.000	.2500	.00667	.0267	149.977	37.499	26.663	999.829	70
72	4.161	.2403	.00633	.0263	158.056	37.984	27.223	1 034.050	72
80	4.875	.2051	.00516	.0252	193.771	39.744	29.357	1 166.781	80
84	5.277	.1895	.00468	.0247	213.865	40.525	30.361	1 230.413	84
90	5.943	.1683	.00405	.0240	247.155	41.587	31.793	1 322.164	90
96	6.693	.1494	.00351	.0235	284.645	42.529	33.137	1 409.291	96
100	7.245	.1380	.00320	.0232	312.230	43.098	33.986	1 464.747	100
104	7.842	.1275	.00292	.0229	342.090	43.624	34.799	1 518.082	104
120	10.765	.0929	.00205	.0220	488.255	45.355	37.711	1 710.411	120
240	115.887	.00863	.00017	.0202	5 744.4	49.569	47.911	2 374.878	240
360	1 247.5	.00080	.00002	.0200	62 326.8	49.960	49.711	2 483.567	360
480	13 429.8	.00007		.0200	671 442.0	49.996	49.964	2 498.027	480

Compound interest factors

4%   4%

	Single Payment		Uniform Payment Series				Uniform Gradient		
	Compound Amount Factor	Present Worth Factor	Sinking Fund Factor	Capital Recovery Factor	Compound Amount Factor	Present Worth Factor	Gradient Uniform Series	Gradient Present Worth	
	Find F Given P F/P	Find P Given F P/F	Find A Given F A/F	Find A Given P A/P	Find F Given A F/A	Find P Given A P/A	Find A Given G A/G	Find P Given G P/G	
n									n
1	1.040	.9615	1.0000	1.0400	1.000	0.962	0	0	1
2	1.082	.9246	.4902	.5302	2.040	1.886	0.490	0.925	2
3	1.125	.8890	.3203	.3603	3.122	2.775	0.974	2.702	3
4	1.170	.8548	.2355	.2755	4.246	3.630	1.451	5.267	4
5	1.217	.8219	.1846	.2246	5.416	4.452	1.922	8.555	5
6	1.265	.7903	.1508	.1908	6.633	5.242	2.386	12.506	6
7	1.316	.7599	.1266	.1666	7.898	6.002	2.843	17.066	7
8	1.369	.7307	.1085	.1485	9.214	6.733	3.294	22.180	8
9	1.423	.7026	.0945	.1345	10.583	7.435	3.739	27.801	9
10	1.480	.6756	.0833	.1233	12.006	8.111	4.177	33.881	10
11	1.539	.6496	.0741	.1141	13.486	8.760	4.609	40.377	11
12	1.601	.6246	.0666	.1066	15.026	9.385	5.034	47.248	12
13	1.665	.6006	.0601	.1001	16.627	9.986	5.453	54.454	13
14	1.732	.5775	.0547	.0947	18.292	10.563	5.866	61.962	14
15	1.801	.5553	.0499	.0899	20.024	11.118	6.272	69.735	15
16	1.873	.5339	.0458	.0858	21.825	11.652	6.672	77.744	16
17	1.948	.5134	.0422	.0822	23.697	12.166	7.066	85.958	17
18	2.029	.4936	.0390	.0790	25.645	12.659	7.453	94.350	18
19	2.107	.4746	.0361	.0761	27.671	13.134	7.834	102.893	19
20	2.191	.4564	.0336	.0736	29.778	13.590	8.209	111.564	20
21	2.279	.4388	.0313	.0713	31.969	14.029	8.578	120.341	21
22	2.370	.4220	.0292	.0692	34.248	14.451	8.941	129.202	22
23	2.465	.4057	.0273	.0673	36.618	14.857	9.297	138.128	23
24	2.563	.3901	.0256	.0656	39.083	15.247	9.648	147.101	24
25	2.666	.3751	.0240	.0640	41.646	15.622	9.993	156.104	25
26	2.772	.3607	.0226	.0626	44.312	15.983	10.331	165.121	26
27	2.883	.3468	.0212	.0612	47.084	16.330	10.664	174.138	27
28	2.999	.3335	.0200	.0600	49.968	16.663	10.991	183.142	28
29	3.119	.3207	.0189	.0589	52.966	16.984	11.312	192.120	29
30	3.243	.3083	.0178	.0578	56.085	17.292	11.627	201.062	30
31	3.373	.2965	.0169	.0569	59.328	17.588	11.937	209.955	31
32	3.508	.2851	.0159	.0559	62.701	17.874	12.241	218.792	32
33	3.648	.2741	.0151	.0551	66.209	18.148	12.540	227.563	33
34	3.794	.2636	.0143	.0543	69.858	18.411	12.832	236.260	34
35	3.946	.2534	.0136	.0536	73.652	18.665	13.120	244.876	35
40	4.801	.2083	.0105	.0505	95.025	19.793	14.476	286.530	40
45	5.841	.1712	.00826	.0483	121.029	20.720	15.705	325.402	45
50	7.107	.1407	.00655	.0466	152.667	21.482	16.812	361.163	50
55	8.646	.1157	.00523	.0452	191.159	22.109	17.807	393.689	55
60	10.520	.0951	.00420	.0442	237.990	22.623	18.697	422.996	60
65	12.799	.0781	.00339	.0434	294.968	23.047	19.491	449.201	65
70	15.572	.0642	.00275	.0427	364.290	23.395	20.196	472.479	70
75	18.945	.0528	.00223	.0422	448.630	23.680	20.821	493.041	75
80	23.050	.0434	.00181	.0418	551.244	23.915	21.372	511.116	80
85	28.044	.0357	.00148	.0415	676.089	24.109	21.857	526.938	85
90	34.119	.0293	.00121	.0412	827.981	24.267	22.283	540.737	90
95	41.511	.0241	.00099	.0410	1 012.8	24.398	22.655	552.730	95
100	50.505	.0198	.00081	.0408	1 237.6	24.505	22.980	563.125	100

Compound interest factors

6%									6%
	Single Payment		Uniform Payment Series				Uniform Gradient		
	Compound Amount Factor	Present Worth Factor	Sinking Fund Factor	Capital Recovery Factor	Compound Amount Factor	Present Worth Factor	Gradient Uniform Series	Gradient Present Worth	
	Find F Given P F/P	Find P Given F P/F	Find A Given F A/F	Find A Given P A/P	Find F Given A F/A	Find P Given A P/A	Find A Given G A/G	Find P Given G P/G	
n									n
1	1.060	.943	1.0000	1.0600	1.000	0.943	0	0	1
2	1.124	.8900	.4854	.5454	2.060	1.833	0.485	0.890	2
3	1.191	.8396	.3141	.3741	3.184	2.673	0.961	2.569	3
4	1.262	.7921	.2286	.2886	4.375	3.465	1.427	4.945	4
5	1.338	.7473	.1774	.2374	5.637	4.212	1.884	7.934	5
6	1.419	.7050	.1434	.2034	6.975	4.917	2.330	11.459	6
7	1.504	.6651	.1191	.1791	8.394	5.582	2.768	15.450	7
8	1.594	.6274	.1010	.1610	9.897	6.210	3.195	19.841	8
9	1.689	.5919	.0870	.1470	11.491	6.802	3.613	24.577	9
10	1.791	.5584	.0759	.1359	13.181	7.360	4.022	29.602	10
11	1.898	.5268	.0668	.1268	14.972	7.887	4.421	34.870	11
12	2.012	.4970	.0593	.1193	16.870	8.384	4.811	40.337	12
13	2.133	.4688	.0530	.1130	18.882	8.853	5.192	45.963	13
14	2.261	.4423	.0476	.1076	21.015	9.295	5.564	51.713	14
15	2.397	.4173	.0430	.1030	23.276	9.712	5.926	57.554	15
16	2.540	.3936	.0390	.0990	25.672	10.106	6.279	63.459	16
17	2.693	.3714	.0354	.0954	28.213	10.477	6.624	69.401	17
18	2.854	.3503	.0324	.0924	30.906	10.828	6.960	75.357	18
19	3.026	.3305	.0296	.0896	33.760	11.158	7.287	81.306	19
20	3.207	.3118	.0272	.0872	36.786	11.470	7.605	87.230	20
21	3.400	.2942	.0250	.0850	39.993	11.764	7.915	93.113	21
22	3.604	.2775	.0230	.0830	43.392	12.042	8.217	98.941	22
23	3.820	.2618	.0213	.0813	46.996	12.303	8.510	104.700	23
24	4.049	.2470	.0197	.0797	50.815	12.550	8.795	110.381	24
25	4.292	.2330	.0182	.0782	54.864	12.783	9.072	115.973	25
26	4.549	.2198	.0169	.0769	59.156	13.003	9.341	121.468	26
27	4.822	.2074	.0157	.0757	63.706	13.211	9.603	126.860	27
28	5.112	.1956	.0146	.0746	68.528	13.406	9.857	132.142	28
29	5.418	.1846	.0136	.0736	73.640	13.591	10.103	137.309	29
30	5.743	.1741	.0126	.0726	79.058	13.765	10.342	142.359	30
31	6.088	.1643	.0118	.0718	84.801	13.929	10.574	147.286	31
32	6.453	.1550	.0110	.0710	90.890	14.084	10.799	152.090	32
33	6.841	.1462	.0103	.0703	97.343	14.230	11.017	156.768	33
34	7.251	.1379	.00960	.0696	104.184	14.368	11.228	161.319	34
35	7.686	.1301	.00897	.0690	111.435	11.498	11.432	165.743	35
40	10.286	.0972	.00646	.0665	154.762	15.046	12.359	185.957	40
45	13.765	.0727	.00470	.0647	212.743	15.456	13.141	203.109	45
50	18.420	.0543	.00344	.0634	290.335	15.762	13.796	217.457	50
55	24.650	.0406	.00254	.0625	394.171	15.991	14.341	229.322	55
60	32.988	.0303	.00188	.0619	533.126	16.161	14.791	239.043	60
65	44.145	.0227	.00139	.0614	719.080	16.289	15.160	246.945	65
70	59.076	.0169	.00103	.0610	967.928	16.385	15.461	253.327	70
75	79.057	.0126	.00077	.0608	1 300.9	16.456	15.706	258.453	75
80	105.796	.00945	.00057	.0606	1 746.6	16.509	15.903	262.549	80
85	141.578	.00706	.00043	.0604	2 343.0	16.549	16.062	265.810	85
90	189.464	.00528	.00032	.0603	3 141.1	16.579	16.189	268.395	90
95	253.545	.00394	.00024	.0602	4 209.1	16.601	16.290	270.437	95
100	339.300	.00295	.00018	.0602	5 638.3	16.618	16.371	272.047	100

Compound interest factors

8%   8%

	Single Payment		Uniform Payment Series				Uniform Gradient		
	Compound Amount Factor	Present Worth Factor	Sinking Fund Factor	Capital Recovery Factor	Compound Amount Factor	Present Worth Factor	Gradient Uniform Series	Gradient Present Worth	
n	Find F Given P F/P	Find P Given F P/F	Find A Given F A/F	Find A Given P A/P	Find F Given A F/A	Find P Given A P/A	Find A Given G A/G	Find P Given G P/G	n
1	1.080	.9259	1.0000	1.0800	1.000	0.926	0	0	1
2	1.166	.8573	.4808	.5608	2.080	1.783	0.481	0.857	2
3	1.260	.7938	.3080	.3880	3.246	2.577	0.949	2.445	3
4	1.360	.7350	.2219	.3019	4.506	3.312	1.404	4.650	4
5	1.469	.6806	.1705	.2505	5.867	3.993	1.846	7.372	5
6	1.587	.6302	.1363	.2163	7.336	4.623	2.276	10.523	6
7	1.714	.5835	.1121	.1921	8.923	5.206	2.694	14.024	7
8	1.851	.5403	.0940	.1740	10.637	5.747	3.099	17.806	8
9	1.999	.5002	.0801	.1601	12.488	6.247	3.491	21.808	9
10	2.159	.4632	.0690	.1490	14.487	6.710	3.871	25.977	10
11	2.332	.4289	.0601	.1401	16.645	7.139	4.240	30.266	11
12	2.518	.3971	.0527	.1327	18.977	7.536	4.596	34.634	12
13	2.720	.3677	.0465	.1265	21.495	7.904	4.940	39.046	13
14	2.937	.3405	.0413	.1213	24.215	8.244	5.273	43.472	14
15	3.172	.3152	.0368	.1168	27.152	8.559	5.594	47.886	15
16	3.426	.2919	.0330	.1130	30.324	8.851	5.905	52.264	16
17	3.700	.2703	.0296	.1096	33.750	9.122	6.204	56.588	17
18	3.996	.2502	.0267	.1067	37.450	9.372	6.492	60.843	18
19	4.316	.2317	.0241	.1041	41.446	9.604	6.770	65.013	19
20	4.661	.2145	.0219	.1019	45.762	9.818	7.037	69.090	20
21	5.034	.1987	.0198	.0998	50.423	10.017	7.294	73.063	21
22	5.437	.1839	.0180	.0980	55.457	10.201	7.541	76.926	22
23	5.871	.1703	.0164	.0964	60.893	10.371	7.779	80.673	23
24	6.341	.1577	.0150	.0950	66.765	10.529	8.007	84.300	24
25	6.848	.1460	.0137	.0937	73.106	10.675	8.225	87.804	25
26	7.396	.1352	.0125	.0925	79.954	10.810	8.435	91.184	26
27	7.988	.1252	.0114	.0914	87.351	10.935	8.636	94.439	27
28	8.627	.1159	.0105	.0905	95.339	11.051	8.829	97.569	28
29	9.317	.1073	.00962	.0896	103.966	11.158	9.013	100.574	29
30	10.063	.0994	.00883	.0888	113.283	11.258	9.190	103.456	30
31	10.868	.0920	.00811	.0881	123.346	11.350	9.358	106.216	31
32	11.737	.0852	.00745	.0875	134.214	11.435	9.520	108.858	32
33	12.676	.0789	.00685	.0869	145.951	11.514	9.674	111.382	33
34	13.690	.0730	.00630	.0863	158.627	11.587	9.821	113.792	34
35	14.785	.0676	.00580	.0858	172.317	11.655	9.961	116.092	35
40	21.725	.0460	.00386	.0839	259.057	11.925	10.570	126.042	40
45	31.920	.0313	.00259	.0826	386.506	12.108	11.045	133.733	45
50	46.902	.0213	.00174	.0817	573.771	12.233	11.411	139.593	50
55	68.914	.0145	.00118	.0812	848.925	12.319	11.690	144.006	55
60	101.257	.00988	.00080	.0808	1 253.2	12.377	11.902	147.300	60
65	148.780	.00672	.00054	.0805	1 847.3	12.416	12.060	149.739	65
70	218.607	.00457	.00037	.0804	2 720.1	12.443	12.178	151.533	70
75	321.205	.00311	.00025	.0802	4 002.6	12.461	12.266	152.845	75
80	471.956	.00212	.00017	.0802	5 887.0	12.474	12.330	153.800	80
85	693.458	.00144	.00012	.0801	8 655.7	12.482	12.377	154.492	85
90	1 018.9	.00098	.00008	.0801	12 724.0	12.488	12.412	154.993	90
95	1 497.1	.00067	.00005	.0801	18 701.6	12.492	12.437	155.352	95
100	2 199.8	.00045	.00004	.0800	27 484.6	12.494	12.455	155.611	100

Compound interest factors

10%   10%

	Single Payment		Uniform Payment Series				Uniform Gradient		
	Compound Amount Factor	Present Worth Factor	Sinking Fund Factor	Capital Recovery Factor	Compound Amount Factor	Present Worth Factor	Gradient Uniform Series	Gradient Present Worth	
	Find F Given P F/P	Find P Given F P/F	Find A Given F A/F	Find A Given P A/P	Find F Given A F/A	Find P Given A P/A	Find A Given G A/G	Find P Given G P/G	
n									n
1	1.100	.9091	1.0000	1.1000	1.000	0.909	0	0	1
2	1.210	.8264	.4762	.5762	2.100	1.736	0.476	0.826	2
3	1.331	.7513	.3021	.4021	3.310	2.487	0.937	2.329	3
4	1.464	.6830	.2155	.3155	4.641	3.170	1.381	4.378	4
5	1.611	.6209	.1638	.2638	6.105	3.791	1.810	6.862	5
6	1.772	.5645	.1296	.2296	7.716	4.355	2.224	9.684	6
7	1.949	.5132	.1054	.2054	9.487	4.868	2.622	12.763	7
8	2.144	.4665	.0874	.1874	11.436	5.335	3.004	16.029	8
9	2.358	.4241	.0736	.1736	13.579	5.759	3.372	19.421	9
10	2.594	.3855	.0627	.1627	15.937	6.145	3.725	22.891	10
11	2.853	.3505	.0540	.1540	18.531	6.495	4.064	26.396	11
12	3.138	.3186	.0468	.1468	21.384	6.814	4.388	29.901	12
13	3.452	.2897	.0408	.1408	24.523	7.103	4.699	33.377	13
14	3.797	.2633	.0357	.1357	27.975	7.367	4.996	36.801	14
15	4.177	.2394	.0315	.1315	31.772	7.606	5.279	40.152	15
16	4.595	.2176	.0278	.1278	35.950	7.824	5.549	43.416	16
17	5.054	.1978	.0247	.1247	40.545	8.022	5.807	46.582	17
18	5.560	.1799	.0219	.1219	45.599	8.201	6.053	49.640	18
19	6.116	.1635	.0195	.1195	51.159	8.365	6.286	52.583	19
20	6.728	.1486	.0175	.1175	57.275	8.514	6.508	55.407	20
21	7.400	.1351	.0156	.1156	64.003	8.649	6.719	58.110	21
22	8.140	.1228	.0140	.1140	71.403	8.772	6.919	60.689	22
23	8.954	.1117	.0126	.1126	79.543	8.883	7.108	63.146	23
24	9.850	.1015	.0113	.1113	88.497	8.985	7.288	65.481	24
25	10.835	.0923	.0102	.1102	98.347	9.077	7.458	67.696	25
26	11.918	.0839	.00916	.1092	109.182	9.161	7.619	69.794	26
27	13.110	.0763	.00826	.1083	121.100	9.237	7.770	71.777	27
28	14.421	.0693	.00745	.1075	134.210	9.307	7.914	73.650	28
29	15.863	.0630	.00673	.1067	148.631	9.370	8.049	75.415	29
30	17.449	.0573	.00608	.1061	164.494	9.427	8.176	77.077	30
31	19.194	.0521	.00550	.1055	181.944	9.479	8.296	78.640	31
32	21.114	.0474	.00497	.1050	201.138	9.526	8.409	80.108	32
33	23.225	.0431	.00450	.1045	222.252	9.569	8.515	81.486	33
34	25.548	.0391	.00407	.1041	245.477	9.609	8.615	82.777	34
35	28.102	.0356	.00369	.1037	271.025	9.644	8.709	83.987	35
40	45.259	.0221	.00226	.1023	442.593	9.779	9.096	88.953	40
45	72.891	.0137	.00139	.1014	718.905	9.863	9.374	92.454	45
50	117.391	.00852	.00086	.1009	1 163.9	9.915	9.570	94.889	50
55	189.059	.00529	.00053	.1005	1 880.6	9.947	9.708	96.562	55
60	304.482	.00328	.00033	.1003	3 034.8	9.967	9.802	97.701	60
65	490.371	.00204	.00020	.1002	4 893.7	9.980	9.867	98.471	65
70	789.748	.00127	.00013	.1001	7 887.5	9.987	9.911	98.987	70
75	1 271.9	.00079	.00008	.1001	12 709.0	9.992	9.941	99.332	75
80	2 048.4	.00049	.00005	.1000	20 474.0	9.995	9.961	99.561	80
85	3 229.0	.00030	.00003	.1000	32 979.7	9.997	9.974	99.712	85
90	5 313.0	.00019	.00002	.1000	53 120.3	9.998	9.983	99.812	90
95	8 556.7	.00012	.00001	.1000	85 556.9	9.999	9.989	99.877	95
100	13 780.6	.00007	.00001	.1000	137 796.3	9.999	9.993	99.920	100

Compound interest factors

12%   12%

	Single Payment		Uniform Payment Series				Uniform Gradient		
	Compound Amount Factor	Present Worth Factor	Sinking Fund Factor	Capital Recovery Factor	Compound Amount Factor	Present Worth Factor	Gradient Uniform Series	Gradient Present Worth	
	Find F Given P F/P	Find P Given F P/F	Find A Given F A/F	Find A Given P A/P	Find F Given A F/A	Find P Given A P/A	Find A Given G A/G	Find P Given G P/G	
n									n
1	1.120	.8929	1.0000	1.1200	1.000	0.893	0	0	1
2	1.254	.7972	.4717	.5917	2.120	1.690	0.472	0.797	2
3	1.405	.7118	.2963	.4163	3.374	2.402	0.925	2.221	3
4	1.574	.6355	.2092	3292	4.779	3.037	1.359	4.127	4
5	1.762	.5674	.1574	.2774	6.353	3.605	1.775	6.397	5
6	1.974	.5066	.1232	.2432	8.115	4.111	2.172	8.930	6
7	2.211	.4523	.0991	.2191	10.089	4.564	2.551	11.644	7
8	2.476	.4039	.0813	.2013	12.300	4.968	2.913	14.471	8
9	2.773	.3606	.0677	.1877	14.776	5.328	3.257	17.356	9
10	3.106	.3220	.0570	.1770	17.549	5.650	3.585	20.254	10
11	3.479	.2875	.0484	.1684	20.655	5.938	3.895	23.129	11
12	3.896	.2567	.0414	.1614	24.133	6.194	4.190	25.952	12
13	4.363	.2292	.0357	.1557	28.029	6.424	4.468	28.702	13
14	4.887	.2046	.0309	.1509	32.393	6.628	4.732	31.362	14
15	5.474	.1827	.0268	.1468	37.280	6.811	4.980	33.920	15
16	6.130	.1631	.0234	.1434	42.753	6.974	5.215	36.367	16
17	6.866	.1456	.0205	.1405	48.884	7.120	5.435	38.697	17
18	7.690	.1300	.0179	.1379	55.750	7.250	5.643	40.908	18
19	8.613	.1161	.0158	.1358	63.440	7.366	5.838	42.998	19
20	9.646	.1037	.0139	.1339	72.052	7.469	6.020	44.968	20
21	10.804	.0926	.0122	.1322	81.699	7.562	6.191	46.819	21
22	12.100	.0826	.0108	.1308	92.503	7.645	6.351	48.554	22
23	13.552	.0738	.00956	.1296	104.603	7.718	6.501	50.178	23
24	15.179	.0659	.00846	.1285	118.155	7.784	6.641	51.693	24
25	17.000	.0588	.00750	.1275	133.334	7.843	6.771	53.105	25
26	19.040	.0525	.00665	.1267	150.334	7.896	6.892	54.418	26
27	21.325	.0469	.00590	.1259	169.374	7.943	7.005	55.637	27
28	23.884	.0419	.00524	.1252	190.699	7.984	7.110	56.767	28
29	26.750	.0374	.00466	.1247	214.583	8.022	7.207	57.814	29
30	29.960	.0334	.00414	.1241	241.333	8.055	7.297	58.782	30
31	33.555	.0298	.00369	.1237	271.293	8.085	7.381	59.676	31
32	37.582	.0266	.00328	.1233	304.848	8.112	7.459	60.501	32
33	42.092	.0238	.00292	.1229	342.429	8.135	7.530	61.261	33
34	47.143	.0212	.00260	.1226	384.521	8.157	7.596	61.961	34
35	52.800	.0189	.00232	.1223	431.663	8.176	7.658	62.605	35
40	93.051	.0107	.00130	.1213	767.091	8.244	7.899	65.116	40
45	163.988	.00610	.00074	.1207	1 358.2	8.283	8.057	66.734	45
50	289.002	.00346	.00042	.1204	2 400.0	8.304	8.160	67.762	50
55	509.321	.00196	.00024	.1202	4 236.0	8.317	8.225	68.408	55
60	897.597	.00111	.00013	.1201	7 471.6	8.324	8.266	68.810	60
65	1 581.9	.00063	.00008	.1201	13 173.9	8.328	8.292	69.058	65
70	2 787.8	.00036	.00004	.1200	23 223.3	8.330	8.308	69.210	70
75	4 913.1	.00020	.00002	.1200	40 933.8	8.332	8.318	69.303	75
80	8 658.5	.00012	.00001	.1200	72 145.7	8.332	8.324	69.359	80
85	15 259.2	.00007	.00001	.1200	127 151.7	8.333	8.328	69.393	85
90	26 891.9	.00004		.1200	224 091.1	8.333	8.330	69.414	90
95	47 392.8	.00002		.1200	394 931.4	8.333	8.331	69.426	95
100	83 522.3	.00001		.1200	696 010.5	8.333	8.332	69.434	100

Compound interest factors

18%   18%

	Single Payment		Uniform Payment Series				Uniform Gradient		
	Compound Amount Factor	Present Worth Factor	Sinking Fund Factor	Capital Recovery Factor	Compound Amount Factor	Present Worth Factor	Gradient Uniform Series	Gradient Present Worth	
n	Find F Given P F/P	Find P Given F P/F	Find A Given F A/F	Find A Given P A/P	Find F Given A F/A	Find P Given A P/A	Find A Given G A/G	Find P Given G P/G	n
1	1.180	.8475	1.0000	1.1800	1.000	0.847	0	0	1
2	1.392	.7182	.4587	.6387	2.180	1.566	0.459	0.718	2
3	1.643	.6086	.2799	.4599	3.572	2.174	0.890	1.935	3
4	1.939	.5158	.1917	.3717	5.215	2.690	1.295	3.483	4
5	2.288	.4371	.1398	.3198	7.154	3.127	1.673	5.231	5
6	2.700	.3704	.1059	.2859	9.442	3.498	2.025	7.083	6
7	3.185	.3139	.0824	.2624	12.142	3.812	2.353	8.967	7
8	3.759	.2660	.0652	.2452	15.327	4.078	2.656	10.829	8
9	4.435	.2255	.0524	.2324	19.086	4.303	2.936	12.633	9
10	5.234	.1911	.0425	.2225	23.521	4.494	3.194	14.352	10
11	6.176	.1619	.0348	.2148	28.755	4.656	3.430	15.972	11
12	7.288	.1372	.0286	.2086	34.931	4.793	3.647	17.481	12
13	8.599	.1163	.0237	.2037	42.219	4.910	3.845	18.877	13
14	10.147	.0985	.0197	.1997	50.818	5.008	4.025	20.158	14
15	11.974	.0835	.0164	.1964	60.965	5.092	4.189	21.327	15
16	14.129	.0708	.0137	.1937	72.939	5.162	4.337	22.389	16
17	16.672	.0600	.0115	.1915	87.068	5.222	4.471	23.348	17
18	19.673	.0508	.00964	.1896	103.740	5.273	4.592	24.212	18
19	23.214	.0431	.00810	.1881	123.413	5.316	4.700	24.988	19
20	27.393	.0365	.00682	.1868	146.628	5.353	4.798	25.681	20
21	32.324	.0309	.00575	.1857	174.021	5.384	4.885	26.330	21
22	38.142	.0262	.00485	.1848	206.345	5.410	4.963	26.851	22
23	45.008	.0222	.00409	.1841	244.487	5.432	5.033	27.339	23
24	53.109	.0188	.00345	.1835	289.494	5.451	5.095	27.772	24
25	62.669	.0160	.00292	.1829	342.603	5.467	5.150	28.155	25
26	73.949	.0135	.00247	.1825	405.272	5.480	5.199	28.494	26
27	87.260	.0115	.00209	.1821	479.221	5.492	5.243	28.791	27
28	102.966	.00971	.00177	.1818	566.480	5.502	5.281	29.054	28
29	121.500	.00823	.00149	.1815	669.447	5.510	5.315	29.284	29
30	143.370	.00697	.00126	.1813	790.947	5.517	5.345	29.486	30
31	169.177	.00591	.00107	.1811	934.317	5.523	5.371	29.664	31
32	199.629	.00501	.00091	.1809	1 103.5	5.528	5.394	29.819	32
33	235.562	.00425	.00077	.1808	1 303.1	5.532	5.415	29.955	33
34	277.963	.00360	.00065	.1806	1 538.7	5.536	5.433	30.074	34
35	327.997	.00305	.00055	.1806	1 816.6	5.539	5.449	30.177	35
40	750.377	.00133	.00024	.1802	4 163.2	5.548	5.502	30.527	40
45	1 716.7	.00058	.00010	.1801	9 531.6	5.552	5.529	30.701	45
50	3 927.3	.00025	.00005	.1800	21 813.0	5.554	5.543	30.786	50
55	8 984.8	.00011	.00002	.1800	49 910.1	5.555	5.549	30.827	55
60	20 555.1	.00005	.00001	.1800	114 189.4	5.555	5.553	30.846	60
65	47 025.1	.00002		.1800	261 244.7	5.555	5.554	30.856	65
70	107 581.9	.00001		.1800	597 671.7	5.556	5.555	30.860	70
75	46 122.1				1 367 339.2	5.556	5.555	30.862	75
100	15 424 131.9				85 689 616.2	5.556	5.555	30.864	100

CHAPTER 18

Final Tips for the Power Exam

OUTLINE

DON'T COMPLICATE 286

BE AWARE OF UNITS 286

USE COMMON SENSE 286

BE PREPARED FOR CODE QUESTIONS 286

LOOK FOR THE FAMILIAR IN THE UNFAMILIAR 287

DON'T TRIP OVER TRICKY QUESTIONS 287
Ideal Devices ■ Balanced and Unbalanced Loads ■ Delta-to-Wye and Wye-to-Delta Conversions ■ Waveforms ■ Nameplate Information ■ Parallel Transformers

SUMMARY 288

This chapter provides a few tips to help you solve some exam problems that may at first appear to be difficult or outside your area of study. These tips come from the author's years of teaching PE exam review courses and hearing the post-exam frustrations and success strategies expressed by candidates. They are also founded on a close analysis of typical sample problems published by different providers of exam prep materials, including National Council of Examiners for Engineering and Surveying (NCEES). The suggestions in this chapter can help you add some precious points to your score.

In reality, our goal is not just to help you *solve* exam problems, but to help you to be able to *recognize the right answer*. It is very possible that a number of questions will not really require calculations, or at least not detailed calculations. Throughout the book, we have tried to present concepts and terminology in a way that will help you to understand power-related topics and their underlying fundamentals. Careful reading and reflection on the material will help you get closer to an answer even if no calculations are required. On the other hand, some problems will definitely test your ability to work with numbers and equations, so you need to be able to use your calculator quickly and accurately.

DON'T COMPLICATE

In a high-stakes exam situation, it is natural to anticipate difficult problems. But be careful not to make a problem more difficult than it really is. Always look for the simplest correct solution method that will solve the problem!

For example, a problem stated as a motor problem might really involve only a simple circuit calculation. One also might expect a simple question or two involving real and reactive power and power factor, or perhaps a problem that can be solved immediately by a simple application of Ohm's law. Knowing the fundamentals reviewed in Chapters 3 through 5 will probably allow you to answer several questions immediately.

One thing we suggest when looking at a problem is to first read all four possible answers. One or more may be patently incorrect.

BE AWARE OF UNITS

Be careful about powers-of-ten prefixes, such as M, m, k, G, and so on. Sometimes, it's best to work in basic units (volts, ohms, amps). In fact, be careful with *all* units, especially when trying to apply an equation that you don't often use to solve a problem. Be sure you can work per-unit problems, as well.

USE COMMON SENSE

Some problems may require an equation, or at least an understanding of numerical dependencies, that you may not have at your fingertips. Often, some common sense reflection can point you in the right direction. For instance, how do you think that capacitance varies with increasing surface area? Think about a simple example that is appropriate to a given case.

BE PREPARED FOR CODE QUESTIONS

The exam specifications indicate that candidates should be prepared for questions on codes and standards, such as NEC and NESC. Because you can bring such references with you to the exam, becoming an expert in the entire code isn't necessary, but you do need to know where to find pertinent information quickly.

Code-related questions often involve one of the following:

- Look-up questions, in which you need to find the correct value from a table; for example, finding ampacity of a conductor. Be sure you carefully note the parameters of the problem (conductors in magnetic or nonmagnetic duct, copper or aluminum conductors, etc.) before selecting an answer. In such problems, both identifying the relevant table and reading it correctly are critical!

- Problems in which common sense can be used to help you arrive at an answer. This type of problem is often stated without any numerical data. If you can rule out one or two of the responses, you will have significantly increased your probability of a correct answer.

LOOK FOR THE FAMILIAR IN THE UNFAMILIAR

Problems involving what might at first seem to be an unfamiliar topic may turn out to be relatively easy. Here are a few examples and possibilities:

- A problem in Supervisory Control and Data Acquisition (SCADA), or relay system design, might be stated in terms of magnitudes of sampled quantities required by the relays, in comparison with actual values. The solution may then only be a matter of knowing that in an instrument (or other) transformer, voltages are related by turns ratios, and currents are related inversely to turns ratios (see Chapters 7 and 15).

- A problem stated in a relaying or system-protection context might really involve only knowledge of the device to be protected. For example, a delta-wye transformer can be differentially protected only if the three-phase connections of the current transformers on the two sides are connected inversely to the connections of the particular windings on the main transformer.

- Some relaying or protection problems might simply involve a common-sense look at where a protective device, such as a surge arrestor, would need to be placed to accomplish a particular objective. For example, to protect a transformer against damage from a lightning strike to the primary, one should place the arrestors at the primary. Chapter 15 provides some insight into this.

DON'T TRIP OVER TRICKY QUESTIONS

Finally, let's look at some specific topics that can have tricky aspects. You may be able to streamline your analysis of such questions by keeping in mind a few basic facts.

Ideal Devices

It is wise to know the characteristics of ideal devices. For example, an inductor can be thought of as a device that tries to maintain its current constant, while a capacitor tries to keep its terminal voltage constant. This follows from the performance characteristics of these devices expressed as simple differential equations. These ideas will help you understand filtered waveforms in rectifier circuits. Further, we know that an ideal diode passes current in only one direction (in the direction of the arrow on the symbol). Knowing this may help answer a question involving, for example, a rectifier circuit. Even more fundamental questions can be answered by knowing, for example, that adding devices across an ideal voltage source (as in power-factor correction) will have no effect on what was there originally.

Balanced and Unbalanced Loads

A problem stated as a balanced problem might ask an obvious question about the neutral; remember that neutral current in a perfectly balanced three-phase device is zero, even if a neutral conductor is present. Alternatively, an unbalanced problem might be very simple. For example, if we have a grounded-wye load without mutual coupling, connected to an ideal grounded three-phase voltage source, and we remove one of the phases, the currents in the remaining phases are unaffected and the neutral current is just the vector sum of the two remaining phase currents. This is obvious from circuit theory, but the normal tendency is to make the problem more complicated.

Delta-to-Wye and Wye-to-Delta Conversions

Sometimes, a quick conversion of a load impedance from delta to wye or vice-versa will allow immediate solution of a problem. Remember that a given balanced delta-connected load can always be replaced by a balanced wye whose impedances are each one-third of a delta impedance (see Chapters 3 and 5).

Waveforms

Power-electronics questions might give several probable-looking waveforms for a voltage or current, only one of which is correct. Often, you can rule out several of the answers immediately based on the waveform appearance (see Chapter 14).

Nameplate Information

A problem may provide several pieces of device nameplate information, only one or two of which are needed to easily solve a problem or answer a question. You need to look at such problems carefully and decide what is relevant to the question. This is particularly true of machines, where many parameters might be given, but only one or two are needed for the particular problem and, to a lesser extent, transformers (see Chapters 7, 10).

Parallel Transformers

Questions involving paralleling of transformers seem to be popular. Remember that you cannot parallel a delta-wye and a wye-wye (or delta-delta) transformer, because there will be improper phase relations that cannot be resolved. Assuming that the data for both transformers is on a common base, transformers in parallel will carry loads that are inversely proportional to their leakage impedances. Often, the smaller unit will overload even if the total kVA load is smaller than the sum of the kVA capabilities of the transformers; knowing this might be enough to answer a question.

SUMMARY

Hopefully, the tips in this chapter have tied together much of the material presented in previous chapters, while also providing you with some specific strategies for dealing with tricky problems on the exam. As a final, general recommendation, remember to look carefully at each problem and at least make an intelligent guess if you are not certain of the answer. Unanswered questions are automatically marked as incorrect—instead, make your best guess in order to improve your chance of passing. At the very least, you will have a 25 percent probability of getting the answer right, and perhaps 50 percent if you have even a small understanding of the topic.

It has been the author's experience that most engineers actually retain or know more technical information than they think they do as they prepare for and take the PE exam. Look at each problem and, if necessary, mark it and put it aside to look at later. Your mind might have found a simple solution technique while you were addressing other problems.

Best wishes for success on the exam!

INDEX

A

ABCD parameters, 112–113
   PI model, T networks, 113
abc sequence ($V_a V_b V_c$ or $V_{ab} V_{bc} V_{ca}$), 65–66
acb sequence, 66
accuracy, 206
ac machines, 125–126
   measuring devices, 214
   types and general characteristics, 126–127
across-the-line starting, 149
active circuit elements, 40
actual value, 79
adequate inspection, 231
admittance, 25, 45–46
all-aluminum-alloy conductors (AAAC), 107
alternating current (ac), 37–38
   circuit models, SSAC analysis, 39
   effective values, 41–43
   impedance, admittance, 45–46
   linear, bilateral, passive elements, 39–41
   nodal problem, 59–62
   phasor, 43–45
   power to single-phase load, 47–51
   resonance, 47
   series and parallel equivalents, 51–53
   sinusoid, 38–39
alternator, 64, 127
aluminum conductor steel-reinforced (ACSR) conductors, 107–108, 109
ammeter, burden associated with, 210–211
amortisseur windings, 126
ampacity, 107
ampere (A), 21
Ampere's circuital law, 11, 16
amplitude (A_m), 38
analog instruments, 206
angles, 5
angular velocity (ω), 38
annual cost, 254–256
ANSI standards, 180
apparent power (AV), 20, 50
approximate unbalance factor (UF_{approx}), 170–171
armature, 67, 120
autotransformer
   connections, 93, 94–96
   starting, 149
availability, 217

average power, 41, 48
AV (volt-ampere), 20

B

backup, 186
   protection, 187
   relaying, 175
balanced load, 64, 287
balanced three-phase circuits, 63
   line-to-line and line-to-neutral voltages, 65
   notation, 66
   phase sequences, 65–66
   transformer connections, 102–104
base quantities, 79, 90
base value, 79
battery voltage (V_s), 35
benefit-cost analysis, 260–261
best alternative, 262–263
b-h curve, 120
bilateral devices, 41
boldface, 5, 20, 66
bolted faults, 174
breakdown torque, 143
breakeven analysis, 261
bridge circuit, 33
bridge network, 34
brushes, 126
buck-boost transformers, 93
bundling, 107
burden voltage, 210–211
burns, 232
bus admittance matrix (Y_{bus}), 59–60, 62
bus impedance matrix (Z_{bus}), 61, 62

C

cage rotor, 138
calculus, 4
candlepower, 221
capacitive susceptance, 111
capacitor (C), 40–41, 52
   -run motors, 155
   -start motors, 155
capitalized cost, 253
CARDINAL conductor, 109
cascade, 112–113
cash flow, 239–240
change-of-base equation, 81–82
characteristic impedance, 114
charge density, 12, 13–14
circuit
   breakers, 174

   elements, representation, 40–41
   laws, 22
   models, 39
   theory, 160–163
circuit-analysis techniques
   mesh, 22–24
   nodal analysis, 24–26
   Thevenin and Norton sources, 28–29
   Thevenin's theorem, 26–28
circular mil, 109
circular trigonometric functions, 2–4
code
   letters, 227
   -related exam questions, 224, 286
coefficient matrix, 6
cogging, 156
color rendition, 220, 221
combination wye and delta connections, 74–76
common sense, 286
commutator, 120–121, 126, 156
compact fluorescent, 220
complex load impedance (Z_L), 88, 90
complex number of array, 7
complex numbers, 5–6
complex power (S), 50–51, 71
composite conductors, 107
compounding, 124
compound interest, 240
   single-payment formulas, 242–244
   symbols, functional notation, 241–242
   tables, 275–284
compound machine, 124
compound windings, 155
conductor
   bundling, 110
   types, configurations, 107–108
constant-horsepower motors, 147
constant power, 146
constant torque, 146
constant-torque motors, 147
continuous compounding, 249–250
controlled Rectifiers, 197
convergence, 162
   tolerance, 162–163
conversion factors, 108
converters, 197
cool white, 220
coordination, 186
copper loss, 123
core form, 86
core permeability (μ), 86

289

cumulative compounding, 124
curl, 12, 15
current direction, 20, 21
current magnitude, 75, 76
current transformers, 186
cycloconverters, 149, 197
cylindrical rotor, 127

D

damper windings, 126, 127
D'Arsonval meter movement, 34
    ac measurements, 209–210
    movement, 207–209
dc component, 210
dc-dc converters, 197
dc motor configurations, 124
dc with ripple, 198
declining-balance depreciation, 265
delta Δ, 67
delta connection, 72–74, 103
    with wye connections, 74–76
delta load, 72
delta-to-wye conversions, 288
delta-wye transformer, 33–34, 180
dependent sources, 30–33
depreciation, 264–268
derived sources, 191
determinant, 7, 8, 9
developed power, 141
diagrams, 21
differential compounding, 124
differential equations, 4
differential protection scheme, 174
differential relaying, 186
differentiation, 4
digital instruments, 206
digital multimeters (DMM), 206
diode, 209
direct axis, 175
direct-axis steady state reactance, 175
direct current (dc), 39
    machinery, concepts, terminology, 120–121
    machines, 119–120
directional-sensing relaying, 186
distance relaying, 186
divergence, 12, 14–15
dot convention, 87
double-subscript, 20–21
DRAKE conductor, 109, 111
driving point impedances, 61
dynamic fields, 16–17

E

economic order quantity, 263
eddy-current, 97
e (electromotive force), 20
effective interest, 248–250
effective values, 41–43
efficacy, 221

efficiency (EFF), 164
    dc machine, 123–124
    transformer, 98–99
electrical codes, standards, 224–226
electrical safety, 225, 230
    insulation by isolation, 231
    reasonable precaution, 230–231
    step and touch potentials, 230
electrical shock, 231–236
electrical speed (n_{elec}), 130–131
electric fields
    charge density, 12, 13–14
    curl, 12, 15
    divergence, 12, 14–15
    gradient, 12, 14–15
    scalar potential as line integral, 12–13
electric flux density, 13
electric-utility generators, 127
electrocution, 232
electromagnet, 120
electromagnetic field theory, 105
electromagnetics, 11
    power engineering relationship, 12
electromotive force (e), 20
electronic starting, 150
empirical equations, 21–22
engineering economics, 238
    problems, 250–251
equipotential grounding systems, 236
equivalence, 241
equivalent uniform annual benefits (EUAB), 254
equivalent uniform annual cost (EUAC), 254
exact values, 114
excitation, 141
exciting current, 98, 193
expected value, 274

F

failure modes and effects analysis (FMEA), 218
FALCON conductor, 109
familiar, 287
Faraday-Maxwell law, 16
Faraday, Michael, 16
Faraday's law, 11, 87
fault, 173, 174–175
    current, multiple sources of, 183–184
    protection, 174–175
fault studies
    generator modeling, 175–176
    short-circuit simulation studies rationale, 178–179
    subtransient model, 178
    transient behavior of synchronous generator subjected to fault, 176–178
    unbalanced faults, 184–185
Ferranti effect, 115

fibrillation, 233–235
field current If, 134–137
field pole, 120
field theory, 11, 12, 105
field winding, 120
first harmonic, 193
fixed input, 250
fixed output, 251
flat start, 161
flicker, 191
fluorescent lamp, 220
flux density, 13, 16
footcandles, 221–222
fractional-horsepower machines, 154
Franklin, Benjamin, 231
frequency (f), 38
    changers, 149–150
    changes and induction motors, 147
    of induced rotor currents, 139
    transformers and, 102
friction and windage loss, 123–124
full load
    performance calculations, 115–117
    torque, 143
    voltage, 122
full wave rectifier, 199–200
fuses, 174
future worth analysis, 253–254

G

Gauss-Seidel iterative procedure, 163
Gauss-Seidel power-flow algorithm, 161
generator, 127
    modeling, 175–176
    operation, 120, 121
    separately excited, 121–124
geometric mean distance (GMD), 110–111
geometric mean radius (GMR), 109, 110
gradient, 12, 14–15
grounded conductor, 191
ground fault circuit interrupters (GFCIs), 235
grounding, 191
ground mats, 230
ground wires, 107

H

half-wave diode rectifier, 197
half-wave-rectified sinusoidal signal, 209
half-year convention, 266–267
halogen lamp, 220
harmonics, 148, 192–194
    induction motors and, 148
    phase sequence, 196–197
hertz (Hz), 38
high intensity discharge (HID) lamp, 220
high-pressure sodium lamp, 220
hyperbolic trigonometric functions, 3–4, 114
hysteresis, 97

curve, 120
motors, 127

I

ideal devices, 287
idealize, 40
ideal transformer, 86
identity matrix (U), 7
IEEE P37.2/D3.1 Draft Standard for Electrical Power System Device Function Numbers, Acronyms, and Contact Designations, 187
illumination engineering, 219–221
impedance (Z), 6, 45–46
　angle, 45
　complex power and, 50–51
　matching, 88
　referral, 88–89
impulses, 192
incandescent lamp, 220
inductance, 109
induction motor, 126
　equivalent circuit of, 140–142
　examples, 152–153
　frequency changers, 149–150
　harmonics on, 148
　machine parameters, 144
　multi-speed, 147
　NEMA designs for, 142
　polyphase, 138–140
　power-factor correction, 150–152
　selection and mechanical load characteristics, 146–147
　single-phase synchronous, 154–155
　speed-torque characteristics, 142–144
　starting methods for, 149–150
　torque equations, conversions, 145–146
　voltage and frequency changes on, 147
　voltage unbalance on, 148
　with higher starting torque, 144–145
inductive reactance, 111
inductors (I), 40–41
infinite life, 253
inflation, 270–272
input/output unfixed, 251
instantaneous power, 41
instantaneous unit, 186
Institute of Electrical and Electronics Engineers (IEEE), 170
　standard device numbers, 187
instrumentation, 206
instrument transformers, 214–215
insulation by isolation, 231
integrals, 4
integration, 4
internal or generated voltage (Ef), 132
interpolate, 206
inverse, 7
inverse phasor transform, 44
inverse transformation matrix, 166

inverters, 197
iron loss, 124
isolation transformer, 235
iterative form, 161

J

joule (J), 41

K

kilowatt-hours (KWH), 41
Kirchhoff laws, 19, 22
Kirchhoff's current law (KCL), 22, 24, 31
Kirchhoff's voltage law (KVL), 22, 24

L

lagging power factor, 49
　load, 55
laminating, 98
leading power factor, 49
leakage
　branch, 96
　flux, 86, 87
　reactance, 97
lighting
　fundamentals, 219–221
　guidelines, indoor, 221–222
linear circuit, 6, 40
line flows calculation, 163–164
line starting, 149
line-to-line voltage, 65, 68, 70
line-to-neutral voltage, 65, 71
load, 180
　convention, 39
　flow, 157
logarithms, 2
long-line equations, 114
long-line model, 112
lossless model, 115
loss of load probability (LOLP), 219
low-pressure sodium lamp, 220
low voltage winding, 88
lumen, 221
lumped-parameter model, 40, 106–107

M

machine types, 155–156
macroshocks, 232–233
magnetic field, 15–16
　intensity, 16
magnetic flux density, 16
magnetic permeability, 16
magnetic saturation, 121–124
magnetization, 97
magnetomotive force (F, mmf), 20
magnitude, 75, 76, 80
matrices, 6–8
maximum "let go" current, 232
maximum (pullout) torque, 143
Maxwell's equations, 16
measurement, 108, 206

medium-length line model, 112
mercury-vapor lamp, 220
mesh analysis, 22–24
metal halide lamp, 220
meters
　D'Arsonval meter movement, 207–209
　waveforms and, 206–207
microshocks, 233–235
minima-maxima, 262
minimum attractive rate of return (MARR), 250–251, 252, 257–260
modes, 1
modified accelerated cost recovery system (MACRS), 265–268
motor, 127
　operation, 120, 122
moving vane, 214
multiplier, 208
multi-speed motors, 147
multi-winding connections, 92–93

N

nameplate, 89
　information, 288
National Council of Examiners for Engineering and Surveying (NCEES), 205–206, 223, 285
National Electrical Code (NEC), 223, 224, 225–226, 286
　Chapter 3: wiring methods and materials, 226–227
　Chapter 4: equipment for general use, 227
　contents of, 227–228
National Electrical Safety Code (NESC), 223, 224, 225–226, 231, 286
　contents of, 229
National Fire Protection Association (NFPA), 226
NEC 2008 Handbook, 226
negative sequence, 66, 165
　components, 166
　heating, 184
　network, 184–185
NEMA (National Electrical Manufacturers Association) designs, 142–145
neon lights, 220
neutral (conductor), 69–70, 191
Newton-Raphson method, 163
nodal analysis, 24–26
no-load performance calculations, 115–117
no load voltage, 122
nominal interest, 248–250
non-annual compounding, 248–249
nonlinear scales, 206
non-rotating loads, 180
normalized representation, 79
Norton conductance G_{eq}, 28
Norton current (I_n), 28, 31–32
Norton equivalent, 28, 30

norton sources, 28–29
notation, 20–21, 264
    single and double subscript, 20–21

O

Ohm's law, 19, 22, 40
    phasor form, 45
ohm Ω, 21
open-circuit test, 97, 100, 101
open faults, 174
operate-fail-repair-operate cycles, 217
optimization, 262–263
overcurrents, 173
overexcited mode, 134
overmotored, 147, 151
overvoltage, 192

P

parallax error, 206
parallel resonance, 47
parallel RLC, 46
parallel transformers, 288
parameter values, per-phase model, 108–111
part-winding starting, 150
passive circuit elements, 40
passive system, 160
peak-to-peak voltage ripple, 201
PE Power Electrical Engineering Sample Exam, 19
percent quantity, 80
percent voltage regulation, 122
periodic compounding, 242
period (T), 38
per-phase equivalent, 67–68
per-unit quantities, 79
    advantages of, 83
    numerical examples, 83
    single-phase ac circuit, 80–82
    three-phase circuits, 82
    transformer, 89–92
per-unit slip (s), 139
phase angle, 43, 44
phase sequence, harmonics, 196–197
phase-to-neutral voltage, 65
phase-to-phase voltage, 65
phase unbalance factor, 165, 170–171
phasor, 39, 43
phasor diagram, 57, 58, 59, 70, 73
    transformer, 100
phasor transform, 43
pictorial symbols, 21
pi model, 113
pi (π) model, 106
polarity, 20, 21
    conventions, 87–88
    two-winding ideal transformer, 86
positive sequence, 65–66, 165
    components, 166
power

reactive, complex, apparent, 49–51
series and parallel equivalents, 51–53
single-phase load, 47–49
units associated with, 50
power electronics, 189–190, 197–198
    full-wave rectifier, 199–200
    half-wave diode rectifier, 197
    single-phase full wave rectifier with capacitive filter, 200–201
    three-phase rectifier, 201
    three-phase rectifier with capacitive filter, 202–204
power factor, 49
    angle, 49, 133
    correction, 53–59, 150–152
power flow, 157
    fundamentals of, 158–159
power (P), 30, 41
    angle, 133
    concepts, 41
    transfer, maximum, 30–33
    triangle, 135
power quality, 189–191
    grounding, 191
    harmonics, 192–194
    harmonics and phase sequence, 196–197
    history of, 192
    terminology, 191–192
    total harmonic distortion, 195–196
power relations, 88–89
power system
    protection, 185–187
    reliability, 217–219
precision, 206
present worth analysis, 251–253
primary winding, 88
probability, 272–273
propagation constant, 114
pu voltage, 91

Q

quadrature power, 49

R

radian frequency (ω), 38
radians, 3
rate of return
    analysis, 256–260
    inflation and, 271–272
reactance, 45, 46
reactive power (Q), 49, 71, 88
reactor starting, 149
real power, 41, 48, 88
reasonable precaution, 230–231
reclose, 175
rectified, 199
rectifiers, 197
reduced-voltage starting, 144, 149
regions, 90
reliability, 173, 217

reluctance motor, 126
repeatability, 206
resistance, 108–109
resistors (R), 40–41
resistor starting, 149
resonance, 47
resonant frequency, 47
respiratory arrest, 233
revolving field, 128–130
right-hand rule, 67
ripple, 200
risk, 273
risk analysis, 272–274
rms (root-mean-square)
    phasor, 43
    value, 40, 41–43
rotary solenoids, 156
rotating field, 128–130
rotating phasor, 43
rotor, 67, 120, 128

S

safety, 225
sag, 191
salient-pole rotor, 127
scalar potential, 12–13
scalar quantities, 80
secondary winding, 88
second harmonic, 193
seconds (s), 21
self-excited dc generator, 124
separately excited generator, 121–124
sequence
    components, 166
    currents, 169
series branch, 96, 97
series resonance, 47
series RLC, 45
series self-excited generator, 124
shaded-pole construction, 155
shading coils, 156
shell form, 86
shock protection, prevention, 235–236
short-circuit, 173
    simulation studies, 178–179
    test, 100, 101
    three-phase study, 179–183
short-line model, 112
shunt
    branch, 97, 98
    capacitance, 109–110
    fault, 173
    resistance, 208
    self-excited generator, 124
siemens (S), 21
silicon controlled rectifiers (SCRs), 197
simple interest, 240
simplicity, 286
single-phase ac circuits, per-unit application, 80–82

single-phase full wave rectifier with capacitive filter, 200–201
single-phase synchronous motors, 154–155
single-phase-to-ground fault, 179
single-subscript, 20–21
sinking-fund depreciation, 265
sinusoid, 38–39
six-pulse rectifier, 202
skin effect, 108
slip, 139
    speed, 139
soft starting, 150
solenoid, 156
source transformation, 28
speed, 126
    regulation, 122, 155
spikes, 192
Standard Handbook for Electrical Engineers, The, 148
star-delta starting, 150
starting torque, 143, 144–145
static wires, 107
stator, 67, 120, 128
steady-state alternating current (SSAC), 37, 39, 157
steady-state period, 177
steady-state range, 177
steady-state reactance (X_d), 177
stepper motors, 156
step potential, 230
straight-line depreciation, 264
subscript, 20–21
subtransient period, 177
subtransient range, 177
subtransient reactance (X_d), 177, 178
sum-of-years'-digits depreciation, 264, 266
superposition, 181
surge-impedance, 115
surge-impedance loading (SIL), 115
surges, 192
susceptance, 45, 46
swell, 192
swing bus, 159
symbols, 21
symmetrical components, 166, 174
    method of, 165–170, 184
    transformation matrix, 166
symmetric network, 113
symmetric waveform, 194
synchronous condensers, 126
synchronous generator, subjected to fault, 176–178
synchronous impedance, 132
synchronous machine, 126, 127–128, 180
    changing field current, 134–137
    electrical vs. mechanical speed, 130–131
    equivalent circuit of, 132–134
    revolving field, 128–130
    torque, 138
synchronous reactance, 64, 132
synchronous speed (n_s), 128, 139

T

table of bases, 90
tax consequences, 269–270
tertiary winding, 93
Tesla, Nikola, 154
thermocouple, 214
Thevenin equivalent, 68, 180–181
Thevenin impedance, 183, 184
Thevenin resistance (R_{th}) 28, 31, 32
Thevenin source, 28, 29
Thevenin's theorem, 26–28, 29, 30
Thevenin voltage (V_{th}), 28, 31
third harmonic, 193
three-phase bridge, 202
three-phase circuits, per-unit calculations, 82
three-phase connections
    combination wye and delta connections, 74–76
    delta connections, 72–74
    per-phase equivalent, 67–68
    wye and equations for total P, Q, and S, 69–72
three-phase faults, 179
    short-circuit, 179–183
three-phase rectifier, 201
    with capacitive filter, 202–204
three-phase synchronous generator, 66–67
threshold of perception, 232
time-overcurrent protection, 186
time value of money, 240
t-model, 96, 101, 113
torque, 138, 142–144
    angle, 130, 133
    equations, conversions, 145–146
    proportional to speed, 146
    proportional to square of speed, 146
total harmonic distortion (THD), 195–196
total power, 71–72
touch potential, 230
transfer impedances, 61
transformer, 180
    balanced three-phase connections, 102–104
    buck-boost, 93
    electromagnetic fundamentals, 85–89
    losses, impedances, models, 96–100
    multi-winding connections, 92–93
    per-unit calculations, 81
    per-unit calculations, base quantities, 89–92
    phasor diagram, 100
    ratings, 89
    tests, model parameters, 100–102
    voltage regulation, 92
transient period, 177
transient range, 177
transient reactance (X_d), 177
transient stability analysis, 177
translational motion, 156
transmission line, 180
transmission-line theory, 105
transmission parameters, 112
transposed line, 108
tricky questions, 287–288
triplen, 194
true-rms-reading, 207
true rms voltage, 210
turns ratio, 88
two-port network, 112
two-wattmeter connection, 216–217
two-winding ideal transformer, polarity considerations, 86–87

U

unbalance, 170
unbalanced faults, 184–185
unbalanced loads, 287
underexcited mode, 134, 137
undervoltage, 192
uniform gradient, 246–248
uniform payment series formulas, 244–246
units, 1, 21, 108, 286
    associated with power, 50
unity (U), 7
universal machine, 119, 126
universal motor, 155
unreliability, 217

V

(v) voltage, 20
valuation, 264–268
variable-torque motors, 147
VARmeter, 216
V curve, 134
vector (V), 6, 7
    magnitude of, 5–6
    of knowns, 6
    of unknowns, 6
    quantities, 80
voltage regulation (VREG), 92, 99, 117, 122, 123
voltage (V, v), 20
    changes and induction motors, 147
    flicker, 191
    ratio, 116
    unbalance and motors, 148
volt-ampere (AV), 20
volts (V), 21

W

warm white, 220
watt, VAR measurement and, 215–217
waveform, 288
    considerations, 211–214
    meters and, 206–207
wavelength, 114–115
Wheatstone bridge, 33, 34–35
winding loss, 97, 98
winding polarity, 88
wiring, 226–227
wound rotor, 138
wye connections, 69–72, 103
   with delta connections, 74–76
wye-delta starting, 150
wye-to-delta conversions, 288
wye (Y), 67
   load, 72

Z

zero sequence, 165
   components, 166
   network, 185
zones, 90

Notes

Notes

Notes

Notes